12

Experimental Meson Spectroscopy—1974

(Boston)

AIP Conference Proceedings

Series Editor: Hugh C. Wolfe

Number 21

Particles and Fields Subseries No. 8

Experimental Meson Spectroscopy—1974

(Boston)

Editor

David A. Garelick

Northeastern University

American Institute of Physics

New York   1974

LC Catalog Card No. 74-82628
ISBN   0-88318-120-7
AEC CONF—740408

American Institute of Physics
335 East 45th Street
New York, N.Y. 10017

Printed in the United States of America

FOREWORD

The Fourth International Conference on Experimental Meson Spectroscopy was held at Northeastern University in Boston on April 26 and 27, 1974. Over 220 physicists attended, representing most of the experimental groups, world wide, working on particle properties, and including quite a few theorists. The primary aim of those working on Conference organization was to maximize the interchange of ideas and discussion among those attending, and to minimize distractions. Information transfer was assisted, for example, by having copies of the most important materials presented by each speaker distributed to the attendees prior to the talks. This necessitated copying multi page papers in several hundred copies in the hours before the sessions. Most Conferees found the papers very useful. It is my purpose here to thank those who labored behind the scenes on this and many similar projects which were designed to make the Conference an easy place to exchange ideas and to provide a non-distracting and, if possible, enjoyable background.

The Organizing Committee of EMS '74 was comprised of W. Faissler (Northeastern), M. Friedman (Northeastern), M. Gettner (Northeastern), U. Kruse (Illinois), D. Leith (SLAC), L. Montanet (CERN), A. Rosenfeld (Berkeley), W. Selove (Pennsylvania), and R. Weinstein (Northeastern). This group made preparations for EMS '74 starting about 18 months prior to the Conference and developed the invitational lists. They were responsible for the balance of the Conference and for its size. A group of "Friends of EMS '74" at Northeastern did the detailed orga-nizing. In addition to the N.U. members of the Organizing Committee mentioned above, the Friends consisted of R. Aaron, R. Arnowitt, B. Cairns, M. Fraade, D. Garelick, E. Saletan, G. Srivastava and M. Vaughn. This group worked hard and effectively on most Conference services, including the audio visual aspects of the Conference, the Zerox service mentioned above, the banquet, shuttle, ladies program, and much much more. I do not diminish my thanks to them when I express special thanks to Dr. G. Srivastava for her organization and operation of the Secretariat and its many services, and for the solution of numerous personal problems of the Conferees. I wish to express special thanks also to B. Cairns, the Conference Secretary, for the long hours and multiple responsibilities undertaken by her on behalf of the Conference. Operation of the Secretariat, and liaison with the hotels was also ably assisted by G. Tang, H. Schneider and E. Pothier of the N.U. Physics Department. Among the graduate students and technical staff who helped, special thanks are due to G. Blanar, C. Boyer, J. Mietus, M. Ronan, G. Simonelli, and M. Tautz for their work with the audio visual and with the shuttles.

I had always thought Conference fees tended to be excessive. I have now found that one is cured of this illusion by running a Conference. I thank those whose financial sponsorship of EMS'74 made its success possible: the U.S. Atomic Energy Commission, the U.S. National Science Foundation, Northeastern University, the Digital Equipment Corporation, EG&G/ORTEC, and Lecroy Research Systems.

June 1974

Roy Weinstein
Organizing Committee Chairman

PREFACE AND EDITOR'S ACKNOWLEDGMENTS

One of the goals of the Conference was to provide, through the Proceedings, an accurate and up to the date picture of meson spectroscopy. Contributors to the Proceedings were given a maximum of seven weeks following the Conference to update and submit their papers. Many of the papers found herein contain additional materials which were not presented at the Conference due to necessary time limitations. The Proceedings contain, in addition to all of the papers presented by the Conference speakers, some comments by session chairmen and others and papers by T. Ferbel and A. Levey, et al. which were thought appropriate for the Proceedings.

I want to thank the other members of the Program Committee, all of whom worked hard for several months defining the topics, selecting the Conference speakers, and reviewing submitted materials. The Program Committee consisted of R. Diebold (ANL), T. Ferbel (Rochester), A. Barbaro-Galtieri (Berkeley), D. Garelick (Northeastern), F. Gilman (SLAC), and Kwan-Wu Lai (BNL). I also want to thank Roy Weinstein, Chairman of the Organizing Committee for his leadership and hard work so vital to the life of EMS '74; and in addition thank Prof. M. Glaubman, Chairman of the Physics Department, and Dean R. Ketchum of Northeastern University for their support and encouragement and for arranging assistance for the Program Committee and editing of the Proceedings. The overall cooperation and help of many others at Northeastern and the support of Northeastern University and the Alfred P. Sloan Foundation are also gratefully acknowledged. Particular thanks are due to Maxine Fraade for her excellent help with the Program Committee's work and with editing the Proceedings.

June 1974

D. Garelick
Program Committee Chairman and Editor

ABBREVIATED INDEX OF SPEAKERS AND FIRST AUTHORS

TABLE OF CONTENTS

Second Afternoon Session (April 27)   Chairman: R. Arnowitt

REVIEW OF NON-STRANGE MESONS

M. G. Bowler
Nuclear Physics Laboratory,
University of Oxford,
Oxford, England

ABSTRACT

The present situation in non-strange meson spectroscopy
is briefly reviewed.

I have 35 minutes to review the field of non-strange mesons.
Every individual topic of great interest has been separately
allotted a comparable length of time, and this is a field in which
I have done very little work. I can only conclude that my role is
intended to be that of the amateur or second-rate comedian employed
to induce in a studio audience a proper frame of mind before the
real stars appear and a show goes on the air.

Let us take the naive non-relativistic quark model as a guide
to our expectations in meson spectroscopy: it seems to work, and
(subject to the review talk of D. Cohen) there is not, and never
has been, any evidence for exotic states which cannot be represented
formally as a bound $q\bar{q}$ pair. This framework highlights the current
problems and future difficulties in meson spectroscopy.

Take a simple harmonic oscillator potential for the $q\bar{q}$ bound
states. Levels are then equally spaced in an appropriate energy
parameter and become increasingly degenerate in orbital angular
momentum ℓ as the level number n increases. In addition we have
both triplet and singlet $q\bar{q}$ configurations in each (n, ℓ) state.
The resulting picture is shown in Fig. 1 up to a level number of
5 (h,f,p). Triplet states are indicated by T, singlets by S. This
looks like a sequence of trajectories, the leading trajectory
accompanied by a family of lower lying daughters, reaching right
down to spin zero. These are not quite the familiar daughters of
dual models however: every column at a given mass has the same
parity. In particular channels you can skip every other daughter
and every other column.

Specializing to non-strange mesons, in Fig. 2 I have exhibited
separately the I = 0 and I = 1 states, and labelled not only (n, ℓ)
but also the G-parity. States accessible in $\pi\pi$ scattering, as
examples of specific channels, are circled and against each entry
(two states for each entry for I = 0) I have given the candidates.
The present parlous state of meson spectroscopy at once becomes
apparent. In I = 1 we have the ρ, A_2 and g on the leading triplet
trajectory. On the leading singlet trajectory we have the π and
very likely the B. In I = 0 we have (ω, ϕ), (f, f') on the leading
triplet trajectory, and (η, η') on the leading singlet trajectory
(assuming η' is 0^- and not 2^-). I regard this list as exhausting
the non-strange meson states which are firmly nailed down: I
shall consider them no further. Note that we only have candidates

qq̄ angular momentum
configuration

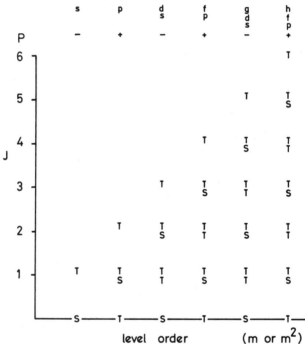

level order   (m or m^2)

Fig.1. qq̄ states in an S.H.O. potential

for the first three columns, taking us up to ∿ 1.7 GeV/c^2 in mass.

The outstanding class of problems at the moment seems to me to be the sorting out of the dubious or missing members of these first three columns, and I address myself now to these items. Focus first on the 0$^+$ states in the second column of Fig. 2, which contains among other things the ε of ill repute. The I = 0 state is accessible in ππ scattering, which is indeed where most of our knowledge has come from, through the analyses of Protopopescu and CERN/Munich. Their data taken in conjunction with the π°π° mass spectrum have yielded S-wave I = 0 ππ phase shifts which now seem quite well determined[1], (Fig. 3) rising slowly to ∿ 90°, whipping on to ∿ 180° as the KK̄ threshold is passed, and then rising slowly through 270° at ∿ 1.25 GeV/c^2. We thus have two I = 0,0$^+$ candidates, as we expect, at ∿ 1 and 1.25 GeV/c^2. The former is the S*, the latter perhaps deserves a new name. There seems no good reason to keep the old ε at ∿ 700 MeV/c^2. In I = 1 CERN/Munich have a possible inelastic p wave resonance under the g,

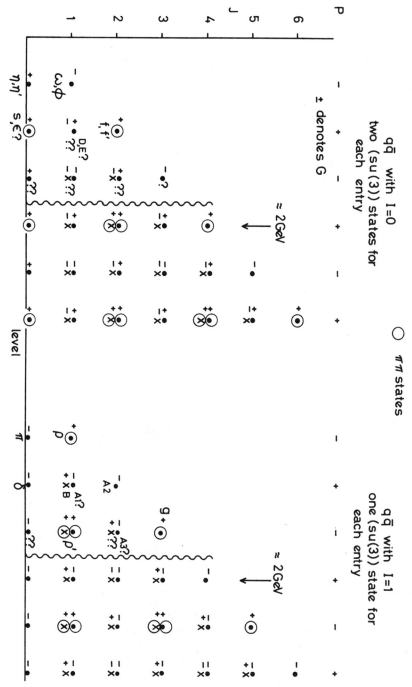

Fig.2. I = 0 and I = 1 q̄q states, with candidates. States accessible through ππ scattering are circled

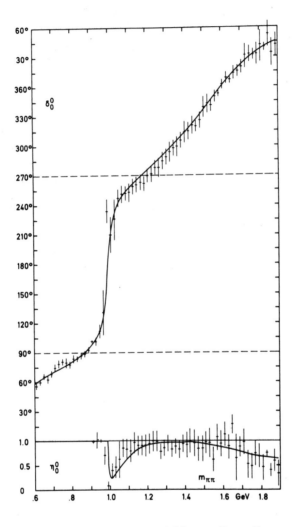

Fig.3. The I = 0 S-wave $\pi\pi$ phase shifts. From Hyams et al, ref. 1

to be identified with the ρ' otherwise claimed in electromagnetic processes.

The only 0^+ candidate in I = 1 is known variously as π_N or δ. There is a very strong I = 1 $K\bar{K}$ enhancement at threshold, which if due to a bound state corresponds to a pole at $\sim$ 960 MeV/c^2, $J^P = 0^+$, and there is evidence for a $\pi\eta$ state at this mass. Much of the evidence for δ comes from $p\bar{p}$ annihilations where it appears as a decay product of the D, which I shall discuss later. It would be nice to have really good data in conventional one particle exchange channels, perhaps.

$$K^-p \to \delta^-Y^{*+}, \quad \pi p \to \delta p \quad \text{followed by } \delta \to \underset{\pi\eta}{K\bar{K}}$$

for a coupled channel analysis. But I think it will stay with us. (Binnie will review the region around 1 GeV/c^2 where many probably spurious effects have been seen from time to time). The non-strange 0$^+$ mesons, S*,ε,δ are thus in fairly good shape.

The second column of Fig. 2 however also contains the biggest scandal of today, namely the 1$^+$ mesons. The I^G = 1$^+$ state (B → ωπ) is the best established and should have associated with it at least an octet. It has long been considered to be 1$^+$ and recent analyses of large samples of new data such as the Ascoli analysis carried out by Chaloupka[2] and the analysis of Chung et al[3] now make this look solid. It is not diffractive (dσ/dt ∿ e^{4t},σ ∿ p$_{in}$$^{-1.5}$) and seems a well established meson. To accompany the B we expect two 1$^+$I^G = 0$^-$ mesons somewhere in the range 1-1.5 GeV/c^2 with decays into ρπ, K$\bar{K}$π. There is no sign of these objects (H,H').

Staying with I = 0 we also expect two 1$^+$I^G = 0$^+$ states. These are usually identified with the D and E mesons, seen almost exclusively in p$\bar{p}$ annihilation and decaying into ηππ, K$\bar{K}$π. The D' looks quite good, decaying through δπ, and with properties which seem experiment invariant. J^P is almost certainly 1$^+$, although some model dependence is still left in. I find evidence for even an E bump unconvincing; something is probably going on but no one knows what.

This brings us to the question of the A1, J^{P}I^G = 1$^+$1$^-$. The only candidate for such a meson is the diffractively produced enhancement at ∿ 1.1 GeV/c^2, dominated by S-wave ρπ. The Ascoli analysis, by assuming a knowledge of the phase variation in specified 2π subsystems, has allowed the extraction of the relative phase variation of 3π states and it is well known to you all that while the A2 behaves very prettily, the A1 phase does not vary through more than ∿ 30° over the whole A1 range. This feature has been confirmed several times over on different sets of data and I know of no counter examples[4]. Furthermore, the Illinois people have constructed a detailed Deck model[5] and subjected it to their analysis: it accounts very nicely, qualitatively at least, for rather detailed features of the data. (Lorella Jones will tell us about this). The idea that the A1 is a diffractively produced resonance is now very sick and indeed the existence of an A1 is in doubt. We would certainly like the A1 to exist: if it does it is hardly conceivable that it does not couple to ρπ. Maybe it just isn't produced diffractively? There are two problems (i) the 1$^+$ρπ phase should still have the characteristics of a resonance, if Watson's theorem is OK (ii) why isn't it produced non-diffractively at the same sort of level as the A2-which perhaps has certain elements of diffractive behaviour in its production[6]. Two years ago, at Philadelphia, Fox listed 5 possibilities for rescuing the diffractive A1[7].

(1) That the A1 does not exist - which he rejected because of the D.

(2) That the 0$^-$S wave had a daughter. The position now is that 1$^+$ρπ S wave exhibits no variation ≳ 30° with respect to any of the

6

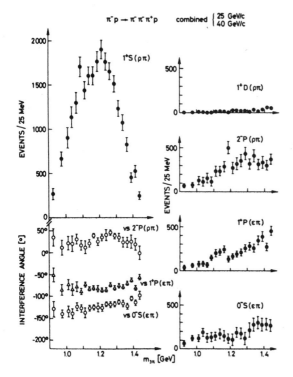

$\pi^- p \to \pi^- \pi^- \pi^+ p$ combined $\begin{cases} 25 \text{ GeV/c} \\ 40 \text{ GeV/c} \end{cases}$

reference waves. [Fig.4]

(3) The A1 is not narrow at $\sim$ 1.07, but broad at a higher mass. There never really has been any evidence for a narrow A1, but this seems implausible. With a model constructed in the same spirit as the Söding model, you can fit the 1^+S mass spectrum quite nicely, with an A1 mass $\sim$1.2 and a width $\sim$ 300 MeV/c^2,[8] but you can't fit the phase this way, with the $1^+\rho\pi$ S wave elastic. Where could the inelasticity be?

(4) Final state interaction theory is inadequate. I have no comment other than that the ρ is not narrow in the A1 region: the $\rho\pi$ system may not be a good approximation to a 2 body problem.[9]

Fig.4. Intensities and relative phases of different partial waves in the A1 region. From Antipov et al, ref. 4.

(5) The analysis is unreliable for a delicate question like the A1 phase. Against this is the pretty behaviour of the A2. For it perhaps two things. (i) The waves the A1 is beaten against are either minority waves or contain the dubious ε: they might be wrong (ii) The isobar model assumptions, built into all analyses of this kind, could be inadequate. From the results of numerical calculations which I have seen, this seems implausible. I don't believe that corrections to the isobar model can destroy an A1 resonant phase variation. I hope we will be hearing about this from Lasinski and Ascoli.

Perhaps the A1 does exist, is not diffractively produced and couples strongly to D-wave $\rho\pi$(inelastic with respect to the S-wave), which the Deck mechanism doesn't generate. For D/S $\sim$ 1 the 1^+S phase variation is then suppressed to $\sim$ 40° only, (excursions of only $\sim$ 20°)[10]. But we still have the problem of why we don't see an effect due to non-diffractive production - a problem emphasised by Kane in a different connection (charge exchange)[11].

If the 1^+ states don't appear soon all our ideas about the structure of the meson spectrum will be in real trouble.

The A3, the 2^- fπ S wave enhancement, may have the same disease as the Al. The Ascoli corporation[12] and Morrison's empire[13] claim no phase variation, in contrast to Purdue[14]. As yet I don't think the data I have seen are good enough to rule out a phase variation characteristic of resonance. But the A3 does not escape the long shadow which is reaching over the diffractive production of resonances. This is a pity, because it is our only candidate for the I = 1 2^- states under the g, and we have a candidate for a 1^- state under the g, the ρ'.

What else do we have in the region below 2 GeV/c^2? Nothing solid. There seems to be something in $\rho\pi$ $I^G = 0^-$ at $\sim$ 1675 (it shows up very nicely in some unpublished data from Berkeley) but J^P is not known: 1^- or 3^- are favoured. The F$_1$ in K$\bar{K}\pi$ is not terribly good - something is probably going on but we don't know what. Finally, the Ascoli analysis of the CERN/Serpukhov data has thrown up a possible 2^+ at $\sim$ 1.8 GeV/c^2 in p-wave fπ.[15]

What do we need to sort out the mess below 2 GeV/c^2? We won't know until it has been done, but certain things are indicated. We have to look at the 10μb level and below, and it seems to me two obvious wants are $\pi^+\pi^-\pi^0$ and K$\bar{K}\pi$, produced by OPE, at a level of at least several hundred events per μb, followed by sophisticated analyses of the general Ascoli type.

At higher masses, getting into the 2 GeV/c^2 region, there is no solid evidence for anything. Kalogeropoulos will review N$\bar{N}$ studies and try and convince you the region is stiff with narrow resonances. p$\bar{p} \to \pi^+\pi^-$ and its inverse are certainly not inconsistent with higher mass resonances [Fig. 5] but no firm conclusions can be

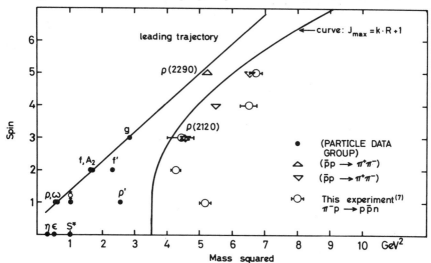

Fig.5. Possible resonances in the p$\bar{p} \rightleftarrows \pi\pi$ system. The figure is due to the CERN/Munich group

drawn. If our current ideas are right, as the mass goes up meson states get more degenerate, closer together and will be highly inelastic in any channel. Studies of $\pi^+\pi^-$ and $\pi^+\pi^0$ scattering show strong diffraction peaks at masses above ~ 1.7 GeV/c^2 [16]. The extraction of high mass resonances from the data is going to be much more difficult than our already not noticeably successful efforts, even if they exist!

That concludes my review of non-strange mesons. I hope that you are now ready for the real speakers, and that if you have not been moved to laughter, you have at least been roused to anger by my ill-informed, idiosyncratic and cavalier treatment of your favourite topics!

REFERENCES

1. For the present status of $\pi\pi$ scattering see $\pi\pi$ scattering 1973 (Tallahassee Conference) Ed. P. K. Williams and V. Hagopian, in particular the papers by:
P. Estabrooks et al p. 37
G. Grayer et al p. 117
B. Hyams et al p. 206
see also the talk by W. Männer in the proceedings.
2. V. Chaloupka CERN/D. Ph II/PHYS 73-33.
3. S. U. Chung et al BNL 18340 (paper 119 submitted to this conference).
4. Yu. M. Antipov et al Nucl. Phys. B63 153 (1973).
G. Thompson et al COO-1428-219 (paper 35 submitted to this conference).
Aachen-Berlin-Bonn-CERN-Heidelberg :paper 80 submitted to this conference.
see also the talks by G. Ascoli and T. Lasinski in these proceedings.
5. G. Ascoli et al Phys. Rev. D8 3849 (1973)
G. Ascoli et al ILL-(TH)-73-9
see also the talk by L. Jones in these proceedings.
6. Yu. M. Antipov et al Nucl. Phys. B63 153 (1973).
7. G. Fox p. 298 in Experimental Meson Spectroscopy 1972 (Third Philadelphia Conference) Ed. A. H. Rosenfeld and K. W. Lai.
8. M. G. Bowler and M. A. V. Game, unpublished.
9. This point was emphasized to me by R. Aaron and M. Vaughn in conversation.
10. M. G. Bowler, unpublished.
11. G. Kane UM HE 74-3 (invited talk, ANL Symposium on Resonance Production, March 1974: paper 121 submitted to this conference).
12. G. Ascoli et al Phys. Rev. D7 278 (1973).
Yu. M. Antipov et al Nucl. Phys. B63 153 (1973).
13. Aachen-Berlin-Bonn-CERN-Heidelberg :paper 80 submitted to this conference.
14. G. Thompson et al Phys. Rev. Lett. 32 331 (1974).

15. Yu. M. Antipov et al Nucl. Phys. B63 153 (1973).
16. W. D. Walker ππ scattering 1973 (Tallahassee Conference) Ed.
    P. K. Williams and V. Hagopian p. 80.
    N. N. Biswas et al Phys. Rev. Lett. 18 273 (1967).

COMMENT ON PAPER OF M. G. BOWLER

A. H. Rosenfeld
Lawrence Berkeley Laboratory, Berkeley, Calif. 94720

Twice Bowler seems to have suggested that one can decide whether or not a resonance exists, and guess its mass if it exists, by looking at separate plots of the modulus and phase of an amplitude, and in particular looking for the energy at which the phase passes through 90°, as opposed to looking at an Argand diagram. I would like to show two figures as counter-examples.

Bowler's Fig. 3 shows the I = 0, S-wave $\pi\pi$ phase shift. He comments that there seems to be no reason to keep the old epsilon at about 700 MeV, but that we see resonant behaviour at about 1000 MeV, and then again at about 1250 MeV, where the phase crosses 270°. My Fig. 1 is the Argand plot that accompanies the phase shift. One sees from the plot that nothing very interesting happens near 1250 MeV; in particular, there is no resonant-like increase in the speed. Moreover, when either Protopopescu or Hyams et al. look for poles (in addition to the clear S^*), they find possible poles near the epsilon, and at 1050 and 1540 MeV.

Next I want to show that for inelastic resonances phases need not go near 90° at all. I have in mind his discussion of the A_1, although I freely admit that the A_1 should not be all that inelastic and is a real puzzle. Nevertheless I continue my campaign for inspection of Argand plots. Remember that we are dealing with an inelastic reaction (Pomeron + $\pi \rightarrow \rho + \pi$) known to be dominated by "background" due to "Deck effect", or "diffraction", or "double Regge exchange", or whatever you want to call the bump in the A_1 region. So we are trying to see a small resonant signal on top of a large and energy-dependent but non-resonant background. In such cases we know from $N\pi$ partial-wave analysis that δ does not go through 90° or even do anything characteristic. I remind you by presenting in Fig. 2 two $N\pi$ Argand plots -- for D_{15} and F_{35} -- and the corresponding plots for elasticity η and phase δ. The δ plots alone don't give a clue to these resonances, although they are clear on the Argand plots.

A final example of why one should make two-dimensional Argand plots instead of phase vs. mass (even for production experiments) comes from our recent experience in partial-wave analysis of the inelastic reaction $N\pi \rightarrow N\pi\pi$. Here we can recognize one or two resonances on almost every Argand plot, but very few of them show interesting behaviour in phase alone. In the A_1 analyses that Bowler has discussed, the initial state (Pomeron + π) is even more complicated than our $N\pi$ initial state because there can be more independent amplitudes, and even the Argand plots will be very difficult to interpret. In that case, I submit that mere reliance on phase plots is probably hopeless.

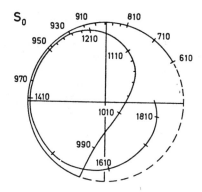

Fig. 1. Argand plot for the I = 0, S-wave ππ phase shift.

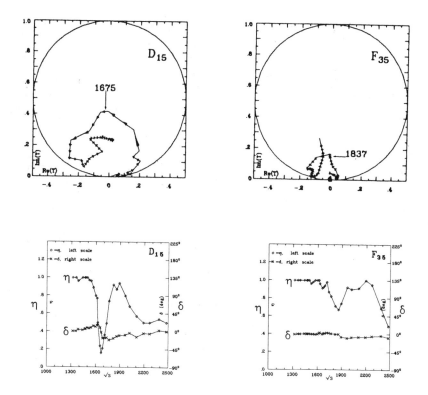

Fig. 2. Two illustrative Nπ Argand plots, for D_{15} and F_{35}, and the corresponding plots for elasticity η and phase δ.

STUDIES OF MESONS NEAR THRESHOLD

Imperial College-Southampton Group[*]
presented by

D.M. Binnie
Physics Dept., Imperial College,
London, England.

ABSTRACT

A technique is described whereby narrow mesons produced in the reaction $\pi^- + p \rightarrow$ meson + nucleon can be studied within a few MeV of the threshold. A summary is given of recently published results on the η, ω, η', S and ϕ, and some new data showing an unexpected behaviour of the ω and η' cross sections is presented.

TECHNIQUE

I will start by outlining the ideas and techniques used in the Imperial College-Southampton missing-mass spectrometer [1]. Consider the reaction $\pi^- + p \rightarrow m^o + n$ where m^o is a neutral meson of mass M. Usually m^o is narrow, e.g. η, ω, η' or ϕ. The restriction on the width Γ is made because most of our studies have been within a few MeV of the meson threshold. In this situation the neutron is contained within a small cone around 0° and has a velocity close to that of the $\pi^- p$ centre of mass. (Fig. 1).

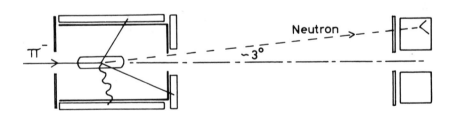

Fig. 1. (Schematic) A π^- meson interacts in a hydrogen target. A neutron produced at a small angle interacts in a neutron counter and its time-of-flight is measured over 6.15m. Secondary π's and γ's are detected by an arrangement of 66 counters nearly surrounding the target.

[*]D.M. Binnie, I.F. Burton, L. Camilleri, N.C. Debenham, A. Duane, D.A. Garbutt, J.R. Holmes, W.G. Jones, J. Keyne, M. Lewis, J.G. McEwen, I. Siotis and P.N. Upadhyay

Let P^* be the final state momentum in the c.m. [2]. As we are very close to threshold we expect a cross section of the form $\sigma \propto P^{*2\ell+1}$ where ℓ is the final state orbital angular momentum. Thus unless the S wave is forbidden, for low enough P^* we must have $\sigma \propto P^*$. This is just the final density-of-states factor. Let us assume the cross section to be of this form. The counting rate is the product of the cross section and the acceptance. The latter is very high as most of the neutrons are swept into the neutron counters. If the S wave is important, the counting rate itself will be high in spite of the low cross section.

Good mass resolution can be obtained near threshold, because both $\partial M/\partial P_n$ and $\partial M/\partial \theta_n \rightarrow 0$ as $P^* \rightarrow 0$. P_n and θ_n are the neutron momentum and angle in the laboratory frame. We can see why this is if we consider the energy balance in the c.m. As P^* is low we can use the non-relativistic expression:

$$W^* = m_n + M + \tfrac{1}{2}P^{*2} \ (1/m_n + 1/M)$$

where W^* is the invariant π^-p mass. The right hand term is small. For example, with $P^* = 50$ MeV/c and $m_n = M \sim 1000$ MeV it is only 2.5 MeV. Therefore an accurate measurement of P^* is not necessary. Both P_n and θ_n depend <u>linearly</u> on P^* (P_n depends on $P^* \cos \theta^*$ and θ_n on $P^* \sin \theta^*$ where θ^* is the c.m. production angle), and the accuracy with which these must be determined therefore tends to zero with P^*. W^* is found from a careful determination of the incident π^- momentum. ($\partial W^*/\partial P\pi \sim 0.5$).

Although the mass resolution is high near threshold, the mass scale and the phase space factor are very non-linear for a given $P\pi$. For example at the ω half the resonance could be below threshold. However we have seen that P_n and θ_n effectively determine P^*. Therefore we predetermine P^* by fixing P_n and θ_n and vary the missing mass by varying P_π. The properties of any meson; mass, width, cross section and decay, can then be examined at a given P^*. There is no discontinuity at 'threshold'. The mass scale is almost linear with P_π.

A valuable feature of a momentum scan in P_π at a fixed P^* is that the background counting rate is usually nearly constant. S channel resonances must modulate the cross section, but fortunately for the meson studies, most non-strange baryons are wide. Also most spurious signals in the neutron counters, for example from neutrons scattered from the floor and walls, will be little affected by relatively small changes in beam momentum. This can be contrasted with a study of the time-of-flight spectrum where spurious signals could well show a marked structure.

The decay counters surrounding the hydrogen target could be used to select simple final states such as $\gamma\gamma$, $\pi^+\pi^-$, $\pi^0\pi^0$, $\pi^+\pi^-\pi^0$. Generally the separation was incomplete and the value of the decay system varied with the meson under study. A 2γ selection was particularly useful for the η and η´, especially as the $2\pi^o$ final state is forbidden. Selection of the two main ω decay channels was also easy, but a detailed study of ω → 2π was not possible.

Most of the recent analysis has been done with neutrons. Protons were also detected, but early expectations of narrow, charged states have not been fulfilled [3].

There have been two main periods of data collection, which we refer to as 1970 and 1972/73 [4]. The 1970 data were mainly concerned with the A_2, but were also analysed for neutral mesons. The 1972/73 data were for specific experiments, mainly to pursue problems and possibilities raised by the earlier results.

RESULTS

The simplest way to study the meson mass spectrum with the spectrometer is to measure the yield of neutrons that lie within a predetermined time-of-flight interval around the c.m. velocity as a function of the incident momentum. The resulting spectrum [5] is shown again in Fig. 2 and is based on 1970 data. The ω, η´ or X^o

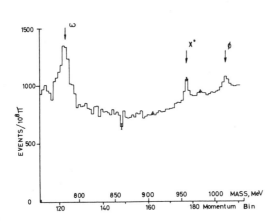

and φ are clearly visible. We now identify the small blip between the X^o and φ with the S^* at the K^+K^- threshold. Meson masses can be extracted directly from the knowledge of the incident momentum. The missing-mass resolution is dominated by the hodoscopes used in the momentum spectrometer, the ionisation loss in the target and the size of the neutron counters. It can therefore be estimated quite accurately and in this data is about 5 MeV f.w.h.m. The results for masses and widths are given in Table I.

Fig. 2. Missing mass spectrum from 750 to 1040 MeV. Neutrons with velocities close to that of the π^-p system are detected near 0°. The beam momentum is increased by steps of 0.5%.

Detected Study of the ω (1972 data)

Detailed Study of the ω (1972 data)

Although the analysis and interpretation of the ω data is not yet complete, the general form of the results is clear. The original motivation for this experiment was to test the reliability of the spectrometer as a tool for the detailed study of a resonance

near threshold. The problem is that some form of final state interaction with the neutron may occur so that we are not really
investigating a well defined meson state. The ω is almost ideal for
such a study; not only is its cross section large close to threshold,
but its width of about 10 MeV is well within the capabilities of the
spectrometer, is narrow enough for background subtraction to be
relatively easy yet is wide enough for the ω decay to occur close to
the neutron. The average flight path of the ω in the c.m. before
decay is about $40/P^*$ fermis (P^* in MeV/c). Finally the ω has two
prominent decay channels, both of which can be readily identified by
the decay counters. A second interest in the experiment was in a
precision determination of the mass and width, assuming of course
that any variation observed in these quantities with P^* was such as
to give confidence that it was the true, 'asymptotic' values which
were being determined.

An interesting technical feature was that to attain a very low
value of P^o a neutron counter was set at 0° to the beam. There was
no room to insert a sweeping magnet and essentially all the non-
interacting beam entered the counter. This was found to produce
extra background in adjacent neutron counters and to avoid this
problem data at 0° had to be collected separately. Much of the extra
background consisted of apparently random pulses in the counter, i.e.
pulses not directly correlated with the time at which beam particles
entered the counter. At least some of these pulses were attributed
to the production by fast π^- of slow π^+ mesons which stopped and
subsequently gave rise to positrons, $\pi \to \mu \to e$. The neutron flight
path was reduced from 6.15 to 4.15m, to improve the signal/back-
ground, and the beam intensity was reduced. Satisfactory agreement
was found with the data from the other neutron counters in the region
of overlap. The counter at 0° allowed a study to be made down to a
mean P^o of 20 MeV/c.

Little evidence was found for any systematic variation in mass,
width or the branching ratio $\omega \to \pi^o \gamma / \omega \to 3\pi$ with P^*. The 1970 data
had shown a curious drop in the cross section below $P^* \sim 100$ MeV/c
and this was confirmed and can now be examined in much more detail.
In Fig. 3 is plotted σ/P^* as a function of P^*. For each P^* region
examined the size of the ω is described in terms of an equivalent
total cross section assuming an isotropic production angular
distribution (above $P^* \sim 60$ MeV/c only neutrons produced with θ^*
close to 0° or 180° are detected).

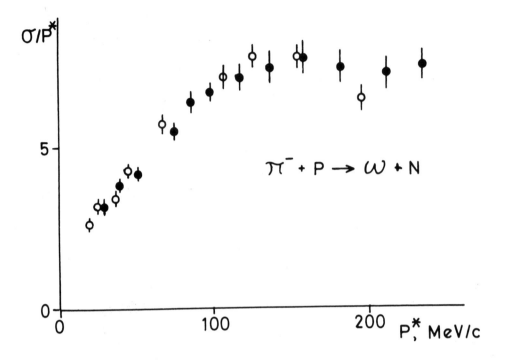

Fig. 3. Values of σ/P^* in μb/MeV/c plotted against the final state c.m. momentum P^* in the reaction $\pi^- p \to \omega n$. The closed and open circles refer to neutrons emitted forward and backward in the c.m. A simple S wave behaviour should show a constant value of σ/P^*.

The cross sections deduced from neutrons forward and backward in the c.m. are the same within the errors. This, taken in conjunction with the evidence from deuterium bubble chambers results[6] on $\pi^+ n \to \omega p$ argues strongly for isotropic production. The near constancy of the ω width and branching ratio with P^* suggest that the ω decay is not the main factor. The large value of σ/P^* may be further evidence for an (on our scale, wide) ωp resonance in the S wave[7]. This would not of course explain the drop in σ/P^* at low P^*. It might be possible to explain the latter as a result of a narrow P_{11} nucleon state[8]. Analysis on these matters is in progress.

S^* Production (1970 and 1972/73 data)

Our analysis of the S^* has been published[9] and we have done no further work on this meson. In outline a sudden drop in the yield of neutrons was found in certain missing-mass plots at a mass within 1.5 MeV of the threshold mass for $K^+ K^-$ production. See Fig. 4 taken from Ref. 9. The drop showed very clearly in the $\pi^+ \pi^-$ channel. Subsequently an examination of the $2\pi^o$ channel showed a very clear bump above a relatively flat background. (Fig. 4(c)). The data were

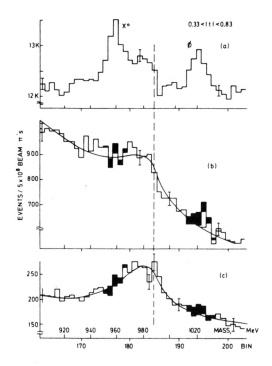

interpreted in terms of the production of the S* with subsequent decay into ππ or KK, and were described in terms of an S wave Breit Wigner distribution. Very detailed work on the S* has been done at much higher momenta with extensive analyses in terms of ππ scattering [10]. The evidence from the present experiment suggests that at threshold one π exchange is not the dominant production mechanism and that possibly some S channel resonance is responsible for the relatively large cross section.

Fig. 4. Missing-mass spectra in the t range indicated (mass bins about 3.6 MeV, about 5×10^8 incident π's per bin). (a) With no decay selection, an abrupt but small drop is seen between the X^0 and the φ. (b), (c) $\pi^+\pi^-$ and $2\pi^0$ selections, respectively. The shaded events are the predicted contributions to the spectra from the X^0 and φ. The dashed line indicates the K^+K^- threshold. The curves are fits to the S* together with a quadratic background over a range from 840 to 1070 MeV. Note the suppressed origins.

Study of the η' (1972 data)

The main aim of this experiment was to make a further attempt to measure the η' width using the missing-mass technique. An additional aim was to search for a 2γ decay mode. The problem was that several possible mesons had been reported in the region close to the η'. The results on the mass, and width and the identification of the 2γ decay mode have already been published [11] (see also Table I below). Here we would like to report some new data on the cross section near threshold. The cross sections involved are much smaller than at the ω which not only makes the evidence weaker statistically, but also raises problems of systematic errors in the background subtraction. In Ref. 1 a tendency for dσ/dt (∝ σ/P as defined above) to increase close to threshold was noted. The 1972 data confirm this effect at somewhat lower P and with better statistics and higher resolution. The data, which are preliminary, are shown in Fig. 5. Other

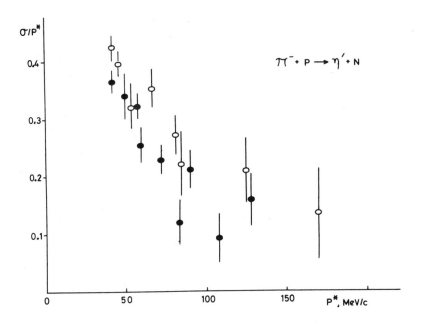

Fig. 5. Values of σ/P^* in µb/MeV/c plotted against P^* in the reaction $\pi^- p \to \eta' n$. The closed and open circles refer to neutrons emitted forward and backward in the c.m. The behaviour should be contrasted with a similar plot for ω production (Fig. 3).

measurements of the cross section exist, notably the comprehensive study by Rader et al. in the reaction $\pi^+ + n \to \eta' + p$. These measurements start at the upper end of the P^* scale shown here and the situation in this region is confused. But I want to draw attention to the striking tendency for σ/P^* to increase as P^* is reduced below 100 MeV/c, at least down to $P^* \sim 40$ MeV/c. Again a simple S wave should give a constant; clearly higher waves cannot be directly responsible for the increase at low P^*. The variation is somewhat similar to that observed several years ago by Berley et al. [13] in the channel $K^- + p \to \eta + \Lambda$. There the behaviour was attributed to the formation of the narrow $Y^*(1670)$. At present there is no suitable N^* near the η' threshold. The effect, taken together with that at the ω, seems rather too much of a coincidence to be explained simply in terms of an 'arbitrary' juxtaposition of meson thresholds and a possible spectrum of narrow N^*'s and may indicate a deeper relationship. It is interesting that ϕ meson production, while small and therefore also hard to determine well, showed a tendency for σ/P^* to rise at low P^* whereas the η showed very little effect although a small ($\sim 10\%$) fall could have been present [1].

Masses and Widths of Mesons

In Table I we show a compilation of masses and widths determined by us with for comparison the Particle Data Group values for 1973 [14].

Table I. Masses and Widths of Mesons

	Mass MeV			Width, MeV		
	1970	1972/73	P.D.G.73	1970	1972/73	P.D.G.73
η	548.1±0.4	547.45±.25	548.8±0.6	check on resolution		(∿ 1 keV)
ω	782.3±0.6	782.4±0.5*	783.8±0.3	12.4±2.2	10.2±0.6*	9.8±0.5
η´	957.1±0.6	957.46±.33	958.1±0.4	< 1.9	< 0.8	< 2
φ	1019.4±0.7	−	1019.6±0.3	4.5±1.1	−	4.2±0.2

(* preliminary value)

The measurements of masses are interesting in that, given the masses of the π, p and n, they are obtained directly in terms of the absolute beam momentum. The latter, apart from a correction for ionisation loss in the beam counters and hydrogen target, is determined from 'floating wire' measurements and is therefore expressed in terms of standard 'weights' and the potential drop across a standard resistor. There is some tendency for the masses obtained to be low (less noticeable when comparison is made with the 1972 Particle Data Group tables). Meanwhile we see no reason to modify our mass scale. We may be able to make an independent check using the reaction $\pi^+ + p \rightarrow K^+ + \Sigma^-$ (K^+ detected).

Experiments at 2 GeV/c (1972 data)

The Iowa-Argonne-Purdue Group (IAP) have presented [15] evidence for several new, narrow mesons. The apparatus was rather similar to ours (ICS); neutrons were detected near 0° and there was a decay array to detect charged particles and γ rays. We therefore performed an experiment which was as similar as we could reasonably make it to IAP. We were fortunate in being able to accumulate much more data than IAP. Thus for a given cross section we could expect about 20 times the number of events. The mass resolution was inferior (13.5 MeV in ICS, 8.6 MeV in IAP, both f.w.h.m.), but this would have only a marginal effect. Our results have recently been published [16]. Like IAP we saw the η´ but we found no evidence for the M(940), δ(962) or M(1033). The decay configurations by which these mesons were selected were rather complicated and we refer the reader to the original publications. Unlike IAP we did find evidence for a 2π effect which might be relevant to the evidence for the lower mass mesons in IAP as it would distort the 'background'.

POSSIBLE DEVELOPMENTS

It is hoped to continue the present investigations using a new spectrometer. This will permit a further improvement in the mass resolution and a more systematic study in meson cross sections close to threshold. In addition it is hoped to investigate the behaviour in the elastic channel across the meson threshold. The high resolution available will be very valuable and should permit a detailed examination to be made of the cusps in the differential cross sections [17]. The form of the effect anticipated is sketched in Fig. 6[18] The data points are taken from some (unpublished) measurements [18] on π p elastic scattering at 180°. Such studies could help to unscramble production mechanisms. They might also serve as a valuable check on phase shift solutions. It is hoped that they could cast some light on the curious behaviour of some of the production cross sections reported here.

CUSP AT η. ILLUSTRATION ONLY.

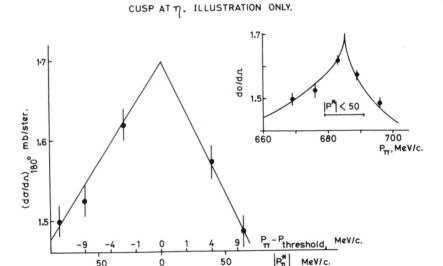

Fig. 6._ The type of behaviour expected in elastic scattering π p → π p across the threshold for η production. Straight lines are expected if the cross sections are plotted against the η momentum in the c.m. Inset. The cusp appears in a plot against beam momentum. The lines are for illustration only and are not fits to the data.

REFERENCES AND FOOTNOTES

1. D.M. Binnie et al., Phys. Rev. $\underline{D8}$, 2789 (1973).

2. Note the approximate relationship $P^* = 20(P_\pi - P_{threshold})^{\frac{1}{2}}$ where all quantities are in MeV/c.

3. G.R. Kalbfleisch et al., Nucl. Phys. $\underline{B69}$, 279 (1973), also our own experience.

4. D.M. Binnie et al., Phys. Lett. $\underline{36B}$, 257 (1971).
   D.M. Binnie et al., Phys. Lett. $\underline{36B}$, 537 (1971). Note that the possible structure reported in this paper was later found to have been enhanced by an error in the analysis program.

5. D.M. Binnie et al., Phys. Lett. $\underline{39B}$, 275 (1972).

6. J.S. Danburg et al., Phys. Rev. $\underline{D2}$, 2564 (1970).

7. V. Davidson et al., Phys. Rev. Lett. $\underline{32}$, 855 (1974).

8. R.J. Cashmore et al., L.B.L. report 2634, to be published.

9. D.M. Binnie et al., Phys. Rev. Lett. $\underline{31}$, 1534 (1973).

10. See the review by Dr. W. Manner, this volume.

11. A. Duane et al., Phys. Rev. Lett. $\underline{32}$, 425 (1974).

12. R.K. Rader et al., Phys. Rev. $\underline{D6}$, 3059 (1972).

13. D. Berley et al., Phys. Rev. Lett. $\underline{15}$, 641 (1965).

14. Particle Data Group, Rev. Mod. Phys. $\underline{45}$, S1 (1973).

15. D.L. Cheshire et al., Phys. Rev. Lett. $\underline{28}$, 520 (1972).
    R.W. Jacobel et al., Phys. Rev. Lett. $\underline{29}$, 671 (1972).
    A.F. Garfinkel et al., Phys. Rev. Lett. $\underline{29}$, 1477 (1972).

16. D.M. Binnie et al., Phys. Rev. Lett. $\underline{32}$, 392 (1974).

17. See for example E.P. Wigner, Phys. Rev. $\underline{73}$, 1002 (1948)
    A.I. Baz,   Sov. Phys. JETP $\underline{6(33)}$ 709 (1958).

18. Rapporteur talk by I. Butterworth, Aix-en-Provence Conference, Supplément au Journal de Physique $\underline{34}$, 180 (1973).

NEW RESULTS IN ππ SCATTERING

W. Männer

Max-Planck-Institut für Physik,
Munich, Germany

ABSTRACT

ππ phase shifts from threshold to 1.8 GeV ππ mass are presented. A value for the I = 0 S-wave scattering length has been obtained and can be compared to other experiments. The ambiguities arising in the phase shifts above 1000 MeV are discussed. K^+K^- and $\bar{p}p$ distributions suggest the existence of high mass states above the g meson.

INTRODUCTION

The subject of this paper is new results in ππ scattering. A large fraction of this is as yet unpublished work of the CERN-Munich group to which I belong. I am grateful to the authors for permission to present these data*.

METHOD OF ANALYSIS

Since ππ scattering cannot be measured directly, each paper on this subject has to start with a discussion of the methods employed to extract ππ phase shifts.

The possibility to determine ππ scattering from existing experiments is given by the Chew-Low equation[1]

$$\lim_{t \to m_\pi^2} \frac{\partial^2 \sigma}{\partial M_{\pi\pi}^2 \partial t} = \frac{M_{\pi\pi}^2 \cdot q_\pi}{4\pi m_p^2 p_{lab}^2} \frac{g^2}{4\pi} \frac{-t}{(m_\pi^2 - t)^2} \sigma_{\pi\pi} , \tag{1}$$

where

m_p, m_π, and $M_{\pi\pi}$ are the masses of the proton, π, and ππ, respectively;

p_{lab} is the beam momentum in the lab. system;

$g^2/4\pi = 2 \times 14.6$;

$q_\pi^2 = \frac{1}{4} M_{\pi\pi}^2 - m_\pi^2$.

In principle, this formula gives a model-independent way to extract ππ scattering. Frequently the statement was made that other exchanges and even background will give vanishing contributions after extrapolation to the pion pole. However, I want to emphasize that this is

---

*The members of the group are: W. Blum, V. Chabaud, H. Dietl, G. Grayer, G. Hentschel, B. Hyams, C. Jones, W. Koch, G. Lutz, G. Lütjens, J. Meissburger, R. Richter, U. Stierlin and P. Weilhammer.

only true for infinite statistics and if no systematic errors are present. In reality the Chew-Low extrapolation will give correct answers only if pion exchange dominates in the physical region.

If we want to extract more information from the data we have to make an amplitude analysis which relies on further assumptions:

i) Factorization

All amplitudes factorize into a part varying rapidly with $M_{\pi\pi}$ determined by the phase shifts and a factor describing the t-dependence which varies slowly with $M_{\pi\pi}$.

In the absence of polarization measurements we cannot determine all amplitudes. As suggested by the authors of the "Poor Man's Absorption Model" [2] all analyses make the assumption of:

ii) s-channel nucleon flip dominance

Experimental tests of this assumption can be made. At present we know that possible non-flip terms are small at low $|t|$, but the errors are large.

In the region $t \approx t_{min}$, small non-flip terms are added according to the absorption model in order to obtain the t-dependence at threshold demanded by angular momentum conservation. With these assumptions the phase angle between the m = 0 amplitude and the helicity-one unnatural exchange contribution can be determined from the data. This angle was found to be consistent with zero (phase coherence) at small $|t| < 0.15$ GeV^2 by Estabrooks et al.[3] for the ρ region and by Ochs[4] for $\pi\pi$ masses $600 < M_{\pi\pi} < 1900$ MeV.

The ideal procedure to extract $\pi\pi$ phase shifts would therefore be to make an amplitude analysis in small $M_{\pi\pi}$,t bins according to Estabrooks and Martin[5] with or without imposing phase coherence. However, the statistics available at present are insufficient. We therefore propose to determine the moments in small $M_{\pi\pi}$,t bins and fit several bins to a parametrization in $M_{\pi\pi}$ and t.

We can introduce a minor modification and use combinations of moments that contain unnatural exchange contributions only, which eliminates the need to parametrize A_2 exchange. For S- and P-waves only this leads to:

$$\sqrt{4\pi}\left\{\langle Y_0^0\rangle + 2\sqrt{\tfrac{5}{6}}\langle Y_2^0\rangle\right\} = \rho_{00}^{00} + \rho_{00}^{11} + 2(\rho_{11}^{11} - \rho_{1-1}^{11})$$

$$\sqrt{4\pi}\left\{\langle Y_2^0\rangle - \sqrt{\tfrac{2}{3}}\langle Y_2^2\rangle\right\} = \sqrt{\tfrac{4}{5}}\,\rho_{00}^{11} - \sqrt{\tfrac{4}{5}}\,(\rho_{11}^{11} - \rho_{1-1}^{11}) .$$

$$(2)$$

The expressions for $\langle Y_1^0\rangle$, $\langle Y_1^1\rangle$, and $\langle Y_2^1\rangle$ are unchanged.

$\pi\pi$ PHASE SHIFTS FOR $M_{\pi\pi} < 1$ GeV

The status of $\pi\pi$ phase shifts at last year's Tallahassee Conference is shown in Fig. 1 for the $I = 2$ S- and D-waves[6]. The $I = 0$ S-wave phase shifts obtained by Protopopescu et al.[7] and by different analyses of our 17.2 GeV experiment are shown in Fig. 2 [8]. In the meantime we have determined new values for δ_S^0 at low $\pi\pi$ mass from our experiment $\pi^- p \to \pi^+ \pi^- n$ at 7 GeV, and they are shown in Fig. 3. The error bars indicate the range of values one can obtain from different analyses. An error of 15% due to uncertainties in normalization and background correction should be added to all phase shifts obtained from our 7 GeV experiment shown in Figs. 3-6. It has been found that the D-wave plays an important role in the calculation of the S-wave although it is badly determined by the data. This fact has also been noticed by Estabrooks and Martin in a new analysis of our 17.2 GeV data[9]. A fit of the D-wave leads to $\delta_D \approx 2°$ at 560 MeV and a high S-wave phase shift. If we constrain the D-wave to zero we obtain the lower value of δ_S at 560 MeV.

Basdevant[10] has recommended that one uses the value of δ_D he obtained from dispersion relations, which is $\delta_D^0 \approx 1°$ at 560 MeV leading to a δ_S^0 close to the middle of our range.

At lower $M_{\pi\pi}$ the uncertainty introduced in δ_S^0 by the D-wave is smaller; here the difference in δ_S^0 is mainly due to two different calculations fitting either the unnatural combinations of Eq. (2) or the moments directly to a parametrization according to P.K. Williams[2].

Figure 4 shows a comparison of our phase shifts with those obtained from K_{e4} experiments.

In Figure 5 we plot $q \cot \delta_S^0$ against q^2, leading to an $I = 0$ S-wave scattering length

$$a_S^0 = (0.44 \pm 0.1) \ m_\pi^{-1} \ .$$

Other values have been obtained recently by Tryon[11] from an analysis of K_{e4} data:

$$a_S^0 = (0.26 \pm 0.08) \ m_\pi^{-1} \ ;$$

by Bunyatov[12] from experiments:

$$\pi^+ p \to \pi^+ \pi^+ n \quad \text{at 230 MeV (7 events)}$$

$$\pi^- p \to \pi^+ \pi^- n \quad \text{at 200-260 MeV (486 events)}$$

$$a_S^0 = (0.18 \pm 0.02) \ m_\pi^{-1}$$

$$a_S^2 = -(0.07 \pm 0.01) \ m_\pi^{-1} \ ;$$

and by Jones, Allison and Saxon[13]

$$-0.06 \; m_\pi^{-1} < a_S^0 < 0.03 \; m_\pi^{-1} \; .$$

Evidently the experimental disagreement is large, but Weinberg's[14] current algebra prediction $a_S^0 = 0.16 \; m_\pi^{-1}$ sits beautifully in the middle.

Figure 6 compares our results for the P-wave phase shifts with dispersion relation calculations by Basdevant et al.[10]. Our 7 GeV data are in agreement with our 17.2 GeV data, but in disagreement with the prediction. It is not yet clear if our data are in contradiction with analyticity, or with some approximations made in the calculations.

HIGH MASS $\pi\pi$ PHASE SHIFTS $1 < M_{\pi\pi} < 1.8$ GeV

In the high mass region we encounter again the ambiguity problem, which can be formulated with the help of Barrelet-zeros[15]. For maximal spin $L_{max} = 3$ we can write the $\pi\pi$ cross-section (θ is the scattering angle)

$$\sigma_{\pi\pi} \propto (\cos \theta - z_1)(\cos \theta - z_2)(\cos \theta - z_3) \times$$
$$\times (\cos \theta - z_1^*)(\cos \theta - z_2^*)(\cos \theta - z_3^*) \; . \tag{3}$$

The experiment measures the cross-section, and if we want to determine the amplitudes, we can take z_ℓ or z_ℓ^* ($\ell = 1, L_{max}$); therefore we have $2^{L_{max}} = 8$ ambiguities for $L_{max} = 3$.

One half of these ambiguities can be eliminated, if the higher phase shift (δ_D or δ_F) is small in a region where some of the lower phase shifts are large.

We expect therefore two solutions in the mass region of the f meson and four solutions in the g-meson region.

In principle the $m \geq 1$ moments do not have these ambiguities; in practice however they change z_ℓ slightly, but do not help to resolve the ambiguities.

In addition to these discrete ambiguities the over-all phase is unknown. The only way to determine it is to demand Breit-Wigner shapes for the f and g mesons in their respective mass regions and to interpolate smoothly in between.

The method applied uses the $I = 2$ phase shifts obtained from our experiment $\pi^+ p \rightarrow \pi^+ \pi^+ n$ at 12.5 GeV [6]. Analogous to Ochs[4] we fit the t-channel amplitudes

$$g_0^\ell, \; g_-^\ell = g_1^\ell - g_{-1}^\ell, \; |g_+^\ell| = |g_-^\ell| \tag{4}$$

simultaneously to 5 $M_{\pi\pi}$ bins of 40 MeV width, 19 t bins for $|t| < 0.16$ GeV^2 and 13 moments $\langle Y_\ell^m \rangle$ ($\ell \leq 6$, $m \leq 1$). Phase shifts and elasticities are binned in 40 MeV bins for S- and P-waves as well as for D-waves at $M_{\pi\pi} > 1.6$ GeV. For $M_{\pi\pi} < 1.6$ GeV the D-wave was para-metrized as a Breit-Wigner with background (as in Ref. 4). The F-wave was assumed to be a simple Breit-Wigner at all masses.

The imaginary parts of the zeros Im z_ℓ resulting from these fits are shown in Fig. 7. The error bars give the range of values found for different fits. We classify the different solutions according to the sign of Im z_ℓ in the region $M_{\pi\pi} \sim 1.5$ GeV, assuming that Im z_1 changes sign around $M_{\pi\pi} = 1.22$ GeV. Solutions where Im z_1 does not change sign at this mass cannot be ruled out and can be obtained by connecting so-lutions with different Im z_1 and equal Im z_2, Im z_3 at $M_{\pi\pi} = 1.22$ GeV.

Below $M_{\pi\pi} = 1.4$ GeV two solutions are possible; at 1.4 GeV Im z_2 goes close to zero, stays negative or crosses over introducing a further ambiguity which then leads to four solutions. At $M_{\pi\pi} \approx 1.78$ GeV new ambiguities arise which cannot be studied with out data.

We show the various solutions in the following figures. The solu-tion (---) [i.e. all Im z_ℓ (1.5) negative] (Fig. 8) shows apart from the f and g mesons, a slowly rising S-wave going through 90° under the f meson and a rapid drop of the elasticity under the g meson which could not be described with a simple resonance form. The P-wave looks like the tail of the ρ meson.

The next solution (-+-) (Fig. 9) corresponds to Im z_2 starting nega-tive and crossing zero at $M_{\pi\pi} = 1.4$ GeV. This solution shows an in-elastic S-wave resonance under the g meson and a rise of the D-wave in the same region.

Solution (+--) (Fig. 10) has an Im z_1 starting negative at $M_{\pi\pi} = 1$ GeV and crossing zero at 1.22 GeV. In the region $M_{\pi\pi} \approx 1.1$ GeV it is identical to the solution found by Protopopescu et al.[7] two years ago. Around 1.4 GeV the S-wave becomes unphysical; therefore we think we can rule out this solution, which shows an inelastic P-wave resonance. It is however, possible to take this solution below 1.22 GeV and con-nect it with solution (---) above this $\pi\pi$ mass.

We have seen preliminary data of the reaction $\pi^- p \to \pi^0 \pi^0 n$ from the Karlsruhe-Pisa group[16], which probably will be able to resolve the ambiguity around 1.1 GeV.

At 1.4 GeV Im z_2 can also cross zero, which leads to solution ++- (Fig. 11). The S-wave elasticity is still larger than one at three points below 1.4 GeV, however in the fit it was possible to constrain the elasticity to values smaller than one leading to worse, but not un-acceptable, values of χ^2. We therefore feel that this solution is some-what less probable than the two other possibilities, but cannot be ruled out at present. This solution has a very rapidly-rising S-wave going through 90° and an inelastic broad P-wave resonance under the g meson.

All these ambiguities can, in principle, be resolved by studying the reaction $\pi^- p \to \pi^0 \pi^0 n$. Unfortunately the data available are not sufficient.

Looking at all solutions we see that Im $z_2(1.5) < 0$ has no S-wave resonance under the g meson, Im $z_2(1.5) > 0$ has one. Similarly Im $z_1(1.5) < 0$ has no P-wave resonance under the g meson, whereas Im $z_1(1.5) > 0$ has one (but with bad χ^2).

If we compare our present solutions with the phase shifts published one year ago[4], we see that these correspond almost exactly to our new solution (---).

As a consequence of the energy-dependent parametrization, the drop in the S-wave elasticity could not be fitted, leading to an Im z_1 crossing zero around 1.5 GeV.

In Figures 12, 13, and 14 we show projections of the fit to solution (---) with the measured moments in the interval $0.01 < |t| < 0.16$ GeV^2. We note the considerable improvement of the new fits compared to the old energy-dependent fits, particularly for $N, Y_4^0 \rangle$ as shown in Fig. 15.

OUTLOOK

In Fig. 16 we show the moments $N\langle Y_7^0 \rangle$ and $N\langle Y_8^0 \rangle$ obtained from brand-new and highly preliminary data of our experiment

$$\pi^- p \to K^+ K^- n$$

at 18.4 GeV. The curve represents the pattern expected for the interference of the g meson with a Breit-Wigner function having mass $M = 2000$ MeV and width $\Gamma = 250$ MeV. There can be no doubt that the spin-4 wave is very important at $M_{\pi\pi} > 1.9$ GeV; whether it really resonates will be the object of further studies.

Indications of high-lying resonances can also be seen in Fig. 17, which shows results from experiments[17-21] $\pi^- p \to \bar{p} p n$ and $\bar{p} p \to \pi^+ \pi^-$. Plotted are the positions found with Breit-Wigner fits to the data. The resonances seem to lie not on the leading trajectory and although each group finds different resonances, they all agree that there is a lot to be discovered in the future.

REFERENCES

1) G.F. Chew and F.E. Low, Phys. Rev. 113, 1640 (1959).

2) G.L. Kane and M. Ross, Phys. Rev. 177, 2353 (1969).
   P.K. Williams, Phys. Rev. D1, 1313 (1970).
   G.C. Fox, in Proc. Conf. on Phenomenology in Particle Physics,
   Calif. Inst. Tech., Pasadena, 1971 (Calif. Inst. Tech.,
   Pasadena, 1971), p. 703.

3) P. Estabrooks, A.D. Martin, G. Grayer, B. Hyams, C. Jones,
   P. Weilhammer, W. Blum, H. Dietl, W. Koch, E. Lorenz, G. Lütjens,
   W. Männer, J. Meissburger and U. Stierlin, Proc. Int. Conf. on
   ππ Scattering and Associated Topics, Florida State Univ.,
   Tallahassee, 1973 (AIP Conf. Proc. No. 13, Particles and Fields
   Subseries No. 5), p. 37.

4) W. Ochs, Thesis, Ludwig-Maximilians-Universität, Munich (1973).
   B. Hyams, C. Jones, P. Weilhammer, W. Blum, H. Dietl, G. Grayer,
   W. Koch, E. Lorenz, G. Lütjens, J. Meissburger, W. Ochs,
   U. Stierlin and F. Wagner, Proc. Int. Conf. on ππ Scattering
   and Associated Topics, Florida State Univ., Tallahassee, 1973
   (AIP Conf. Proc. No. 13, Particles and Fields Subseries No. 5),
   p. 206 and Nuclear Phys. B64, 134 (1973).

5) P. Estabrooks and A.D. Martin, Phys. Letters 41 B, 350 (1972).

6) G. Grayer, B. Hyams, C. Jones, P. Weilhammer, W. Blum, H. Dietl,
   W. Koch, E. Lorenz, G. Lütjens, W. Männer, J. Meissburger,
   U. Stierlin and W. Hoogland, Proc. Int. Conf. on ππ Scattering
   and Associated Topics, Florida State Univ., Tallahassee, 1973
   (AIP Conf. Proc. No. 13, Particles and Fields Subseries No. 5),
   p. 337.
   W. Hoogland, G. Grayer, B. Hyams, C. Jones, P. Weilhammer, W. Blum,
   H. Dietl, W. Koch, E. Lorenz, G. Lütjens, W. Männer,
   J. Meissburger and U. Stierlin, Nuclear Phys. B69, 266 (1974).

7) S.D. Protopopescu, M. Alston-Garnjost, A. Barbaro-Galtieri,
   S.M. Flatté, J.H. Friedman, T.A. Lasinski, G.R. Lynch, M.S. Rabin
   and F.T. Solmitz, Proc 3rd Int. Conf. on Experimental Meson
   Spectroscopy, Univ. Pennsylvania, Philadelphia, 1972 (AIP Conf.
   Proc. No. 8, Particles and Fields Subseries No. 3), p. 17.

8) G. Grayer, B. Hyams, C. Jones, P. Schlein, P. Weilhammer, W. Blum,
   H. Dietl, W. Koch, E. Lorenz, G. Lütjens, W. Männer,
   J. Meissburger and U. Stierlin, High statistics study of the
   reaction $\pi^- p \rightarrow \pi^+ \pi^- n$: Apparatus, method of analysis, and general
   features of results at 17 GeV/c, to be published in Nuclear
   Phys. B.

9) P. Estabrooks and A.D. Martin, University of Durham Preprint,
   April 1974.

10) J.L. Basdevant, C.D. Froggatt and J.L. Petersen, Proc. 2nd
     Int. Conf. on Elementary Particles, Aix-en-Provence, 1973
     (J. Phys. $\underline{34}$ Colloque C-1, Suppl. to No. 10), p. 220 and
     Construction of phenomenological $\pi\pi$ amplitudes, Preprint,
     Laboratoire de Physique Théorique et Hautes Energies, Paris,
     September 1973.

11) E.P. Tryon, $\pi\pi$ parameters from K_{e_4} data and a rigorous represent-
     ation for $\pi\pi$ amplitudes, submitted to Phys. Rev. D (Feb. 1974).

12) S.A. Bunyatov, Investigation of the $\pi^+ p \rightarrow \pi^{\pm}\pi^+ n$ reactions near
     threshold, paper submitted to this conference.

13) J.A. Jones, W.M. Allison and D.H. Saxon, Measurement of the
     S-wave I = 0 $\pi\pi$ scattering length, paper submitted to this
     conference.

14) S. Weinberg, Phys. Rev. Letters $\underline{17}$, 616 (1966).

15) E. Barrelet, Nuovo Cimento $\underline{8}$ A, 331 (1972).

16) H. Müller, private communication.

17) N. Barash-Schmidt, A. Barbaro-Galtieri, C. Bricman, V. Chaloupka,
     D.M. Chew, R.L. Kelly, T.A. Lasinski, A. Rittenberg, M. Roos,
     A.H. Rosenfeld, P. Söding, T.G. Trippe and F. Uchiyama, Phys.
     Letters $\underline{50}$ B, 1 (1974).

18) H. Nicholson, B.C. Barish, J. Pine, A.V. Tollestrup, J.K. Yoh,
     C. Delorme, F. Lobkowicz, A.C. Melissinos, Y. Nagashima,
     A.S. Carrole and R.H. Phillips, Phys. Rev. Letters $\underline{23}$, 603
     (1969) and CALT-68-332 (1972).

19) A. Donnachie and P.R. Thomas, Daresbury Report DNPL/P149 (1973),
     and Nuovo Cimento Letters $\underline{7}$, 285 (1973).

20) B. Hyams, C. Jones, P. Weilhammer, W. Blum, H. Dietl, G. Grayer,
     W. Koch, E. Lorenz, G. Lütjens, W. Männer, J. Meissburger and
     U. Stierlin, Nuclear Phys. $\underline{B73}$, 202 (1974).

21) E. Eisenhandler, W.R. Gibson, C. Hojvat, P.I.P. Kalmus,
     L.C.Y. Lee Chi Kwong, T.W. Pritchard, E.C. Usher, D.T. Williams,
     M. Harrison, W.H. Range, M.A.R. Kemp, A.D. Rush, J.N. Woulds,
     G.T.S. Arnison, A. Astbury, D.P. Jones and A.S.L. Parsons,
     Phys. Letters $\underline{47}$ B, 536 (1973).

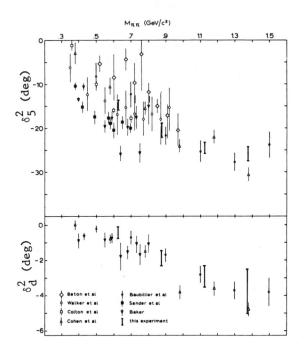

Fig. 1.   I = 2 S- and D-wave phase shifts.

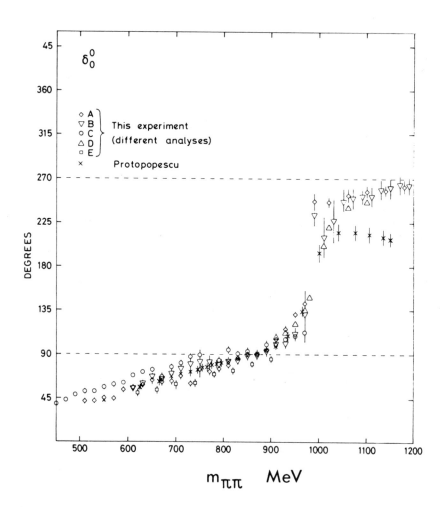

Fig. 2. I = 0 S-wave phase shift.

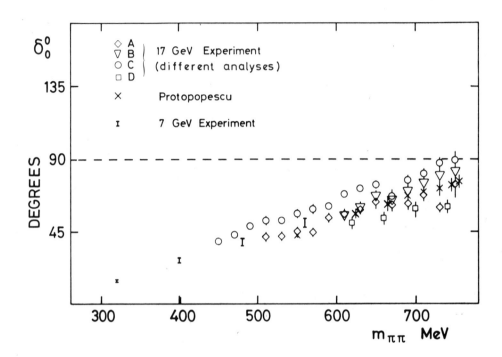

Fig. 3. I = 0 S-wave phase shifts at low mass.

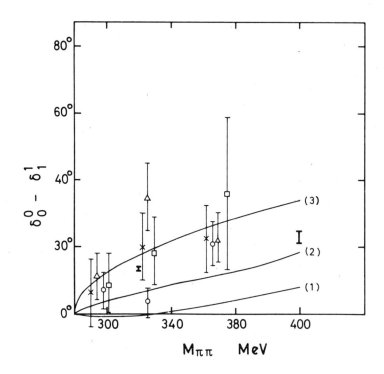

Fig. 4. K_{e4} results for $\delta_0^0 - \delta_1^1$ from Ref. 10
I values from our 7 GeV experiment.

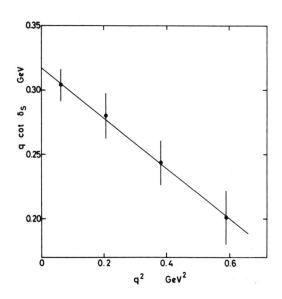

Fig. 5. $q \cot \delta_S^0$ versus q^2
for our 7 GeV data. The
error bars indicate the
range of values obtained
from various fits.

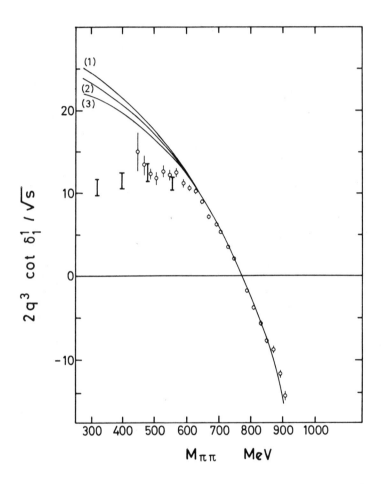

Fig. 6. $2q^3 \cot \delta_1^1/\sqrt{S}$ in units of m_π, curves
from Basdevant (Ref. 10).
ϕ: 17.2 GeV data.
I: 7 GeV data.

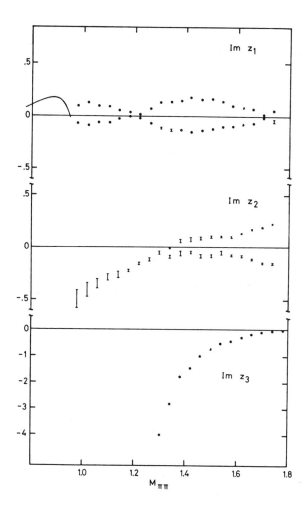

Fig. 7.   Imaginary parts of Barrelet zeros·
as obtained from fits. Error bars indicate
range of values for different solutions.

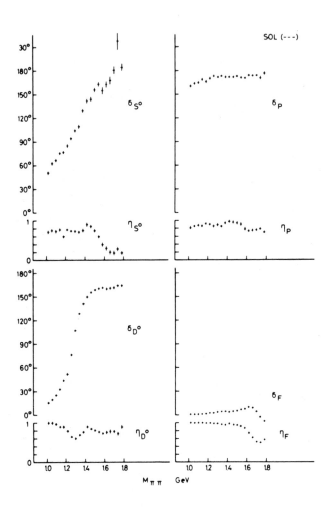

Fig. 8. Solution (---).

37

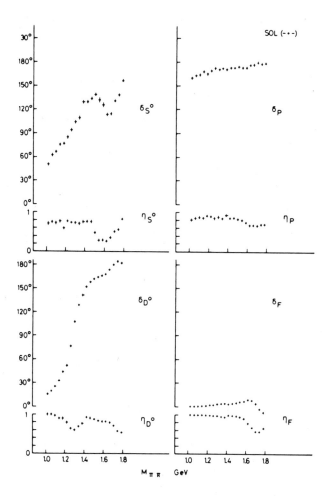

Fig. 9. Solution (-+-).

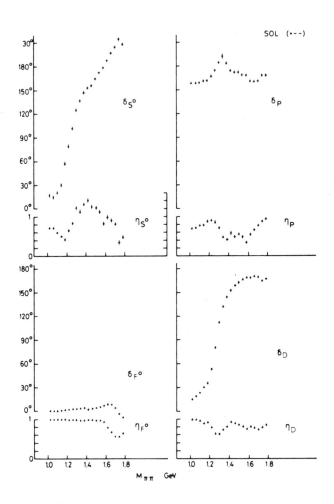

Fig. 10. Solution (+--).

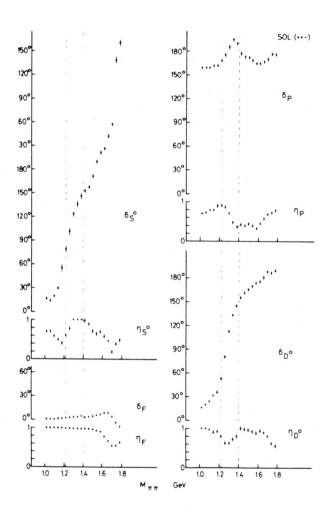

Fig. 11. Solution (++−).

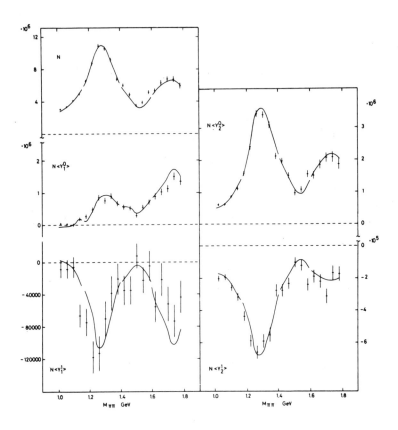

Fig. 12. Unnormalized moments in $0.01 < |t| < 0.16$ with projection of fit for solution (---).

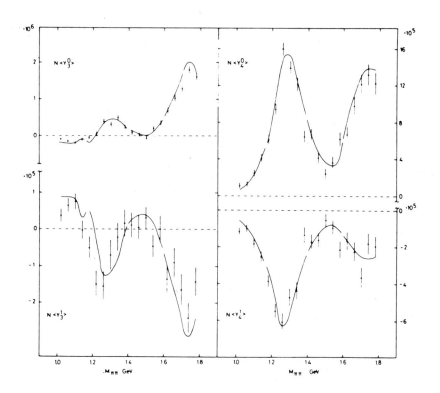

Fig. 13. Unnormalized moments in 0.01 < |t| < 0.16 with projection of fit for solution (---).

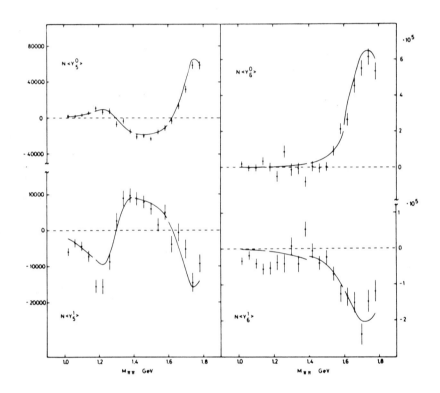

Fig. 14. Unnormalized moments in 0.01 < |t| < 0.16
with projection of fit for solution (---).

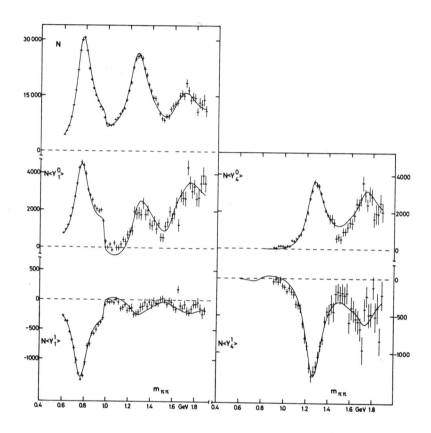

Fig. 15. Unnormalized moments with fitted curve from Ref. 4.

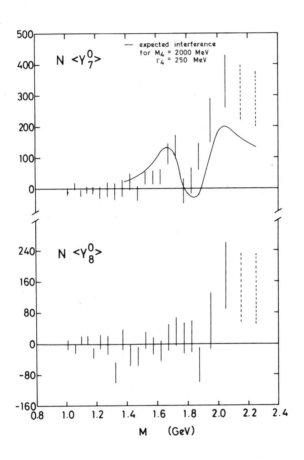

Fig. 16. Unnormalized moments $N\langle Y_7^0\rangle$ and $N\langle Y_8^0\rangle$ from $\pi^- p \to K^+ K^- n$ at 18.4 GeV.

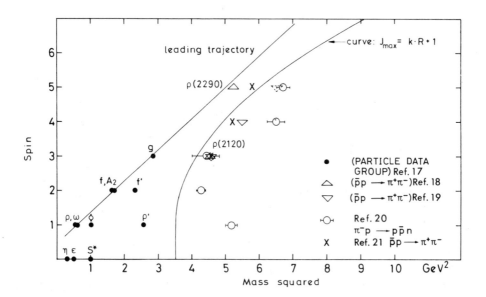

Fig. 17. Possible resonances in $p\bar{p} \leftrightarrow \pi^{+}\pi^{-}$.

AMPLITUDE ANALYSIS OF $(3\pi)^+$ PRODUCTION AT 7 GeV/c*

M. Tabak, E. E. Ronat,[†] A. H. Rosenfeld,
and T. A. Lasinski
Lawrence Berkeley Laboratory
University of California, Berkeley, CA 94720

R. J. Cashmore[‡]
Stanford Linear Accelerator Center
P. O. Box 4349, Stanford, CA 94305

ABSTRACT

An amplitude analysis of the reaction $\pi^+ p \to p(\pi^- \pi^+ \pi^+)$ at 7 GeV/c is in progress. Although our method of analysis and assumptions are quite different from those of the Illinois group, our preliminary results are nevertheless consistent with theirs. We fit coherent amplitudes, not the density matrix, and make different assumptions about decay amplitudes of J^P states into various isobars. This approach, furthermore, allows us to probe a much larger set of partial waves than that quoted by the Illinois group, at the expense of imposing explicit rank conditions on the density matrix.

I. INTRODUCTION AND FORMALISM

We are pursuing a 3π partial wave analysis of the LBL Group A 7-GeV/c $\pi^+ p \to \pi^+ \pi^- \pi^+ p$ data.[1] These data have not heretofore been subjected to such analyses. There are two additional new features of the present analysis which should be noted:

1) our formalism and programs have been developed independently of the Illinois effort,[2] and

2) our fitting parameters are associated with amplitudes instead of density matrix elements.

Our formalism[3] and programs have been used in an extensive analysis[4] of the reactions $\pi N \to \pi\pi N$ in the s-channel resonance region. In spirit, our approach is quite similar to that of the Illinois group.[2] We think of a 3π system with spin J^P and projection M as decaying into an isobar (ϵ, ρ, f) of spin ℓ and a pion with relative orbital angular momentum L. The probability for a given event may be written as

*Work supported in part by the U. S. Atomic Energy Commission.
†Permanent address: Weizmann Inst. of Science, Rehovot, Israel.
‡Now at Oxford University, Oxford, England.

$$P(\text{event}) = \sum_{\substack{\eta=\pm 1}} \left\{ \left| \sum_{\substack{JM \\ L\ell}} A_{++}^{JM\eta L\ell}(M_{3\pi}, t) T_{L\ell}(M_{3\pi}, M_{2\pi}) G^{JM\eta L\ell}(\alpha\beta\gamma\theta_h) \right|^2 \right.$$

$$\left. + \left| \sum_{\substack{JM \\ L\ell}} A_{+-}^{JM\eta L\ell}(M_{3\pi}, t) T_{L\ell}(M_{3\pi}, M_{2\pi}) G^{JM\eta L\ell}(\alpha\beta\gamma\theta_h) \right|^2 \right\}. \qquad (1)$$

The functions $G^{JM\eta L\ell}$ are essentially the real ($\eta = +1$) and imaginary ($\eta = -1$) parts of $G^{JML\ell}$ given in Ref. 3. The angles $\alpha\beta\gamma$ define an Euler rotation from the production coordinate system (s- or t-channel) to a system with z-axis in the plane of the three pions; θ_h is the helicity decay angle of the isobar. The functions $T_{L\ell}$ are of the form

$$T_{L\ell}(M_{3\pi}, m_{2\pi}) \propto Q^L \frac{e^{i\delta_\ell} \sin\delta_\ell}{q^{\ell+1}}, \qquad (2)$$

where δ_ℓ is either approximated by Breit-Wigner behavior or is specified by the actual $\pi\pi$ phase shifts. The discrete quantum number η, which for $M = 0$ is given by

$$\eta = (-1)^{J+L+\ell}, \qquad (3)$$

may be associated with natural and unnatural parity exchange.[5] Our fitting parameters are the complex "amplitudes"[6] $A^{JM\eta L\ell}$ not density matrix elements. For a given η, we see that for N partial waves there are 4N real parameters to be determined; the corresponding number for a density matrix analysis is N^2. For a given number of fitting parameters, we are thus able to investigate the importance of many more waves than in a density matrix approach. A difficulty associated with an amplitude analysis is the uniqueness of solutions; that is, there are generally several solutions with comparable likelihoods. In view of these remarks, the primary emphasis of the present analysis is to determine which waves are the important ones and to investigate how serious is the ambiguity in solutions.

II. ASSUMPTIONS AND FITTING PROCEDURES

In addition to the isobar assumption implicit in Eq. (1), we have also assumed
1) No $\eta = -1$ waves. Preliminary studies of our data indicate that this approximation is good to about 10%. A notable exception is the A_2 for which $\eta = -1$ contributes about 1/8 to the $J^P = 2^+$ cross section for $M_{3\pi} \approx 1300$ MeV.

2) $A^{JML\ell}_{+-} = 0$. Strictly speaking the assumption[6] here is that of spin-coherence,

$$A^{JML\ell}_{+-} = a\, A^{JML\ell}_{++}, \qquad (4)$$

where a is some complex number.

3) Δ^{++} cut, $1160 \leq M_{\pi^+p} \leq 1280$ MeV. The principal effect of this cut is to reduce statistics in the 1600-MeV 3π mass region. No systematic study of the effect of such a cut on our fits has yet been made.

The fitting parameters $A^{JML\ell}$, indexed by $M_{3\pi}$ and t, are determined by likelihood fits to the data. The data were binned into low and high $|t|$ intervals,

$$|t| \leq 0.1 \text{ GeV}^2 ; \quad 0.1 \leq |t| \leq 0.6 \text{ GeV}^2.$$

The $M_{3\pi}$ binning consisted of nine 100-MeV intervals from 950 to 1750 MeV. For low $|t|$ there were typically 2000 to 700 events per mass bin; for high $|t|$, 2000 to 1500 events per mass bin.

Table I. The waves considered in the present analysis. All waves are $\eta = +1$, $M = 0$ unless otherwise indicated. Double underlined waves were generally "strong"; single underlined waves were considerably "weaker" but definitely present.

	$\rho\pi$	$f\pi$	$\epsilon_{LO}\,\pi$	$\epsilon_{HI}\,\pi$
	0^-P	0^-D	0^-S	0^-S
(A_1)	1^+S	1^+P	1^+P	1^+P
	1^+S, M=1		1^+P, M=1	1^+P, M=1
	1^-P, M=1	1^-D, M=1		
	1^+D			
	2^-P	$(A_3)2^-$S	2^-D	2^-D
		2^-S, M=1		
(A_2)	2^+D, M=1	2^+P, M=1		
	3^+D	3^+P	3^+F	3^+F
	3^-F, M=1	3^-D, M=1		

The waves included in the present fit are indicated in Table I. The reference wave was taken to be $0^- S \epsilon_{LO} \pi$; that is, in each mass bin we set

$$\text{Im } A_{++}^{00S \, \epsilon_{LO}} = 0. \tag{5}$$

Consequently, all phases shown here are measured relative to this wave. The starting parameters for these waves were found through the following procedure. First, we randomly generated in each bin 500 sets of the parameters corresponding to the waves of Table I. We next considered only those 20 sets which had the highest likelihood. Each set was optimized by our fitting program. The net results of this procedure were some 4-17 potential solutions per bin. Depending on the mass bin,[7] these fits involved 43 or 53 parameters.

III. RESULTS FOR LOW MASS (< 1500 MeV)

In Fig. 1 we show the $2^+ \rho\pi$ wave for the high $|t|$ interval. Mass, phase, and Argand plots are presented for those solutions which differ by less than 10 points from the highest likelihood solution in each mass bin. The mass and phase plots indicate a rather clean Breit-Wigner-like behavior with mass ~ 1300 MeV and width ~ 150 MeV. Note that in the Argand plot (the radius here corresponds to (events)$^{1/2}$) the most rapid motion is between the 1250 and 1350 MeV bins (D and E). The fits for the high $|t|$ data in 50-MeV bins are shown in Fig. 2. The highest likelihood 100-MeV solutions are also indicated with open circles. A similar behavior for the A_2 is also observed at low $|t|$, but only contributes ~60 events out of ~ 2000 in the 1250-MeV region.

We show our results for the $A_1(\rho\pi)$ in 50- and 100- MeV bins at low and high $|t|$ in Fig. 3. This peak is rather broad, 200-300 MeV, and its position shifts by some 100 MeV between low and high $|t|$. In both cases there is little phase motion, though the high $|t|$ data indicate some small but definite behavior above ~1200 MeV. A better feeling for the significance of this motion is obtained from Fig. 4, where we show the high $|t|$ $A_1(\rho\pi)$ Argand plot. Thus points A through F (925 to 1175 MeV) lie along one radius vector, whereas the higher mass points (G to K) fall off that vector.

For $|t| \leq 0.1$ GeV^2 and $1.0 \leq M_{3\pi} \leq 1.1$ GeV, the linear combinations of $A_1(\rho\pi)$ t-channel density matrix elements corresponding to states of definite η are[5]

$$\rho_{00} = 0.965 \pm 0.086$$

$$\sqrt{2} \text{ Re } \rho_{01} = -0.183 \pm 0.054$$

$$\sqrt{2} \text{ Im } \rho_{01} = 0.005 \pm 0.041$$

$$\rho_{11} - \rho_{1-1} = 0.035 \pm 0.011$$

$$\rho_{11} + \rho_{1-1} = 0 \text{ (Input)}$$

These numbers correspond to the $1^+\rho\pi$, M=1 wave being present to ~ 3.5σ in our low $|t|$ fits. They are quite similar to those of Ref. 8, keeping in mind the somewhat different $|t|$ and mass intervals.

IV. RESULTS FOR HIGH MASS (> 1500 MeV)

As noted earlier, we have not yet systematically studied the effects of our Δ^{++} cut. In addition, as seen in Fig. 5, there is a definite tendency for our reference wave, $0^-\epsilon_{LO}\pi$, to decrease in the high mass region (1550-1750 MeV). For these reasons we are not prepared as yet to make definite conclusions about phases in this region. With these qualifications in mind, we consider next the principal high-mass waves in our analysis.

A composite of these waves is shown in Fig. 6 for the high $|t|$ region. Notice that the mass region 1550 to 1750 MeV, commonly associated with the A_3, is decomposed in our analysis into six different partial waves: $2^-Sf\pi$, $2^-P\rho\pi$, $2^+Pf\pi$ (M =1), $2^-D\epsilon\pi$, $3^+D\rho\pi$, and $3^+F\epsilon\pi$. The 3^+ waves are 9 to 10σ effects in our data; the importance of 3^+ has also been observed by a European collaboration.[9] The $2^+Pf\pi$ is also seen by the CERN-Soviet group,[8] though at a somewhat higher mass and with more intensity.

Our more important waves at low $|t|$ are shown in Fig. 7. Note that the two decay modes for the $A_1(\rho\pi$ and $\epsilon\pi)$ are roughly $90°$ out of phase. A similar feature is present in the high $|t|$ results. While the 3^+ waves are less striking, it should be noted that "Chew-Low" boundary effects are more severe for the low $|t|$ data.

V. CONCLUSIONS AND PROSPECTUS

The results of the present study may be summarized as follows:

1) The principal waves at low mass (≤ 1500 MeV) are $0^-\epsilon\pi$, $1^+\rho\pi$, $1^+\epsilon\pi$, and $2^+\rho\pi$.

2) The $2^+\rho\pi$ wave is quite consistent with Breit-Wigner resonance behavior. There is no evidence in the $1^+\rho\pi, \epsilon\pi$ phases for such an interpretation.

3) Although there are multiple solutions at low mass, they are quantitatively consistent within errors. In addition they are considerably fewer in number than those at high mass.

4) In the A_3 region (1600-1800 MeV), there are at least five important waves present: $2^-f\pi$, $2^-\rho\pi$, $2^-\epsilon\pi$, $3^+\rho\pi$, $3^+\epsilon\pi$. Here there are more solutions and they are farther apart than at low mass.

The consistency of our results with those of the Illinois group[2] indicate the viability of an amplitude approach. However, to fully develop this approach, we are pursuing the following projects:

1) Modification of the analysis to measure the eigenvectors of the density matrix. The assumption of spin coherence, while justified by other analyses, clearly deserves independent verification.

2) Development of statistical tests for comparing competing solutions in a given set of partial waves and for selection of the

minimal set of partial waves required by the data. The present criterion, a difference of 10 points in likelihood, is <u>ad hoc</u>.

In the high mass region certain questions independent of our model and peculiar to our data require investigation:

1) How sensitive are the fits to different Δ^{++} cuts?

2) What other choices can be made for reference waves? The sharp behavior of the $0^-\epsilon\pi$ wave in this region (for high $|t|$) indicates that its use as a reference wave may be unwarranted.

To study the production mechanisms of 3π states, we shall include unnatural parity exchange states in the fits and study more closely the t-dependence of the amplitudes.

REFERENCES

1. M. Alston-Garnjost et al., Phys. Letters <u>33B</u>, 607 (1970); S. D. Protopopescu et al., Phys. Rev. D <u>7</u>, 1279 (1973).
2. G. Ascoli et al., Phys. Rev. D <u>7</u>, 669(1973) and refs. therein. P. V. Brockway, Thesis, U. of Illinois, 1970 (unpublished). J. D. Hansen et al., CERN/D.Ph.II/PHYS 73-34, submitted to Nucl. Phys. B.
3. D. J. Herndon et al., LBL-543 (submitted to Phys. Rev.), 1972.
4. D. J. Herndon et al., LBL-1065 (Rev.), 1974.
5. K. Gottfried and J. D. Jackson, Nuovo Cimento <u>33</u>, 309 (1964); G. Cohen-Tannondji et al., Nuovo Cimento <u>55A</u>, 412 (1968); J. P. Ader et al., Nuovo Cimento <u>56A</u>, 952 (1968).
6. As discussed by Hansen et al. (Ref. 2), we may not in principle uniquely associate A_{++} and A_{+-} with helicity nonflip and flip amplitudes without additional information on the polarization of the protons. This does not, however, invalidate our approach provided additional constraints are imposed on A_{++} and A_{+-} so as to distinguish their contributions in Eq. (1) from the point of view of the fitting program.
7. The waves labeled $\epsilon_{HI}\pi$ were present only in mass bins with $M_{3\pi} > 1200$ MeV. Our ϵ was basically that of the CERN-Munich group, B. Hyams et al., p. 206, $\pi\pi$ Scattering-1973, ed. P. K. Williams and V. Hagopian, AIP, New York, 1973. The distinction between ϵ_{HI} and ϵ_{LO} was as follows:

$$\epsilon_{LO} = \begin{cases} \epsilon_{CM}, & M_{2\pi} \leqslant 2m_K \\ 0, & M_{2\pi} > 2m_K \end{cases} \qquad \epsilon_{HI} = \begin{cases} 0, & M_{2\pi} < 2m_K \\ \epsilon_{CM}, & M_{2\pi} \geqslant 2m_K \end{cases}$$

This division was made to take account of the fact that δ_0^0 passes twice through 90°. For $M_{3\pi} \lesssim 1500$ MeV, the ϵ_{HI} waves contribute no more than ~ 20 events for all solutions. Above this mass these waves are only somewhat larger (~ 30-40 events). There is one solution in the 1650 mass bin which has ~ 100 events in the $2^- \epsilon_{HI}\pi$ wave; we would reject this solution on the basis of continuity in the mass plot.
8. Yu. M. Antipov et al., Nucl. Phys. <u>B63</u>, 141 (1973); <u>B63</u>, 153 (1973).
9. Aachen-Berlin-Bonn-CERN-Heidelberg Coll., CERN/D.Ph.II/ PHYS 74.

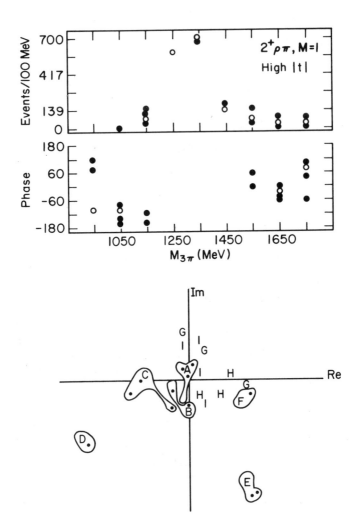

Fig. 1. The A₂ mass, phase and Argand plots for the high |t| interval. Highest likelihood solutions are marked with an open circle in mass and phase plots. In Argand plot they are marked with letters (A = 950 MeV, ···, F = 1450 MeV).

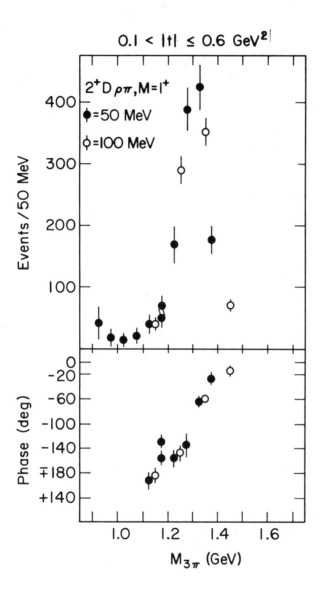

Fig. 2. The A_2 mass and phase plots in 50- and 100-MeV bins for high $|t|$.

54

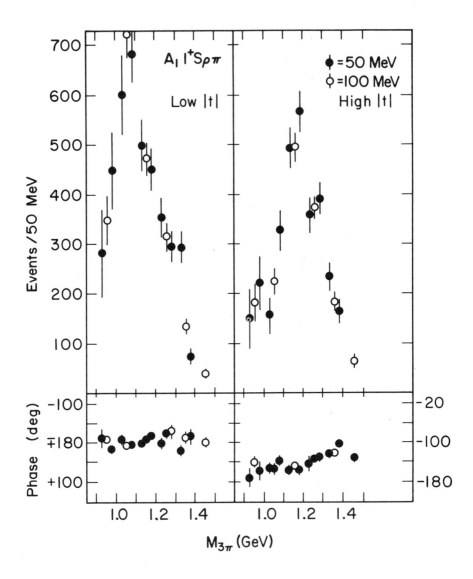

Fig. 3. $1^{+}\rho\pi$ (A$_1$) mass and phases plots in 50- and 100-MeV bins for both low and high $|t|$.

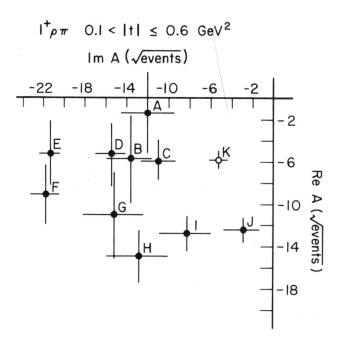

Fig. 4. The A₁ Argand plot for high |t|. A = 925, B = 975,
..., I = 1325, J =1375, K = 1450 MeV.

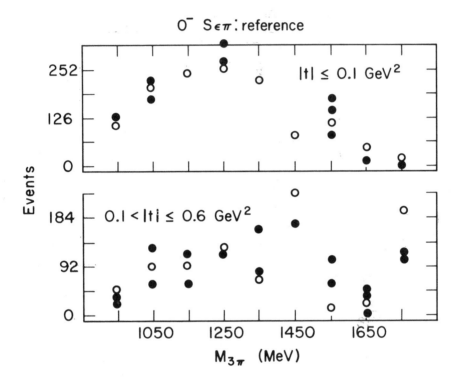

Fig. 5. Mass plots for the reference wave, $0^- \epsilon \pi$, for both low and high $|t|$. The highest likelihood solution in each mass bin is marked by an open circle.

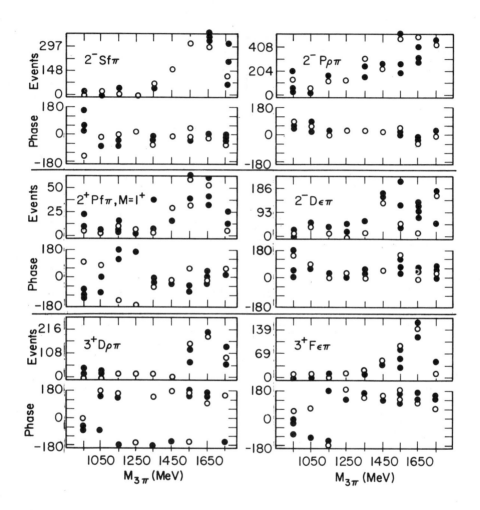

Fig. 6. Additional significant waves for the high |t| interval. An open circle indicates the highest likelihood solution.

58

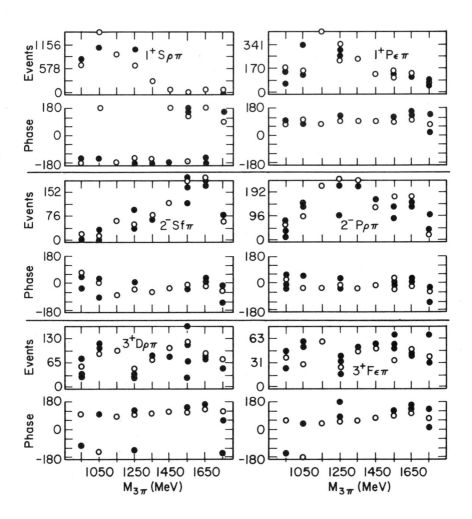

Fig. 7. Important waves for the low |t| interval. An open circle indicates the highest likelihood solution.

ANALYSIS OF 3-π SYSTEM AND UNITARITY*

G. Ascoli and H. W. Wyld
University of Illinois, Urbana, Ill. 61801

ABSTRACT

We present a re-analysis of the 3-π states produced in $\pi^- p \to \pi^+ \pi^- \pi^- p$, using unitarized amplitudes.

During the past three years a considerable effort has been made to perform a complete angular momentum analysis of the 3π system produced in the reaction $\pi^- p \to \pi^+ \pi^- \pi^- p$. Both bubble chamber data,[1] and more recently, a counter experiment at Serpukhov,[2] have been analyzed. One of the results of this analysis was that for both the $A_1 (1^+ S \to \rho\pi)$ and the $A_3 (2^- S \to f\pi)$ bumps the variation with $M_{3\pi}$ of the phase (relative to other - presumably non-resonant states) did not show the behavior expected for a resonance. The results from the Serpukhov experiment[3] are particularly convincing, because of the good statistics.

Some criticism[4] has been voiced regarding these results, particularly on the grounds that in the analysis no account was taken of 2- or 3-particle unitarity. To answer this criticism we have undertaken to repeat the analysis taking unitarity explicitly into account.

In this talk no attempt will be made to go into the details of the calculation. I will try to give a very quick outline of the method and show a comparison of fits using the original method and the new improved (?) method.

The general approach which had been used[5] was to describe the decay distribution (angles + Dalitz variables) of the 3π system by

$$\sum_{JPM} \sum_{J'P'M'} \mathcal{M}_{J'P'M'} \, \rho_{J'P'M',JPM} \, \mathcal{M}^*_{JPM} , \qquad (1)$$

where ρ is a density matrix describing the production of states with quantum numbers $J^P M$ while $\mathcal{M}_{JPM}$ describes the decay of the state $J^P M$ into the final 3π system. In writing the decay amplitudes $\mathcal{M}_{JPM}$ we assume that they can be described by sequential decays corresponding to diagrams of the type:

---

* Research supported in part by the U. S. Atomic Energy Commission and by the National Science Foundation.

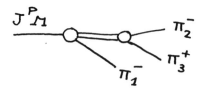

with the double line standing for the propagator of a dipion system of low spin. This means that we write

$$\mathcal{M}_{JPM} = \sum_{LS} C_{LS}^{J} (M_{3\pi}) \, \mathcal{M}_{LS}^{JPM} . \tag{2}$$

The decay amplitude $\mathcal{M}_{LS}^{JPM}$ for a given partial wave (or decay mode) is written down by inspection from the above diagram plus the diagram obtained by interchanging the two identical π's, π_1^- and π_2^-

$$\mathcal{M}_{LS}^{JPM} = \mathcal{R}_{LS}(s_1, M_{3\pi}) \, Z_{LS}^{JM}(\Omega_1 \bar{\Omega}_1) + (1 \leftrightarrow 2) , \tag{3}$$

where

$$Z_{LS}^{JM}(\Omega_1 \bar{\Omega}_1) = \sum_{m\mu} <JM|LSm\mu> \, Y_L^m(\Omega_1) \, Y_S^\mu(\bar{\Omega}_1)$$

gives the angular dependence appropriate to the decay of a state $J^P M$ into a dipion ($M_{2\pi}^2 = s_1$, spin S) and a pion with relative orbital angular momentum L, followed by the decay of the spin S dipion into π_2 and π_3. $\mathcal{R}_{LS}$ contains the vertex factors and the dipion propagator. One notes that the two pieces (corresponding to dipion = $\pi_2^- + \pi_3^+$ and the dipion = $\pi_1^- + \pi_3^+$) do satisfy 2-body unitarity (Watson's theorem) but the symmetrized amplitude does not. Presumably Nature satisfies unitarity by allowing rescattering between π's to occur.

So far I have described our standard analysis. Our alternate scheme, which does satisfy unitarity is briefly as follows:

We begin by looking at the structure of the 3π- 3π scattering amplitude, T. Unitarity implies that it can be written as

$$T = K + iK\rho T \tag{4}$$

where K is a real symmetric matrix. We know very little about K except for this property. We now make the crucial assumption that we can write K as

$$K = k_1 + k_2 + k_3 , \tag{5}$$

where k_1 (k_2, k_3) are π-π scattering amplitudes with π_1 ($\pi_2 \pi_3$) as a

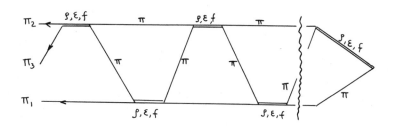

$k_1 =$ [diagram with lines 1, 3, 2] $k_2 =$ [diagram with lines 1, 3, 2]

$k_3 =$ [diagram with lines 1, 2, 3]

Fig. 1. Graphs for K-matrix

[diagram: Graphical representation with labels π_2, π_3, π_1 and markings $\mathcal{S}, \mathcal{E}, f$ and π]

Fig. 2. Graphical representation of the $\mathcal{S}_1$ term.

spectator (see Fig. 1). After some manipulations we arrive at the
result that one should replace $\mathcal{R}$ by $\mathcal{S}$ with $\mathcal{S}$ determined by the
integral equation

$$\mathcal{S}_1 = \mathcal{R}_1 + i\, t_1\, \rho\, A_{12}\, \mathcal{S}_2 \, . \qquad (6)$$

Here $\mathcal{S}_1$ ($\mathcal{S}_2$) corresponds to a diagram in which the last rescattering
occurs between π_2^- and π_3^+ (or π_1^- and π_3^+), t_1 is the $\pi_2^-\pi_3^+$ scattering
amplitude ($e^{i\delta_{\pi\pi}}\sin\delta_{\pi\pi}$ after an appropriate decomposition into
partial waves) and ρ is a phase space factor. A_{12} is a recoupling
coefficient (in both momentum and isospin space). Eqn. 6 is
actually a set of integral equations for the decay amplitudes corresponding to a particular value of J^P (one amplitude for each L/S
value we choose to include). The solution to the integral equation
corresponds to the (infinite set of) diagrams illustrated in Fig. 2.
There is, of course, some arbitrariness in selecting the inhomogenous
term $\mathcal{R}$. After solving the set of integral equations, we use - in

our "unitarized" fits -

$$\mathcal{M}_{LS}^{JPM} = \sum_{L'S'} \mathcal{S}_{L'S'}^{JPLS} (s_1) \ Z_{L'S'}^{JM} (\Omega_1 \bar{\Omega}_1) + (1 \leftrightarrow 2) \qquad (3U)$$

in place of (3). The upper set of indices (LS) refers to the quantum numbers of the inhomogeneous term (see Eqn. 6). We have arbitrarily chosen the inhomogeneous term $\mathcal{R}$ in Eqn. (6) to have the same form as the $\mathcal{R}$ used in Eqn. 3 in our non-unitarized fits.

Before proceeding to a comparison with the data, we illustrate the results obtained by solving the integral equation. Fig. 3 shows the results obtained for the state $J^P = 1^+$ at $M_{3\pi} = 1.1$ GeV. We kept 5 terms: S/L = 0/1 ($1^+P \rightarrow \epsilon\pi$), S/L = 1/0 ($1^+S \rightarrow \rho\pi$), S/L = 1/2 ($1^+D \rightarrow \rho\pi$), S/L = 2/1 ($1^+P \rightarrow f\pi$) and S/L = 2/3 ($1^+F \rightarrow f\pi$). At $M_{3\pi}$=1.1 GeV only the first two are important. The first row in the figure corresponds to an initial (before rescattering) term $1^+P \rightarrow \epsilon\pi$, which feeds (after rescattering) not only the 1^+P term but also the $1^+S\rightarrow\rho\pi$. The second row corresponds to the initial term $1^+S \rightarrow \rho\pi$. The real and imaginary parts of $\mathcal{S}$ are shown as a function of $M_{\pi\pi}$ (actually $M\pi\pi - 2$ $M\pi$).For the diagonal terms we show also the corresponding non-unitarized amplitude, $\mathcal{R}$. We note that the corrections are rather large, in particular there is rather large effect which corresponds to the $1^+(S \rightarrow \rho\pi)$ wave feeding the $1^+(P \rightarrow \epsilon\pi)$ wave.

We would like to emphasize as clearly as possible one point: our unitarized amplitudes do indeed satisfy unitarity. One should not be confused into thinking that they are ipso facto the correct amplitudes. Our assumption (Eqn. 5) could be drastically wrong. To drive home this point Fig. 4 shows the ratio between the integral (over phase space) of the square of the unitarized 2^+D amplitude to the corresponding integral for the non-unitarized 2^+D amplitude. There is no sign of any A_2 effect, as we would expect to find had we used the correct 3π-3π scattering amplitude.

Y. Goradia, T. Lasinski, H. Tabak, and G. Smadja[6] independently have carried out a calculation similar to ours. Although the starting point and the details are seemingly quite different, they appear to arrive to the same integral equation. We have had no opportunity- so far- of comparing their numerical results with ours. They had - at the time of the Conference - made no comparison of their model with 3π data.

Up to this time we have made only a reasonably careful re-analysis of the data up to $M_{3\pi} = 1.4$ GeV, using the combined 11-25GeV $\bar{\pi}p$ data described in Ref. 1. We will probably delay publication until a reanalysis of the larger data sample from the Serpukhov experiment (Ref. 2) is carried through.

Although other, more complicated hypotheses were also tried, we present here only a comparison of the results obtained using a simple hypothesis in which the following partial waves are included:

63

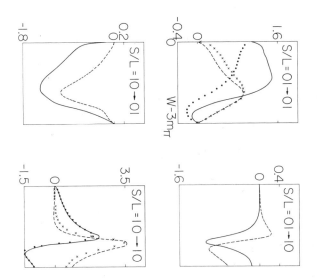

Fig. 3.   Unitarized, S, and non-unitarized, R, amplitudes for $J^P = 1^+$, $M_{3\pi} = 1.1$ GeV.

——— Re S ----- Im S
····· Re R xxxxx Im R

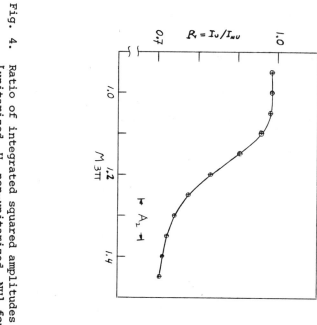

Fig. 4.   Ratio of integrated squared amplitudes [unitarized, U, non-unitarized, NU] for A_2.

$$0^- S \to \epsilon\pi, \quad 0^- P \to \rho\pi$$
$$1^+ P \to \epsilon\pi, \quad 1^+ S \to \rho\pi, \quad J_z = 0 \text{ only}$$
$$2^- P \to \rho\pi \qquad\qquad , \quad J_z = 0 \text{ only}$$
$$2^+ D \to \rho\pi \qquad\qquad , \quad J_z = \pm 1 \text{ only*}$$

At the risk of being repetitive, we note that the above is a list of input states; for the case of unitarized amplitudes, all partial waves listed in Table 1 (for the states 0^-, 1^+, 2^-, 2^+) are included in the final state.

Table 1

S and L Values Included in
Calculation of Unitarized Amplitudes

		$S = 0 (\epsilon)$	$S = 1 (\rho)$	$S = 2 (f)$
$J^P = 0^-$	$L =$	0	1	2
$J^P = 1^+$	$L =$	1	0,2	1,3
$J^P = 2^-$	$L =$	2	1,3	0,2,4
$J^P = 2^+$	$L =$	–	2	1,3

We summarize briefly the results:

(a) When the unitarized amplitudes are used to fit the data, the fits are significantly worse. This is illustrated in Fig. 5. We have verified (by generating Monte-Carlo events corresponding to the unitarized and non-unitarized fits--in the region $M_{3\pi} = 1.2$-1.4 GeV) that this is a real effect. One can account for the decrease in likelihood by comparing Dalitz plot distributions and moments of angular distributions for the Monte Carlo events and for the data. We remark (but omit details) that the inferiority of the U (unitarized) fits relative to the NU (non-unitarized) fits persists when more complicated hypotheses (more partial waves) are used.

(b) The amount of production of various J^P states is essentially un-affected. The same total amounts of $J^P = 0^-$, 1^+, 2^-, 2^+ are obtained in both U and NU fits. (See Figs. 6-8).

(c) For the states $J^P = 0^-$ (Fig. 6) and $J^P = 1^+$ (Fig. 7), while the total number of $J^P = 0^-$ or $J^P = 1^+$ events is unchanged, the de-composition into $\epsilon\pi$ and $\rho\pi$ waves is different for the two fits. For $J^P = 0^-$ (Fig. 6) the number of events ascribed to both 0^- (S $\to \epsilon\pi$) and 0^- (P $\to \rho\pi$) by the U fit is larger. The excess

---

* In the 'natural parity exchange' combination: $|2^+ 1\rangle + |2^+ -1\rangle$.

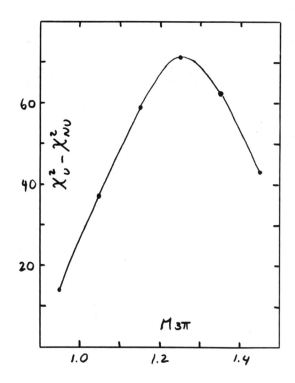

Fig. 5. χ^2 difference between unitarized (U) and non-unitarized (NU) fits to 3π data.

events are cancelled by a strong negative interference between the two waves. Exactly the same remarks can be applied to describe the $J^P = 1^+$ results (Fig. 7). Again the U fits give more $1^+P \to \epsilon\pi$, more $1^+S \to \rho\pi$, and a strong negative interference.

It is quite clear what the data are telling us: the rescattering from $\rho\pi$ to $\epsilon\pi$ produces more $\epsilon\pi$ in the final state than there is $\epsilon\pi$ in the data. The fitting procedure then includes an excess of direct $\epsilon\pi$ production amplitude and cleverly adjusts the relative phases to cancel (as much as possible of) the unitarity correction term. One is strongly tempted to conclude that the rescattering corrections inferred from our particular model are too large. The correct amplitudes, which must, of course, be unitary, nevertheless appear to be better approximated by ignoring the rescattering corrections altogether (NU fits) than by our present attempts at unitarization.

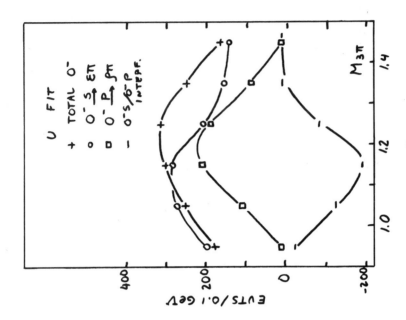

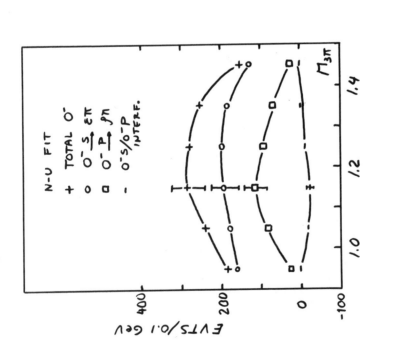

Fig. 6. Fits to 3π data. Events due to 0⁻S, 0⁻P, interference and total 0⁻ are shown at left for non-unitarized (N-U) fits and at right for unitarized (U) fits.

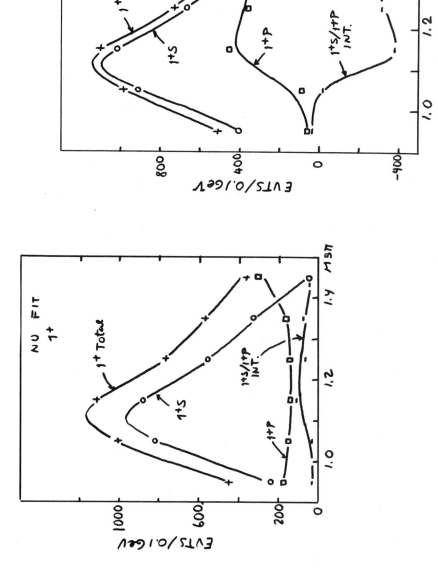

Fig. 7. Fits to 3π data. Events due to 1⁺S, 1⁺P, interference and total 1⁺ are shown at left for non-unitarized (NU) fits and at right for unitarized (U) fits.

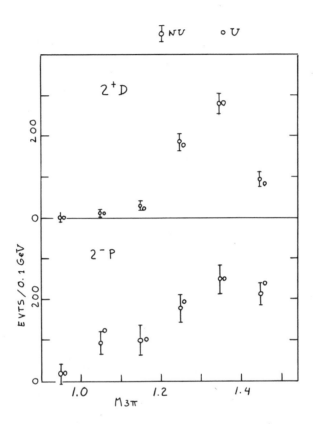

Fig. 8. Fits to 3π data. Events due to 2^+D and 2^-P are shown for non-unitarized (NU) and unitarized (U) fits.

(d) In spite of our strong reservations about the correctness of our rescattering correction, we show (Fig. 9) what happens to the phases of the A_1 ($J^P = 1^+S \to \rho\pi$) wave relative to other waves. We would comment:

(d1) The $M_{3\pi}$ dependence of the relative phases is essentially the same, although the phases themselves have been shifted.

(d2) For this data sample, one can draw significant conclusions only with regard to the $1^+S/1^+P$ relative phase. In both U and NU fits the phases show little sign of change in the region $M_{3\pi} = 0.9$-1.4 GeV.

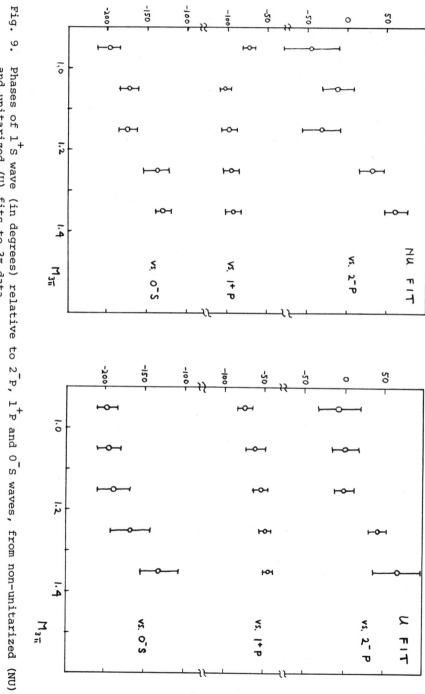

Fig. 9. Phases of 1^+ S wave (in degrees) relative to 2^- P, 1^+ P and 0^- S waves, from non-unitarized (NU) and unitarized (U) fits to 3π data.

REFERENCES

1. G. Ascoli et al., Phys. Rev. D7, 669 (1973).
2. Yu. M. Antipov et al., Nucl. Phys. B63, 153 (1973).
3. See, in particular, Figs. 1C and 9 of Ref. 2.
4. G. Fox, Experimental Meson Spectroscopy-1972, p. 271. AIP Conference Proceedings No. 8.
5. For a more detailed description of the standard (non-unitarized) analysis see Ref. 1 or
    Yu. M. Antipov et al., Nucl. Phys. B63, 141 (1973).
6. Y. Goradia, T. Lasinski, H. Tabak, and G. Smadja. Unitarity and Isobar Model. Paper submitted to this Conference.

COMMENT ON THE AMPLITUDE ANALYSIS OF A_1-MESON PRODUCTION

Michael T. Vaughn

Northeastern University, Boston, Mass. 02115

ABSTRACT

It is suggested that the decay amplitude for $A_1 \rightarrow \rho\pi \rightarrow 3\pi$ may have a strong dependence on the c.m. energy ω of the 3-pion system, such that the decay width is a <u>decreasing</u> function of ω in the A_1 region (this is predicted in some chiral Lagrangian models, for example). This can strongly distort the A_1 propagator from a simple Breit-Wigner form, and, together with a decay width for $A_1 \rightarrow \sigma\pi \rightarrow 3\pi$ which is an <u>increasing</u> function of ω, can account qualitatively for the lack of phase variation of the 1^+ S ρ π amplitude in the analyses of Ascoli and collaborators. This effect is in addition to the rescattering (unitarity) corrections suggested by Aaron and Amado, and reported to this conference by Lasinski.

INTRODUCTION

Ascoli [1] and Lasinski [2] have discussed at this conference the analysis of the reaction

$$\pi + N \rightarrow \pi + \pi + \pi + N \tag{1}$$

in which a partial wave decomposition of the three-pion final state is made in an attempt to isolate resonant structures in the three-pion system. Specifically, the amplitude for the reaction (1) is decomposed into a sum of terms of the type shown schematically in Fig. 1, in which each term has the structure

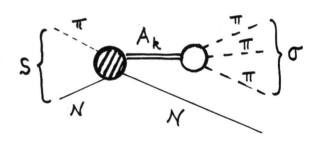

Fig. 1. Schematic diagram for amplitude analysis of reaction (1).

$$T_k = P_k(s) \Delta_k(\sigma) M(A_k \rightarrow 3\pi) . \tag{2}$$

Here $P_k(s)$ is a production amplitude which depends only on the c.m. energy $W = \sqrt{s}$ and the angular momentum state A_k of the three-pion system (dependence on the momentum transfer between initial and final nucleon is suppressed either by restricting the analysis to a small range of momentum transfer, or by effectively averaging over a larger range).

$\Delta_k(\sigma)$ is an effective propagator for the angular momentum state A_k, and $M(A_k \to 3\pi)$ is an amplitude for the state A_k to decay into three pions.

The decay amplitude is expressed in an isobar approximation shown schematically in Fig. 2, where R is one of the several dipion resonances $(\sigma, \sigma', \rho, f^0)$, and the sum of appropriate isobar terms is included for each state A_k.

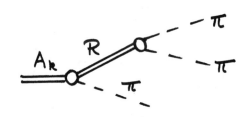

Fig. 2. Diagram for the decay amplitude $M(A_k \to 3\pi)$.

The decay amplitudes introduce phases into the amplitudes T_k; these are reasonably well understood (although the Breit-Wigner approximation used for the σ, σ' propagator could be refined somewhat). The purpose of the amplitude analysis is to extract the residual phase of the effective propagator $\Delta_k(\sigma)$ as a function of σ, since it is this phase which is expected to go through $\pi/2$ at resonance, with a slope related to the width of the resonance.

It is the purpose of this note to emphasize that the structure of the propagator and the corresponding partial wave amplitudes may be rather more complicated than simple Breit-Wigner formulas would imply, if the relevant partial decay widths are rapidly varying functions of the invariant mass-squared σ of the three-pion system. In fact, just such rapid variation of the $A_1 \to \rho\pi$ partial width is expected in some reasonable chiral $SU(2) \otimes SU(2)$ Lagrangian models (or hard-pion current algebra models).

STRUCTURE OF THE PROPAGATOR

The scalar part of the effective propagator for the three-pion state A_k has the general structure

$$\Delta_k^{-1}(\sigma) = \sigma - m_{ok}^2 + \frac{1}{\pi} \int_{9m_\pi^2}^{\infty} \frac{\sqrt{x}\ \Gamma_k(x)}{x - \sigma}\ dx \tag{3}$$

with positive weight function $\Gamma_k(x)$, containing contributions from each decay channel. (It may be necessary to add CDD poles to this to describe more than one resonance, but that is not important in the present context.)

A resonance at $\sigma = m_R^2$ is characterized by

$$\text{Re } \Delta_k^{-1}(m_R^2) = 0 \qquad (4)$$

$$Z \equiv \left[\frac{d}{d\sigma} \text{ Re } \Delta_k^{-1}(\sigma) \right]_{\sigma = m_R^2} > 0 \qquad (5)$$

If $\Gamma_k(x)$ is a slowly varying function of x near $x = m_R^2$, then the propagator is reasonably approximated by a Breit-Wigner amplitude, with width parameter

$$\Gamma_R = Z^{-1} \Gamma_k(m_R^2) \qquad (6)$$

and inclusion of simple centrifugal barrier factors is a reasonable and sometimes necessary refinement.

Partial widths into various decay channels are also obtained from the contributions to $\Gamma_k(\sigma)$ from each decay channel;

$$\Gamma_R^c = Z^{-1} \Gamma_k^c(m_R^2) \qquad (7)$$

where the superscript c labels a particular decay channel.

Suppose now that for a __principal__ decay channel, $\Gamma_k^c(\sigma)$ is a rapidly decreasing function of σ near $\sigma = m_R^2$. Then the peak of the decay spectrum $S_k^c(\sigma)$ in channel c, which is given by

$$S_k^c(\sigma) = \sqrt{\sigma} \; \Gamma_k^c(\sigma) |\Delta_k(\sigma)|^2 \qquad (8)$$

is in fact shifted to a point __above__ $\sigma = m_R^2$, and the peak is somewhat narrower than suggested by Eq. (6). If the phase variation of $\Delta_k(\sigma)$ above m_R^2 is somewhat slower than near m_R^2, a rather slow variation in the phase of $\Delta_k(\sigma)$ over the peak of the spectrum $S_k^c(\sigma)$ can result (such as the slow variation seen in the $A_1 \to \rho\pi$ amplitude observed by Ascoli, et al. [1]).

A note of caution must be injected here: the two conditions set out in the preceding paragraph are inconsistent with the analytic structure of Eq. (3) for a single decay channel. In order to obtain the slow variation of the phase of $\Delta_k(\sigma)$ above $\sigma = m_R^2$, it is necessary to have a second channel whose partial width is rapidly rising not too far above $\sigma = m_R^2$ (but such that the total width has a minimum above $\sigma = m_R^2$).

A_1 DECAY

These conditions can be satisfied, at least qualitatively, in the A_1-decay. The partial width for decay into the ρ-π channel can be a rapidly decreasing function of σ near m_A^2, and, as explained below, there are plausible theoretical models, which predict just this behavior. The partial width for decay into the σ-π channel, which has a somewhat higher threshold, can be expected to rise with a normal P-wave kinematical factor, and provide the mechanism for slowing down the phase variation of the propagator.

It is a rather remarkable fact that the rapid decrease in the partial width for $A_1 \rightarrow \rho\pi$ is predicted by some chiral $SU(2) \otimes SU(2)$ Lagrangian models (or hard-pion current algebra models) [3]. A typical set of results (based on a ρ width of 140 Mev, central A_1 mass of 1070 Mev) is

$\sqrt{\sigma}$	$\Gamma\ (A_1 \rightarrow \rho\pi)$
(Mev)	(Mev)
1000	119
1070	61
1140	13

While the width is perhaps somewhat narrower than the observed peak in the $\rho\pi$ spectrum, the qualitative mass dependence of the width is clearly shown by this solution, and persists with reasonable variations of the parameters in the model. The characteristic feature is that the S-wave amplitude for $A_1 \rightarrow \rho\pi$ has a zero not too far above the A_1 mass, so that the corresponding partial width will be small (both D-wave contributions and effects due to the finite width of the ρ will leave a residual partial width $\sim$ 5-10 Mev). The A_1-σ-π coupling is not uniquely predicted by the model, but it can give a significant partial width above 1100 Mev.

Future analyses should attempt to take the energy variation of the partial widths into account, in addition to the rescattering (unitarity) corrections suggested by Aaron and Amado [4], and reported to this conference by Lasinski [2].

REFERENCES

1. G. Ascoli, these proceedings; see also G. Ascoli, et al., Phys. Rev. D7, 669 (1973).
2. T. Lasinski, these proceedings; see also Y. Goradia, T. Lasinski, M. Tabak and G. Smadja, Berkeley preprint (1974).
3. A general discussion of the models is given by M. T. Vaughn, Phys. Rev. D3, 2738 (1971); the application to A_1-decays is in P. J. Polito and M. T. Vaughn, Nuovo Cimento 7A, 449 (1972).
4. R. Aaron and R. D. Amado, Phys. Rev. Letts. 31, 1157 (1973).

COMMENTS ON HEAVY MESONS

R. H. Dalitz
Oxford University, Oxford, England

At one time, it appeared that missing mass experiments[1] had established the existence of a series of narrow (typically $\Gamma \lesssim 35$ MeV) resonances with high mass values, the states S(1930), T(2200) and U(2360), which continued the leading Regge trajectory beginning with the well-known states ρ(765), A_2(1310), and g(1680). Their narrowness was attributed to the high spin values required for them by their assignment to the leading trajectory, and this was in accord with theoretical notions. The theoretical models predict a great profusion of mesonic states at high mass values. For example, the q-q model requires a series of states with configurations $^{2S+1}_n L_J$, where S denotes the Pauli spin, L the internal orbital angular momentum, J the total angular momentum, and n is the radial quantum number, and with masses M given by $M^2 = M_0^2 + (2n+L-2)\hbar\omega$ for the case of a harmonic interaction with frequency ω. This model predicts 10 I = 1 states and 20 I = 0 states in the neighbourhood of the T mass, their spin values ranging from J = 0 to J = 5. Dual models give various patterns of mesonic levels, as function of M^2 and J, which have the same qualitative behaviour. Two factors which systematically influence the meson decay widths are the centrifugal barriers against decay (which increase with increasing J) and the level spacing (which decreases with increasing n and L). For given mass value, the state on the leading trajectory (highest J) will be the narrowest and its width will decrease with increasing J, for sufficiently large J, according to the estimates by Goldberg[2] for the q-q model and by Chan and Tsou[3] for a particular dual model which illustrates these remarks rather nicely.

After some years, the existence of T(2200) and U(2360) became confirmed from the measurement of $\bar{p}p$ and $\bar{p}d$ total cross sections by Abrams et al.[4], but they were found to be quite broad, their widths being 85 MeV and 140 MeV, respectively. Of course, these T and U bumps could each include a number of resonant states, but no indications have been found in other, more detailed experiments, for the existence of the many other states predicted in this mass region. For example, the angular distributions recently reported[5] for the reactions $\bar{p}p \rightarrow \pi^+\pi^-$ and K^+K^- all vary smoothly with c.m. energy, giving no indications for the existence of resonance states.

It is therefore of interest to draw attention to the existence of two narrow resonances lying near the $\bar{N}N$ threshold, as described by Kalogeropoulos in this Session. An I = 1 multi-pion resonance was found by Gray et al.[6] at 1795 MeV, with width $\Gamma \lesssim 8$ MeV, in their study of the reaction $\bar{p}d \rightarrow p+(N\pi)$ at rest. A narrow resonance, probably with I = 1 and lying at 1932 MeV with $\Gamma = 9^{+4}_{-3}$ MeV, has been reported quite recently by Carroll et al.[7] from their $\bar{p}p$ and $\bar{p}d$ total cross section data for low $\bar{p}$ momentum.

The 1795 MeV resonance lies 83 MeV below the $\bar{N}N$ threshold. Two dynamical interpretations can be mentioned here for this state,

(i) as an "unstable bound state resonance" occurring because of strongly attractive forces in the $\bar{N}N$ channel, sufficient to bind the $\bar{N}N$ system, which then decays through its coupling with the open multi-pion channels. This is quite analogous to the interpretation proposed[8] for the baryonic state $\Lambda(1405)$ in terms of the binding forces in the $\bar{K}N$ channel, with which it is strongly coupled. Calculations by Peaslee[9] have shown that the theoretical $\bar{N}N$ forces may well be sufficiently strong as to bind the $\bar{N}N$ system by 83 MeV, in either s-wave or p-wave, so that this interpretation would require a low spin value ($J \leqslant 2$) for this resonance. A low spin value is required to account for the strong rate observed for the $\bar{p}n$ capture process in the $\bar{p}$-d atom, in any case. With this interpretation, the small decay width reflects the fact that the reduced widths of this state are dominated by the closed $\bar{N}N$ channel; this small width may be an indication that this $\bar{N}N$ bound state is p-wave, since the s-wave $\bar{N}N$ annihilation reaction is known to have great strength.

(ii) as a $\bar{q}$-q bound state, which belongs to an appropriate SU(3) nonet. The narrow width would then argue for a large J value. This conflicts appreciably with its strong formation from the $\bar{p}n$ interactions occurring in a $\bar{p}$-d atom, since the recoil proton cannot carry away large angular momentum.

The 1932 MeV resonance lies 54 MeV above the $\bar{N}N$ threshold. Hence, the "unstable bound state resonance" interpretation is not available for this state unless one appeals to higher thresholds such as $\bar{N}\Delta$ and $\bar{\Delta}N$, or $\bar{\Sigma}\Lambda$ and $\bar{\Lambda}\Sigma$, which are rather far from this mass value. The elasticity of this state is given by $1.5/(2J+1)$, if a single I = 1 state is involved, which argues for a large value for J; on the other hand, the resonance is excited at quite a low c.m. momentum (≈ 225 MeV/c), which argues for a relatively low J. It is difficult to be more quantitative at present. The main point is that there do appear to be rather narrow resonances in the B=S=0 systems at quite high mass values. It will be of great interest to determine the spin values for these states.

For meson spectroscopy, e^+e^- annihilation has one striking advantage, that it excites only states with JP=1-. This process of excitation has already taught us much concerning the ρ, ω and ϕ mesons, as well as contributing significantly to the discovery of the $\rho'(1600)$ vector meson. In the e^+e^- experiments at SPEAR, reviewed by Chinowsky in this Session, the c.m. energy $2E_e$ lies far above these states, and the cross section $\sigma(e^+e^- \to$ hadrons), measured at 100 MeV intervals in E_e, varies smoothly with E_e. If daughter vector mesons are being excited by this process, as envisaged by Renard[10], they must be broad and strongly overlapping. However, it appears quite possible that, in the approximately linear rise now observed for the ratio R = σ(hadrons)$/\sigma(\mu^+\mu^-)$ as function of $(2E_e)^2$, we are seeing the onset of the excitation of a "Very Heavy Vector Meson", the gluon G (of mass M_G), which has been hypothesized by many as the origin of the internal binding between the quark-like partons within hadronic states.

Some of the effects of the gluon may be summarized briefly:

(i) they may act as neutral partons within the nucleons. Such neutral objects are required in the parton interpretation of the data on e-P deep-inelastic scattering.

(ii) they would give a finite size to the charged partons (and to themselves). Chanowitz and Drell[11,12] have pointed out that this effect would cause deviations from scaling in e-P deep-inelastic scattering, by introducing an additional scale-dependent factor $(1-q^2/M_G^2)^{-2}$ into the cross sections. They have suggested that such deviations may already exist in the SLAC data. Recently, indications that these deviations may be as large as 35% for $q^2 = -40(GeV/c)^2$ have been reported[13] from μ-P scattering at NAL; this would correspond to a value $M_G \approx 13$ GeV. We note that the gluon might not account for the whole of the parton charge structure; the parton might still have a point-like core[14], in which case asymptotic scaling could set in for momentum transfers $|q^2| \gg (M_G c)^2$.

(iii) they would contribute an additional factor $F_G(q^2)$ to the electromagnetic form factors for hadrons. The data on the proton magnetic form factor $F_{PM}(q^2)$ does call for such an effect, as was pointed out in 1969 by Massam and Zichichi[15], who found that the empirical $F_{PM}(q^2)$ falls faster for large q^2 than does the form factor calculated from the low-mass vector mesons ϱ, ω and ϕ. They obtained a good fit for $F_{PM}(q^2)$ for the choice $M_G = 7.7 \pm 1.1$ GeV.

(iv) they would contribute a magnetic moment to the partons. West[16] has emphasized that this anomalous magnetic moment may help to account for the early onset of scaling observed for e-P deep inelastic scattering. However, this scaling would then appear rather accidental in character, not a very appealing situation, but still quite possible.

(v) e^+e^- annihilation could well proceed then through a γ-G coupling, in the same way that it goes through the γ-ϱ, γ-ω and γ-ϕ couplings in the low-energy regime, rather than through a direct interaction with a parton pair, as has normally been assumed to be the case. In the energy range $2E_e = M_G \pm \Gamma/2$, the virtual gluon G formed might then decay to a parton pair, but it might equally well decay by the direct emission of hadrons, being a piece of hadronic matter itself. Since the gluon would be envisaged as the carrier of the superstrong interactions, it is reasonable to expect its decay width to be large, with $\Gamma \approx M_G c^2$.

If the gluon mass is indeed of order 10 GeV, then the improved SPEAR which is due to come into operation early in 1975 will be able to take us far towards the gluon resonance peak, and perhaps even beyond it. Even if the gluon resonance is not reached, we would still expect a continual increase in R with increase of E_e in the energy range available, so that it should become possible to rule out this vector gluon hypothesis, if it is without validity. Thus, the topic of e^+e^- annihilation stands very central to the domain of Meson Physics, and indeed to the whole of Hadron Physics.

REFERENCES

1. G. Chikovani, L. Dubal, M. N. Focacci, W. Kienzle, B. Levrat, B. C. Maglic, M. Martin, C. Nef, P. Schubelin, and J. Sequinet, Phys. Letters 22, 233 (1966).

2. H. Goldberg, Phys. Rev. Letters 21, 778 (1968).

3. H. M. Chan and S. T. Tsou, Phys. Rev. D4, 156 (1971).

4. R. J. Abrams, R. L. Cool, G. Giacomelli, T. F. Kycia, A. B. Leontic, K. K. Li, and P.N. Michael, Phys. Rev. D1, 1917 (1970).

5. E. Eisenhandler, W. Gibson, C. Hojvat, P. Kalmus, L. Kwong, T. Pritchard, E. Usher, D. Williams, M. Harrison, W. Range, M. Kemp, A. Rush, J. Woulds, G. Arnison, A. Astbury, D. Jones, and A. Parsons, Phys. Letters 47B, 531 and 536 (1973); 49B, 201 (1974).

6. L. Gray, P. Hagerty, and T. Kalogeropoulos, Phys. Rev. Letters 26, 1491 (1971).

7. A. S. Carroll, I. H. Chiang, T. F. Kycia, K. K. Li, P. O. Mazur, D. N. Michael, P. Mockett, D. C. Rahm, and R. Rubinstein, Phys. Rev. Letters 32, 247 (1974).

8. R. H. Dalitz and S. F. Tuan, Ann. Phys. (N.Y.) 8, 100 (1959); 3, 307 (1960).

9. D. C. Peaslee, Phys. Rev. D9, 272 (1974).

10. F. M. Renard, Vector Mesons in e^+e^- Annihilation, Montpellier preprint PM/73/10.

11. M. Chanowitz and S. Drell, Phys. Rev. Letters 30, 807 (1973).

12. M. Chanowitz and S. Drell, Speculations on the Breakdown of Scaling at 10^{-15} cm, SLAC-PUB-1315(T/E), October 1973.

13. See Physics Today, 27, p.17 April issue (1974).

14. Such a possibility was much discussed for the nucleons in the early days of work with the vector mesons. See, for example, E. Clementel and C. Villi, Nuovo Cimento 4, 1207 (1956).

15. T. Massam and A. Zichichi, Lett. Nuovo Cimento 1, 387 (1969).

16. G. B. West, Evidence for Quark Substructure from Electron-positron Colliding Beam Experiments?, Stanford University preprint ITP-454, January 1974.

REVIEW OF EXOTIC MESONS *

D. Cohen

Columbia University, N.Y., N.Y. 10027

ABSTRACT

Recent data pertaining to exotic mesons is presented. Unambiguous evidence now exists for the occurrence of reactions proceeding via exchanges involving exotic quantum numbers, although the interpretation of such reactions remains unclear. Exotic mesons have been searched for in both meson exchange and baryon exchange reactions, and no evidence has been found for their existence.

I. INTRODUCTION

Exotic mesons were last reviewed at this conference four years ago by Rosner.[1] At that time, there was no confirmed direct evidence for the existence of exotic meson states, although several claims for their observation had been put forth. Upper limits quoted on the cross sections for exotic meson production were typically on the order of 10-15 μb. Evidence was presented at that time that exchanges involving exotic quantum numbers might occur in certain high energy interactions. Most of the reactions studied, however, involved the production of unstable resonances, for example,

$$pn \rightarrow \Delta^-_{FWD} \Delta^{++}, \ ^{[2]} \text{ and } K^-p \rightarrow \pi^+_{FWD} Y^{*-}. \ ^{[3]}$$

These have alternatively been explained as arising from kinematic reflections of processes not involving exotic exchanges.[4] There was thus no totally unambiguous evidence for the existence of exotic exchanges at that time.

I would briefly like to discuss the results on exotic mesons that have been presented since Rosner's review. By an exotic meson I mean, as usual, a mesonic state that cannot be formed from only simple quark-antiquark ($q\bar{q}$) couplings, while exotic baryons refer to those baryonic states that cannot be formed from the couplings of only three quarks (qqq).[5]

* Research supported by the National Science Foundation.

II. EXOTIC EXCHANGES

I would like to begin by discussing new evidence that has recently come to light concerning reactions proceeding via exchanges of exotic quantum numbers. To avoid the ambiguities inherent in exotic exchange reactions leading to unstable particles in the final state that I referred to earlier, I will concentrate on processes leading to stable final-state particles, i.e., final-state particles having no strong decays.

As an example of the evidence that now exists for the presence of so-called exotic exchange reactions, Fig. 1 shows the full angular distributions of K^-p and $\bar{p}p$ elastic scattering, as measured at 5 GeV/c by a CERN-Paris-Stockholm-Orsay Collaboration.[6] Both reactions show prominent peaks in the backward direction. Other similar exotic exchange reactions that have been unambiguously observed include

$$K^-p \to \pi^+_{FWD} \Sigma^-, ^7 \quad K^-p \to K^+_{FWD} \Xi^-, ^7 \quad \pi^-p \to K^+_{FWD} \Sigma^-, ^7$$
$$\bar{p}p \to K^+_{FWD} K^-, ^6 \quad \text{and} \quad \bar{p}p \to \bar{\Sigma}^+_{FWD} \Sigma^-. ^8$$

These forbidden reactions have alternatively been conjectured as arising from intermediate s-channel resonance effects, two-particle exchange, or exotic single-particle exchange.

In an attempt to determine the dominant mechanism by which these reactions proceed, it is instructive to examine the energy dependence of the differential cross sections for processes that can proceed by the exchange of known, non-exotic particles. Three such reactions are examined in Fig. 2:

$$\bar{p}p \to \pi^-_{FWD} \pi^+, \quad \bar{p}p \to \pi^+_{FWD} \pi^-, \quad \text{and} \quad \bar{p}p \to K^-_{FWD} K^+. ^6$$

Each of these reactions is characterized by an initial sharply falling energy dependence that goes approximately as s^{-10} for values of s, the square of the center of mass energy, less than about 5 GeV^2, and then, for values of $s > 5$ GeV^2, by a more gradually falling energy dependence that goes approximately as s^{-2} or s^{-3}. The sharply falling s dependence at low energies is generally attributed to the effects of the dominance of s-channel resonance formation. The more gradual s dependence observed for $s \gtrsim 5$ GeV^2 is thought to arise from the dominance of the single-particle exchange mechanism, with the exact energy variation depending on the nature of the exchanged particle. The energy dependence described above is more or less true of all reactions that can proceed by the exchange of some known, non-exotic particle.

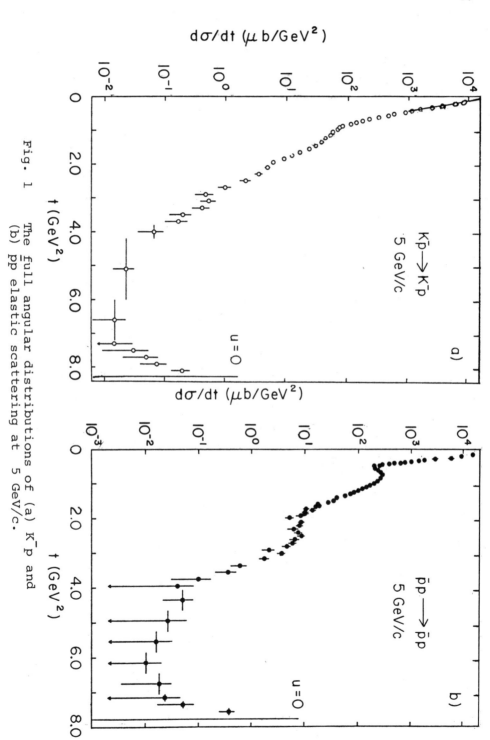

Fig. 1 The full angular distributions of (a) K⁻p and (b) p̄p elastic scattering at 5 GeV/c.

82

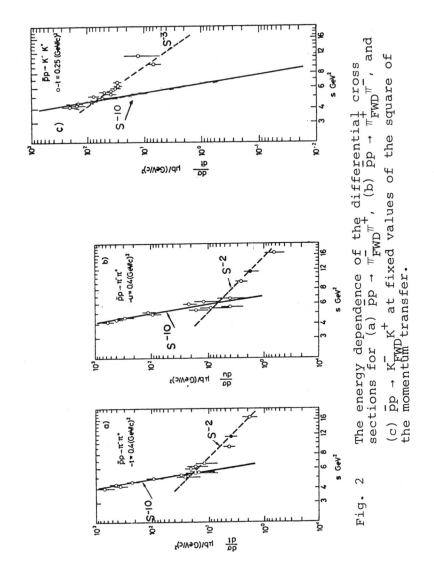

Fig. 2   The energy dependence of the differential cross
sections for (a) $\bar{p}p \to \pi^-_{FWD}\pi^+$, (b) $\bar{p}p \to \pi^+_{FWD}\pi^-$, and
(c) $\bar{p}p \to K^-_{FWD}K^+$ at fixed values of the square of
the momentum transfer.

Let us now turn to the energy dependence of the differential cross section for reactions that cannot proceed by non-exotic exchange. Figure 3 shows the differential cross section at 0° as a function of s for K^-p backward elastic scattering,6
$$K^-p \rightarrow \pi^+_{FWD} \Sigma^-, \quad 7 \text{ and } K^-p \rightarrow K^+_{FWD} \Xi^- . \quad 7$$

Each of these reactions is characterized by an approximate s^{-10} dependence up to the highest center of mass energy at which they have been measured, which is about 10 GeV^2. From the rapidly decreasing energy dependence observed, it appears that intermediate s-channel resonance formation is the dominant mechanism responsible for the occurrence of these forbidden reactions for $s \leqslant$ 10 GeV^2. Along these lines, it is also noteworthy that the differential cross sections at 0° for K^-p backward elastic scattering and $K^-p \rightarrow \pi^+_{FWD} \Sigma^-$ are almost identically equal over the measured energy range, as indicated by the solid line in Fig. 3b.

The energy dependence of three more exotic exchange reactions is examined in Fig. 4. Both $\bar{p}p$ backward elastic scattering6 (Fig. 4a) and $\bar{p}p \rightarrow K^+_{FWD} K^-$, 6 (Fig. 4b) also exhibit an approximate s^{-10} dependence for $s \leqslant$ 10 GeV^2, and hence can also be thought of as arising from s-channel effects. The reaction $\pi^-p \rightarrow K^+_{FWD} \Sigma^-$ 7 (Fig. 4c), however, displays an approximate s^{-10} behavior only up to $s \sim 7$ GeV^2, and then appears to flatten off considerably to an $s^{-1.1}$ behavior above this energy. It thus appears that this reaction displays the more gradual energy dependence expected in a particle exchange picture for $s > 7$ GeV^2. Of course, the two possibilities of either exotic single-particle exchange or two-particle exchange still remain. It thus seems that the π^-p channel experiences weaker s-channel resonant forces than the K^-p and $\bar{p}p$ channels, and hence displays the characteristics expected in a particle exchange picture at a lower energy. This hypothesis of weaker resonant forces in the π^-p channel is also borne out by the fact that, for values of $s < 7$ GeV^2 where, as discussed above, the s-channel resonant picture seems to dominate, the π^-p exotic exchange cross section is at least an order of magnitude smaller than the K^-p and $\bar{p}p$ exotic exchange cross sections. The relatively flat $s^{-1.1}$ dependence observed for $\pi^-p \rightarrow K^+\Sigma^-$ for $s > 7$ GeV^2 is noteworthy in that two-particle exchange models predict a dependence that goes approximately as s^{-3} at high energies.[9] It should be pointed out, however, that the observed dependence is based on only two experimental points, and that the exponent has a statistical error of 0.8 (i.e.,

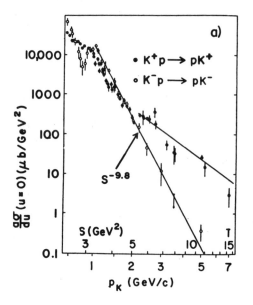

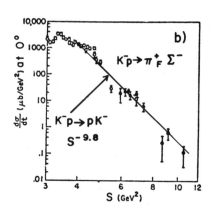

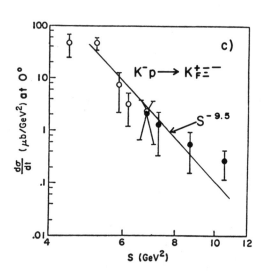

Fig. 3   The energy dependence of the differential cross sections at 0° for (a) $K^{\pm}p$ backward elastic scattering, (b) $K^-p \rightarrow \pi^+_{FWD}\Sigma^-$, and (c) $K^-p \rightarrow K^+_{FWD}\Xi^-$.

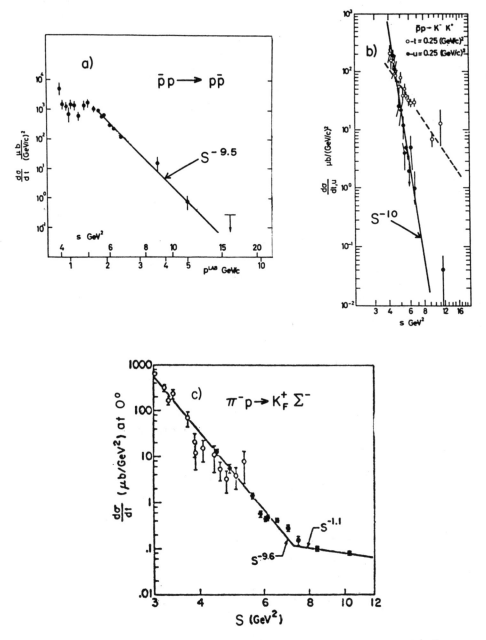

Fig. 4   The energy dependence of the differential cross sections for (a) $\bar{p}p$ backward elastic scattering, (b) $\bar{p}p \rightarrow K^-K^+$, and (c) $\pi^-p \rightarrow K^+_{FWD}\Sigma^-$.   (a) and (c) are at 0° while (b) is at a fixed value of the square of the momentum transfer.

- 1.1 $\pm$ 0.8). It is therefore extremely important to
see if this behavior persists at higher energies. It
would also be very interesting to extend the K$^-$p and $\bar{p}$p
studies to higher energies to see if, and at what energy,
the more gradual energy dependence expected in a particle
exchange picture sets in, and to determine what the
asymptotic energy dependence of these differential cross
sections is.

III. SEARCHES FOR EXOTIC MESON PRODUCTION

I would now like to turn my attention to the results
of experiments that have directly searched for exotic
meson production. The Rochester bubble chamber group
has recently completed the analysis of a high statistics
π^-d experiment at an incident momentum of 7 GeV/c.[10]
They have studied the reaction π^-d → pp + x^{--}, where
x^{--} represents a multi-mesonic system with isotopic
spin 2. They have investigated the x^{--} system in detail
to see whether it contains contributions from any resonant states. The $\rho^-\pi^-$ mass distribution from this experiment is shown in Fig. 5a. The observation of an
enhancement in this mass spectrum at ∼ 1320 MeV has
been claimed.[11] The authors see no such enhancement,
and place an upper limit of 4 μb on resonance production
in this channel assuming a resonance width $\leqslant$ 160 MeV.
They have obtained upper limits ranging from 2 μb to
6 μb for the other decay modes of the x^{--} system examined. Figure 5b shows the x^{--} mass spectrum for the
sum of all the processes studied, and an upper limit of
9 μb is quoted for the overall production cross section
of I = 2 resonances in the x^{--} system.

A similar study has also been carried out utilizing
K$^-$d interactions at 3 GeV/c.[12] A search was made for
strange exotic mesons, non-strange exotic mesons, and
exotic hyperons. The results are shown in Fig. 6. The
authors find no evidence for exotic production, and, in
particular, place an upper limit of 9 μb on the production cross section for the previously reported I = 3/2
K$\pi\pi$ (1170) state, [13] and 6 μb for the $\rho^-\pi^-$ (1320) state.

The above searches were sensitive to the detection
of exotic meson states produced via meson exchange
reactions as shown in Figs. 7a and 7b. Recently, results
have become available that are sensitive to the production of exotic meson states produced by a baryon exchange mechanism. The motivation for looking at baryon
exchange reactions arises from an observation made by
Rosner[14] that the baryon-anti-baryon system requires
exotics in order to satisfy two-component duality. Thus
exotics might be expected to couple more favorably to
a baryon-anti-baryon vertex than to a meson vertex.

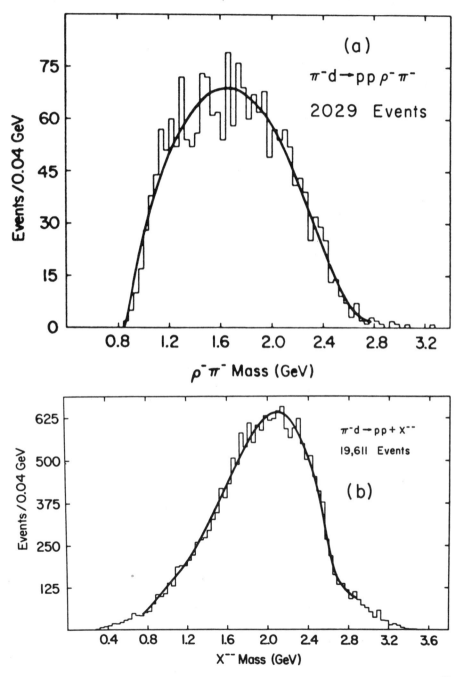

Fig. 5 (a) The $\rho^-\pi^-$ mass spectrum for the reaction $\pi^-d \rightarrow pp\rho^-\pi^-$ at 7 GeV/c; (b) The x^{--} mass spectrum for the reaction $\pi^-d \rightarrow pp + x^{--}$. The curves are the results of low-order polynomial fits to the histograms.

88

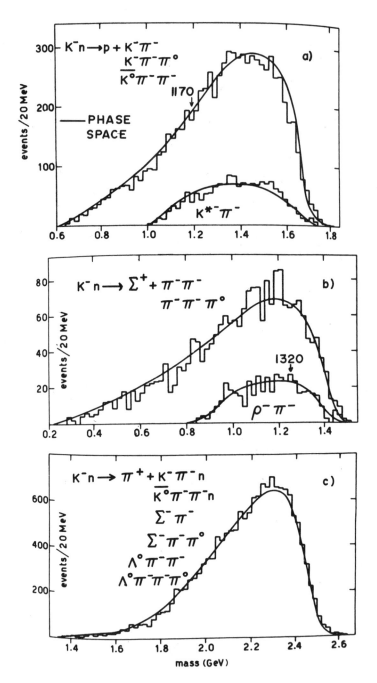

Fig. 6   Invariant mass distributions for (a) strange exotic mesons, (b) non-strange exotic mesons, and (c) exotic hyperons produced in $\bar{K}d$ interactions at 3 GeV/c.

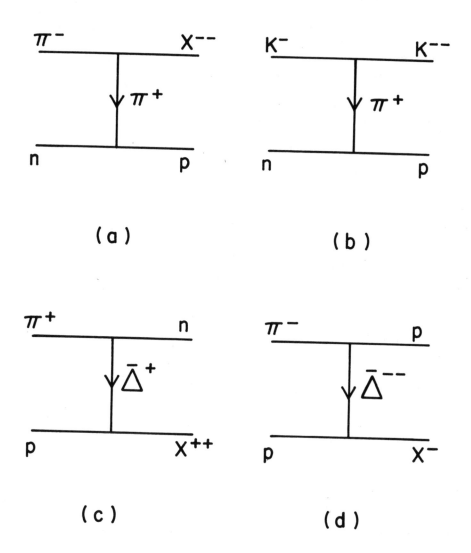

Fig. 7   Particle exchange diagrams for reactions discussed
in the text.

Motivated by these theoretical suggestions, an Indiana-Purdue-SLAC-Vanderbilt Collaboration has recently carried out an experiment in the SLAC rapid cycling triggered 15-in. bubble chamber utilizing a downstream neutron detector to investigate the reaction $\pi^+ p \rightarrow n_{FWD} + x^{++}$ at 8.4 GeV/c. As can be seen from the particle exchange diagram in Fig. 7c, the exotic x^{++} state in this case is produced at a baryon-anti-baryon vertex. Lipkin,[15] in fact, has derived a sum rule relating backward exotic production in this reaction to backward non-exotic meson production in the reaction $\pi^- p \rightarrow p_{FWD} + x^-$ (Fig. 7d). His prediction is that the cross section for the production of backward exotic states in $\pi^+ p \rightarrow n_{FWD} + x^{++}$ should be equal to the cross section for backward non-exotic resonance production in $\pi^- p \rightarrow p_{FWD} + x^-$, if two-component duality is correct.[16] Since there is evidence from the Brookhaven-Carnegie Mellon experiment [17] for backward resonance production at the 2 μb level in $\pi^- p$ interactions at 8 GeV/c, one might then be led to expect backward exotic production in the $\pi^+ p$ channel.

Figure 8 shows the preliminary results from the Indiana-Purdue-SLAC-Vanderbilt experiment [18] for the decay of the x^{++} state into two pions. The sensitivity of this experiment corresponds to $\sim$ 40 ev/μb, while the mass resolution (σ) varies from about 5 MeV at the lowest masses to about 30 MeV at the highest masses. The two categories of events shown in Figs. 8a and 8b, namely the 1-constraint and 3-constraint categories, arise from the fact that the neutron direction is well known for only about one-half of the events, and the events for which the neutron direction is not known make only 1-constraint fits and not, in addition, 3-constraint fits. The smooth curve drawn on Fig. 8a represents a low-order polynomial fit to the observed mass distribution, and indicates that there is no obvious evidence for resonance production in this channel. To give one an idea of the statistical sensitivity of this experiment, the two fluctuations noted in Fig. 8a correspond to 0.25 and 0.16 μb of cross section. The curve in Fig. 8b represents the predictions of transverse momentum damped phase space, and describes the distribution extremely well. Figure 8c shows the four standard deviation upper limit for exotic resonance production in this channel as a function of mass, assuming a resonance width of 100 MeV. This upper limit is relatively flat as a function of mass, and is typically $\sim$ 1 μb. Figure 9 examines four pion decay mode of the exotic x^{++} state. Again, modified phase space describes the data reasonably

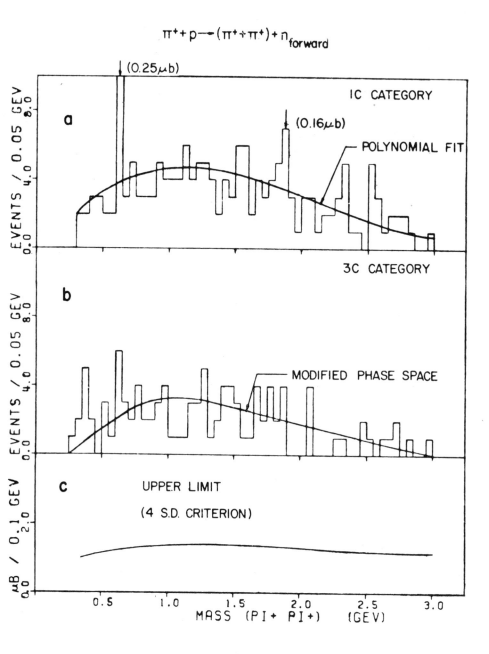

Fig. 8   Results of exotic meson search from the reaction
$\pi^+ p \to n_{FWD} + \pi^+\pi^+$. See text for details.

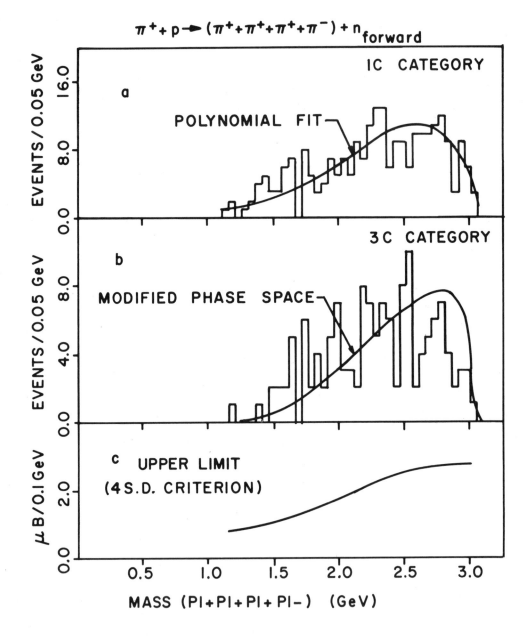

$$\pi^+ + p \longrightarrow (\pi^+ + \pi^+ + \pi^+ + \pi^-) + n_{forward}$$

Fig. 9   Results of exotic meson search from the reaction
$\pi^+ p \rightarrow n_{FWD} + \pi^+\pi^+\pi^+\pi^-$.

well, and the upper limits are ~ 1-2 μb. The six pion
mass distributions are shown in Fig. 10 along with the
upper limits for exotic resonance production in this
channel. The analysis of these data is continuing, and,
in addition, the same group has an approved streamer
chamber experiment at SLAC to investigate backward exotic
and non-exotic resonance production in π^-d interactions
at 8 and 14 GeV/c.[19] They hope to obtain a sensitivity
of about 1000 events/μb at each energy.

A search for backward exotic meson production has
also been carried out in a Columbia-Binghamton experiment
utilizing a preliminary 10 ev/μb sample of the reaction
$\pi^+p \to \Lambda_{FWD} + K^{++}$ at 15 GeV/c. The data are derived from
a million-picture exposure of the SLAC 82-in. bubble
chamber. In this reaction, the exotic K^{++} state is also
produced at a baryon-anti-baryon vertex (see Fig. 11).
No evidence is seen for exotic meson production, and an
upper limit of 5 μb is obtained on exotic resonance production in this channel for masses < 4 GeV. This analysis is also continuing, and will quadruple its statistics in a short time. [20]

IV. CONCLUSIONS

In conclusion, unambiguous evidence now exists for
the occurrence of exotic exchange reactions, unclouded
by possible kinematic interpretations. s-channel effects
seem to be responsible for K^- and $\bar{p}$ induced exotic exchange reactions for $s \leqslant 10$ GeV^2, and for $\pi^-p \to K^+_{FWD} \Sigma^-$
for $s < 7$ GeV^2. For $s > 7$ GeV^2, $\pi^-p \to K^+_{FWD} \Sigma^-$ shows the
more gradual energy dependence expected in a particle
exchange picture, although it is not clear whether two-
particle exchange or exotic single-particle exchange
is the dominant mechanism.

There is no evidence for exotic meson production in
meson exchange reactions, and upper limits vary from ~ 2
to 9 μb depending on the final state examined. There is
also no evidence for exotic meson production in baryon
exchange reactions. Upper limits are ~ 1-2 μb for
$\pi^+p \to n_{FWD} + x^{++}$ and 5 μb for $\pi^+p \to \Lambda_{FWD} + x^{++}$.

I would like to thank R. Panvini and S. Stone for
their cooperation in providing me with the preliminary
results of the Indiana-Purdue-SLAC-Vanderbilt experiment.
I also gratefully acknowledge helpful discussions with
C. Akerlof, C. Baltay and J. Finkelstein.

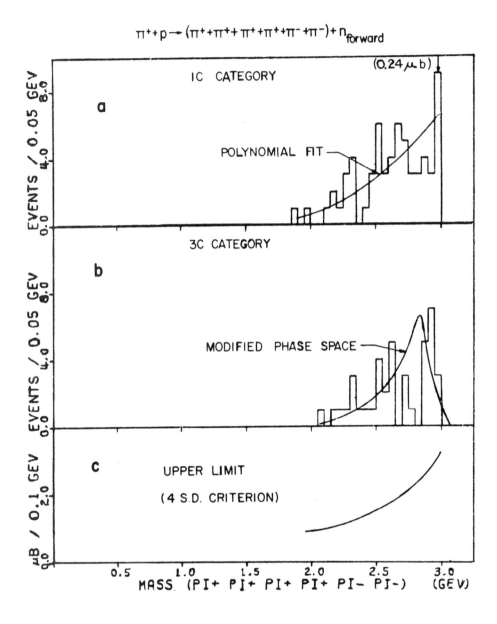

$\pi^+ + p \rightarrow (\pi^+ + \pi^+ + \pi^+ + \pi^+ + \pi^- + \pi^-) + n_{forward}$

Fig. 10   Results of exotic meson search from the reaction
$\pi^+ p \rightarrow n_{FWD} + \pi^+\pi^+\pi^+\pi^+\pi^-\pi^-$.

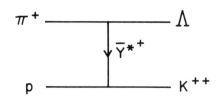

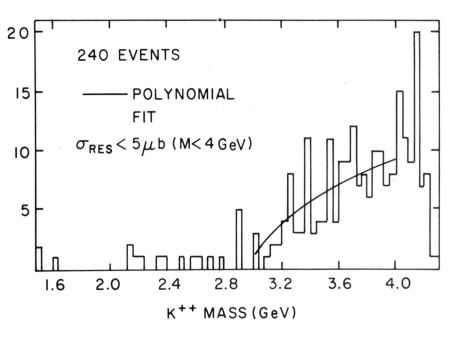

g. 11   K^{++} mass distribution for the reaction $\pi^+ p \to \Lambda_{FWD} + K^{++}$.

REFERENCES

1. J.L. Rosner in Experimental Meson Spectroscopy, edited by C. Baltay and A.H. Rosenfeld (Columbia University Press, New York, 1970), p. 499.
2. G. Yekutielli et al, Phys. Rev. Lett. 25, 184 (1970).
3. P.M. Dauber et al, Phys. Lett. 29B, 609 (1969).
4. See, for example, E.L. Berger, Phys. Rev. Lett. 23, 1139 (1969); A. Dar and A. Gal, Phys. Rev. D1, 2714 (1970); E.L. Berger, R.A. Morrow, Phys. Rev. Lett. 25, 1136 (1970); D. Sivers, F. vonHippel, Phys. Rev. D6, 874 (1972).
5. See, for example, J.J.J. Kokkedee, The Quark Model (W.A. Benjamin Inc., New York, 1969).
6. A. Eide et al, Nucl. Phys. B60, 173 (1973).
7. C.W. Akerlof et al, University of Michigan preprint (1974). The author thanks Prof. Akerlof for discussions regarding these data.
8. H.W. Atherton et al, Phys. Lett. 42B, 522 (1972).
9. C. Michael, Phys. Lett. 29B, 230 (1969).
10. D. Cohen et al, Nucl. Phys. B53, 1 (1973); D. Cohen, Ph.D. Thesis, University of Rochester, 1973 (unpublished), U. of Rochester Report UR-447.
11. R. Vanderhagen et al, Phys. Lett. 24B, 493 (1967); W.M. Katz et al, Phys. Lett. 31B, 329 (1970).
12. G. Giacomelli et al, Phys. Lett. 33B, 373 (1970).
13. A.H. Rosenfeld in Meson Spectroscopy, edited by C. Baltay and A.H. Rosenfeld (W.A. Benjamin Inc., New York, 1968), p. 455. See also Ref. 1.
14. J.L. Rosner, Phys. Rev. Lett. 21, 950 (1968).
15. H.J. Lipkin, Phys. Rev. D7, 237 (1973) and Phys. Rev. D7, 2262 (1973).
16. See also, M. Jacob, J. Weyers, Il Nuovo Cimento LXIX, 521 (1970).
17. E.W. Anderson et al, Phys. Rev. Lett. 22, 102 (1969).
18. Private communication. The author expresses his gratitude to Profs. R. Panvini and S. Stone for communicating these results to him prior to publication, and for numerous enlightening discussions concerning these data.
19. C. Baglin et al, Search for Exotic Mesons Using the SLAC Two-Meter Streamer Chamber, Proposal submitted to SLAC.
20. Other recent exotic meson searches not specifically mentioned in the text include J. Lys and J.W. Chapman, Phys. Rev. D2, 2525 (1970); L. Young et al, Phys. Rev. D5, 2727 (1972); T. Buhl et al, Nucl. Phys. B37, 421 (1972).

DIRECT CHANNEL N̄N PHENOMENA. II[†]

T. E. Kalogeropoulos

Department of Physics, Syracuse University, Syracuse, New York 13210

ABSTRACT

First, the well established s-channel phenomena will be presented and discussed in terms of an over-all picture indicated by the behavior of cross-sections. Second, the observations at low energies will be presented which clearly demonstrate that the annihilations exhibit many direct narrow s-channel phenomena and many others are indirectly suggested.

I. INTRODUCTION

It was my intention to present to this conference some recent work on P̄D annihilations at low energies which firmly establishes the existence of a (P̄N) resonance like structure at $(2m + 20)$ MeV. Moreover, the results of Carrol et al.[1] are confirmed and consequently a second resonance like bump exists in P̄N annihilations at $(2m + 56)$ MeV. I was asked not only to report on these observations but to comment as well on the phenomena which in my opinion are relevant to the title of this talk. A talk was presented with the same title last September in Moscow.[2] I have tried to avoid repetition and therefore both of them are somewhat complimentary. I would also like to recommend the Proceedings of the II Nucleon-Antinucleon Symposium[3] particularly my talk and Bizzarri's representing an alternate view on the subject.

Although my talk is a very personal view on the subject, I nevertheless, have avoided as much as possible drawing any speculative conclusions. I do avoid in particular calling these phenomena resonances in order to eliminate discussions on their relevance to the usual resonance classification schemes. It is my opinion that such considerations are premature and must wait until some firm spin-parity observations are made.

II. TOTAL CROSS-SECTIONS AS AN OVER VIEW

Very early it was noticed that the total N̄N cross-sections behave as $1/P_L$. This fact was particularly emphasized at low momenta and was viewed as an example of the $1/v$-law or the dominance of S-waves. This view is now totally unjustified because several observations show that the absorption proceeds from high waves in the region of the "$1/v$" behavior. In Fig. 1   the behavior of P̄P, P̄D cross-sections are shown in terms of the inverse of the laboratory projectile momentum; for comparison the PP total cross-section is

---

† Work supported by The National Science Foundation

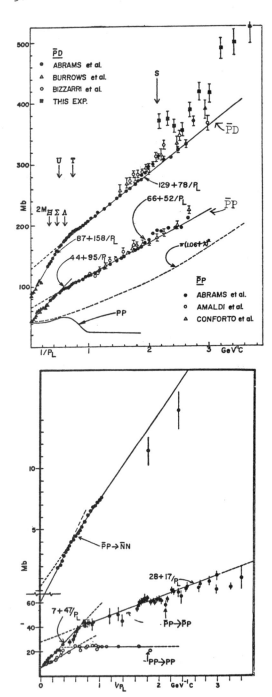

Fig. 1

Total cross-sections as a function of the inverse of the projectile momentum. Curves are eye fits to straight lines.

Fig. 2

Charge exchange cross-sections are those of Cutts et al. (Ref. 6)

also shown. The most dramatic characteristics are:

(i)   the break at 1.92 GeV/c ($\sqrt{s}$ = 2.4 GeV, s = 5.8 GeV^2),

(ii) the linear dependences below and above the break.

Table I summarizes the results implied by Fig. 1 .

Table I. Cross-Sections in (mb); P_L in GeV/c

	Break, $(GeV/c)^{-1}$	High Energy	Intermediate[2]
$\sigma_{\bar{P}D}{}^{T}$	.53	$87 + 158/P_L$	$129 + 78/P_L$
$\sigma_{\bar{P}P}{}^{T}$	.51	$44 + 95/P_L$	$66 + 52/P_L$
$(\sigma_{\bar{P}N}{}^{T})$ [1]	.54	$43 + 63/P_L$	$63 + 26/P_L$
$\sigma_{\bar{P}P}{}^{El}$	.70	$7 + 47/P_L$	$28 + 17/P_L$

[1] $\sigma_{\bar{P}N}{}^{T} \equiv \sigma_{\bar{P}D}{}^{T} - \sigma_{\bar{P}P}{}^{T}$. This is suggested by recent observations below 1 GeV/c. See Reference 16.

[2] This region extends down to ~ 0.5 GeV/c.

The following conclusions can be drawn:

(i)   The energy independent terms do not indicate any I-spin dependence.

(ii) The energy dependent terms are significantly dependent on I-spin. If decomposed in terms of I-spin, then the ratio of the I = 0 to I = 1 parts are 2 and 3 for the high and intermediate regions respectively. Although it is substantial the I-spin dependence in all energy regions, its overall contribution is only significant in the intermediate region because it contributes equally or more to the total cross-section.

(iii) At very high energies $\bar{N}N$ and NN cross-sections become identical to the extent that the energy dependent part of the $\bar{N}N$ cross-sections becomes small in comparison to the constant term of ~ 44 mb.

(if) In view of the large contributions of the I-spin dependent part to the total cross-section at

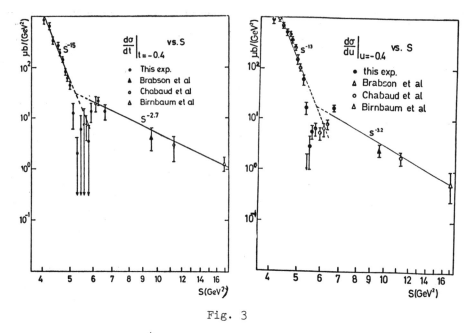

Fig. 3

Differential $\overline{P}P \rightarrow \pi^- \pi^+$ cross-sections from Reference 7.

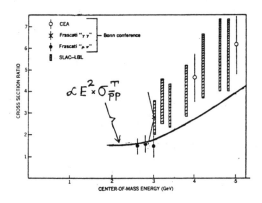

Fig. 4

Comparison of $\overline{P}P$ total cross-section to $e^+ e^- \rightarrow$ hadrons. (Data from Phys. Today March, 1974, p. 17).

intermediate energies this region is a favored one in searching for s-channel resonances.

(v) The unexplored low energy region is indeed interesting.

The following evident questions are born out of these observations:

(1) <u>Why does the $1/P_L$ provide such good parametrizations?</u>

We can not answer the question with any degree of confidence. We would like however to point out that this general behavior characterizes all $\bar{N}N$ reactions for which good data exists. As an illustration of this point in Figs. 2, 3 the behavior of two body reactions as diverse in frequency and possibly mechanisms as $\bar{P}P \to \bar{P}P$, $\bar{N}N$, and $\to \pi^+\pi^-$ are shown. A $1/P$ dependence can be obtained from the usual cross-section formula written in terms of partial waves by assuming that only a narrow range of waves contributes around a high value. Crudely, $\sigma \sim (2\ell + 1)/P^2 \simeq 2\ell/P^2 \sim RP/P^2 \to 1/P$ and the $1/P$ dependence seems natural. Moreover and in this spirit, the well defined break at ~ 2 GeV/c can be viewed as the introduction of a new radius of interaction.

(2) <u>Why the break?</u>

The break occurs in the region of the broad bumps which have been observed long ago[4], the so called T and U bumps. It has been suspected that these bumps might be due to threshold effects but there is no such experimental support. We would like to point out that this transition characterizes not only $\bar{N}N$ but e^+e^- hadrons as well (Fig. 4). Furthermore, we observe that the break is in the region of the $\bar{N}N \to \Lambda\Lambda$, $\bar{\Sigma}\Sigma$ $\bar{\Xi}\Xi$ thresholds. Whether these observations are accidental or relevant to this question cannot be answered at the present time.

III. S-CHANNEL PHENOMENA IN
THE INTERMEDIATE AND HIGH ENERGY REGIONS

A. <u>The status of the T and U bumps.</u> The bumps T(2193, 98), I = 1, U(2359, 164), I = 0 (with possibly a second with I = 1), which were first observed in $\bar{P}P$ and $\bar{P}D$ total cross-sections by Abrams et al.[4], have also been observed recently by Alspector et al.[5] in the elastic, inelastic, and annihilation channels (Fig. 5) and by Cutts et al.[6] in $\bar{P}P \to \bar{N}N$ (Fig. 6). In spite of the many searches to see these structures in specific annihilation channels no clear evidence exists. In conclusion these bumps are very real but their nature is uncertain due to their proximity to the transition region and the inelastic thresholds (e.g. $\bar{N}N \to \bar{N}N\pi$ etc.)

B. <u>The $\pi^+\pi^-/K^+K^-$ Channels.</u> There are at last some excellent data[7] (Fig. 7) from 0.9 to 2.3 GeV/c. The data show rapidly and regularly varying angular distributions - particularly $\pi^+\pi^-$-exhibiting a

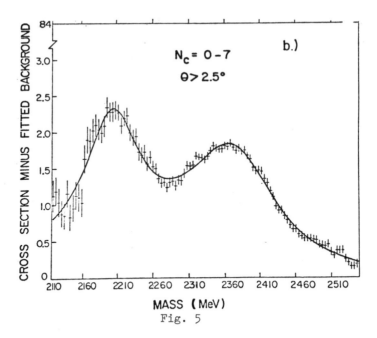

Fig. 5

Annihilation cross-sections from the Rutgers Annihilation Spectrometer

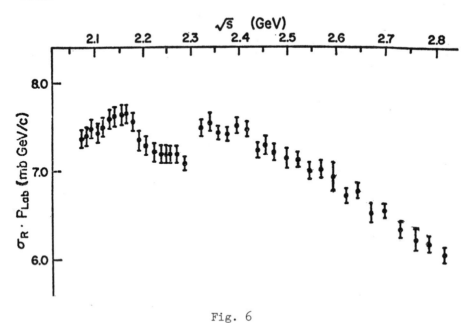

Fig. 6

$\bar{P}P \rightarrow \bar{N}N$ Cross-section from Reference 6

$\overline{P}P \to \pi^- \pi^+$

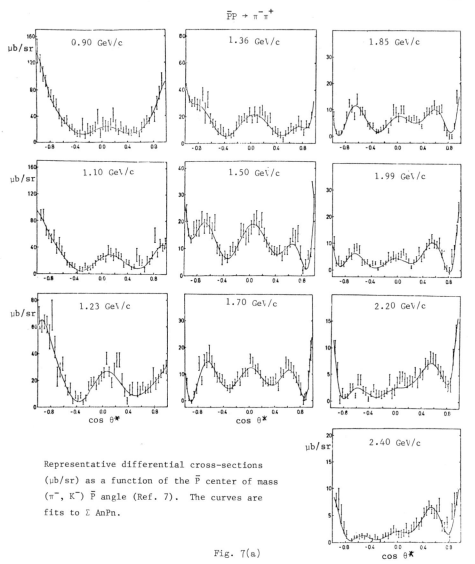

Representative differential cross-sections
(μb/sr) as a function of the $\overline{P}$ center of mass
(π^-, K^-) $\overline{P}$ angle (Ref. 7). The curves are
fits to Σ AnPn.

Fig. 7(a)

$\bar{P}P \to K^- K^+$

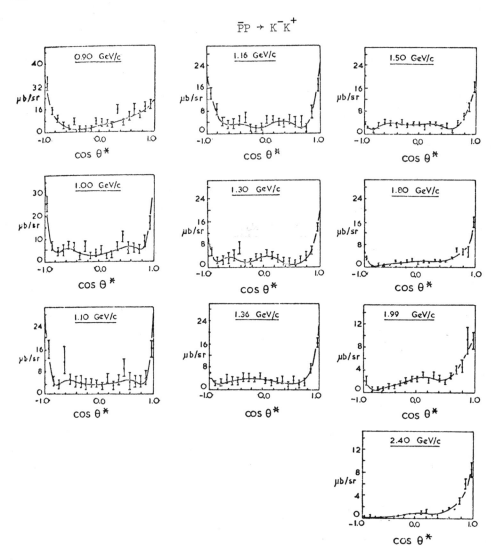

Fig. 7(b)

$$\bar{P}P \rightarrow \pi^- \pi^+$$

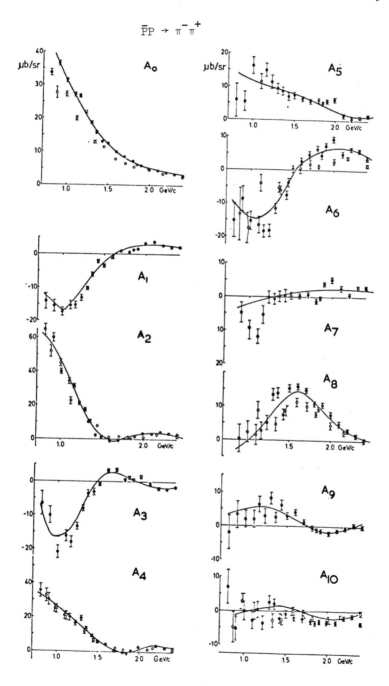

Fig. 8(a)

Legendre coefficients of the differential cross-sections (Fig. 7).

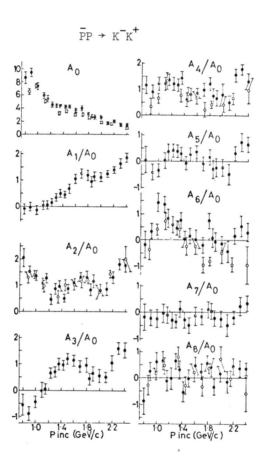

Fig. 8(b)

number of minima and maxima with their number increasing mono-
tonically with energy (Fig. 8   ). Attempts have been made to
fit them in terms of background and resonance terms.[8] In spite of
the introduction of many waves (up to $J = 5$) there are no good fits
(best fits have $\chi^2 \sim 5/DF$).

The cross-sections vs. energy do not show any bump structures
except for the break at ~ 2 GeV (Fig. 3   ). In such a situation
it seems to me, that the search for resonances, presupposes the
knowledge of the background and since the energy dependence of the
cross-sections has the same behavior as all $\bar{N}N$ reactions, the
understanding of the "background" is part of the whole problem and
cannot be parametrized.

IV. LOW ENERGY

Through the years of observing annihilations at rest and low
energies ($P_L \leq .5$ GeV/c) in hydrogen and deuterium I became aware
of their unique problems and opportunities. For many years the
emphasis was on final state interactions and very little attention
was given to "initial state interactions". The following, puzzling
observations in chronological sequence, have forced me to seriously
reconsider this "traditional" emphasis.

(i)   Absence of interactions between the spectator and
the products (π, K) of the annihilation in deuterium
at rest[2,3]. I cannot understand this essentially
geometrical puzzle except by getting the spectator
away from the pion source which implies that the
emission of the pions (annihilation time) is not
instantenous, $\sim (10^{-24} - 10^{-23})$ sec. Calculations
based on the known πN interactions, the measured
pion spectra and the deuteron wave function show
that these observations cannot be reconciled with
annihilation time smaller than 5×10^{-22} sec or
$\Gamma > 1.5$ MeV. Preliminary $\bar{P}D$ in flight data show
again a similar effect: antiprotons which first
scatter on the proton do not annihilate with the
spectator neutron. (See IV B and V).

(ii)   The reaction $\bar{P}D \rightarrow (\pi^-\pi^0) + P_s$ at rest goes with
invisible (≤ 100 MeV/c) but does not go with
visible spectators[9]. This means that the annihilation into 2π from $I = 1$ is very sensitive to
the total energy: it is produced within the
energy interval $(2m_p - 10)$ MeV $\leq \sqrt{s} < 2 m_p$.

(iii) The observation of $\bar{P}P \rightarrow 2\pi^0$ at rest[10] (see also
Ref. 9) implies that 40% of all $\bar{P}P \rightarrow 2\pi$ events
come from P(ODD) $\bar{N}N$ orbitals in strong disagreement with the traditional view that

annihilations at rest go dominantly from
S-states. Furthermore, this result is
dramatically, factor of ~ 40, different
from the implication of the observed small
$(\bar{P}P \to K_S K_S)/\bar{P}P \to K_S K_L)$ branching ratio.
Comparison of the $\pi\pi$, $\bar{K}^\circ K^\circ$ results and
independent of their implications to the
overall $(\bar{P}P)_{ODD}/(\bar{P}P)_{EVEN}$ ratio, demonstrate
that the 2π, $\bar{K}K$ to $\bar{N}N$ coupling constants are
very different. This implies that important
dynamical effects are present in annihilations
(compare f°, $f^{\circ\,\prime}$ to 2π, $\bar{K}K$).

(iv)   There has been some evidence[11] for a narrow
(< 7 MeV) $\bar{P}N$ bound state at -83 MeV decaying
into an even number of pions.

(v)   Presence of high angular momentum states in
$\bar{P}P \to \pi^+\pi^-$ at low energies.[12] We now have[13]
good data on $\bar{P}P \to 2\pi^\circ$ which do not only
show the presence of high angular momentum
states at low energies but in addition abnormal
behavior of the cross-section and strong spin-
orbit effects. (See IV C).

(vi)   Independent theoretical investigations performed
by Prof. I. Shapiro and collaborators at ITEP
which are based on the known non-relativistic
N-N potentials[14] produce many non-relativistic
bound and resonant $\bar{N}N$ states.

Although we are not yet certain on whether the further classification of the low energy region into Bound ($\sqrt{s} < 2m_p$) and
Resonant ($\sqrt{s} > 2m_p$) regions carries physical significance, it is
nevertheless convenient from the experimental point of view. The
resonance region is accessible through $\bar{N}$ interactions in H_2 and D_2
but the bound region is only accessible in deuterium (complex
nuclei) experiments. Generally, the $\bar{N}N$ interaction is expected to
depend on the total energy and the relative momentum which
constraints the favored angular momenta. In hydrogen the total
energy defines as well the relative momentum, one variable
problem, but in deuterium this is not true. For non-relativistic
$\vec{P}(\vec{k})$ and spectator $(\vec{q})$ momenta

$$Q \equiv \sqrt{s} - 2m_p = \frac{(\vec{k}+\vec{q})^2}{4m} - B_D - \frac{q^2}{2m} \qquad (1)$$

where $(\vec{k}+\vec{q})/m$ is the $\bar{N}N$ relative momentum. Therefore the same Q
can be reached from a range of relative momenta. This fact will
have important implications which should be kept in mind in
evaluating any observations. Consider, for example, a high spin
resonance above threshold. Its production in hydrogen experiments

will be highly suppressed by the factor $(KR)^{2J}$ but in deuterium by properly selecting the beam and spectator momenta this centrifugal barrier suppression can be eliminated. In the following we shall present the most recent H_2 and D_2 experimental results which beyond doubt establish this region as the prime one in searching for s-channel phenomena (resonances).

A. Cross-Sections

The high statistics H_2, D_2 total cross-section measurements obtained by Carrol et al.[1] have unearthed a bump at ~ .45 GeV/c (Fig. 9). Several bubble chamber measurements existed in this range but did not have the statistical significance required for the identification of this ~ 10 mb structure with $(Q, \Gamma) = (56 \pm 2, 9 {}^{+4}_{-3})$ MeV. In spite of the great statistical significance it has been considered prudent not to be classified among the established resonances[15].

Carrol et al. favor $I = 1$ with a warning that the standard Glauber corrections are large and uncertain. Strangely however, recent data by Bizzarri et al.[16] in this energy region show identical $\bar{P}P$ annihilation cross sections (Fig. 10) as measured in H_2 and D_2! This is another interesting facet of the $\bar{N}N$ interactions at low momenta.

We have measured low energy, ~ (.26 - .56) GeV/c, cross-sections in D_2 from a bubble chamber exposure[17]. A well defined beam, stopping near the chamber exit, allows us to select in-flight events without contamination from at rest (Fig. 15a). The total cross-sections are shown in comparison with other existing measurements in Fig. 1. The bubble chamber cross-sections are generally higher than the counter data of Carrol et al. but apart from a (few percent) normalization the bubble chamber data support the Carrol et al. bump (see also Fig . 11).

Far more significantly, however, our data do not lie on the $A + B/P_L$ extrapolation but are much higher indicating a possible new regime or the tail of another bump lying below ~ .3 GeV/c. This enhancement is associated with $(\bar{P}N)$ and $(\bar{P}P)$ annihilations but not with the quasi-elastic events (Fig. 11 _). We would like to point out that these enhancements, if due to $\bar{P}$-nucleon s-channel resonances, must be associated with low momenta spectators and this is implied by the narrowness of the peak.

B. $\bar{P}D$ Stripping

Antiproton cross-sections at low momenta are very difficult to measure because of their rapid ionization losses and contaminations from stopping annihilations. We thought of transpassing this problem by studying annihilations of the type

$$\bar{P} + D \rightarrow N_{\pi^-} + P_{vis} \qquad (2)$$

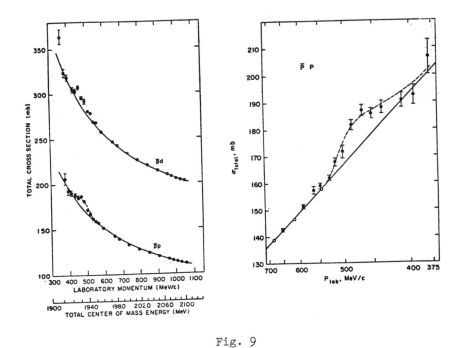

Fig. 9

Total cross-section of Carrol et al. (Reference 1). Curves are A + B/P fits.

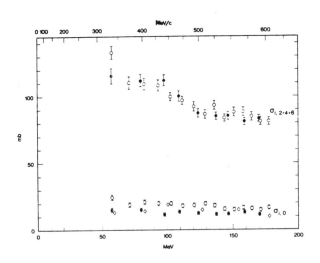

Fig. 10

P̄P annihilation cross-sections as measured in H_2 (o) and D_2 (.) for 2-, 4-, 6-prong events and zero-prong.

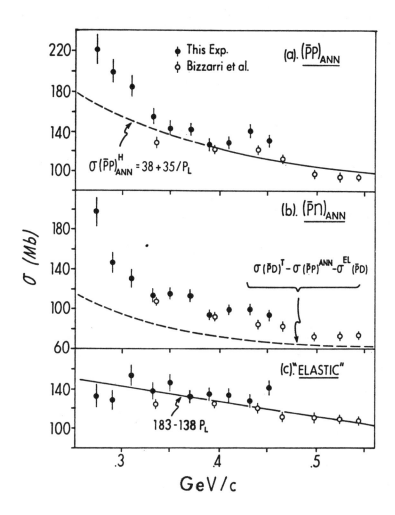

Fig. 11

Antiproton cross-sections in deuterium at low energies. (a) annihilation with the proton (b) on the neutron and (c) P̄D interaction without annihilation.

in the exposure discussed previously. Low energies (above and below threshold) are thus achieved because of the spectator. As it has been discussed already (see IV) this is not identical (for better or worse!) to the free nucleon target case. This problem can be looked upon as the deuterium stripping by the antiproton and is completely analogous to the stripping of deuterium by nuclear targets. The differential cross-section is[14]

$$\frac{d^2\sigma_{\bar{P}D}}{dq\,d\Omega} = \frac{q^2|F_D(q)|^2}{(2\pi)^3} \cdot \frac{|\vec{k} + \vec{q}|}{k} \cdot \sigma_{\bar{P}N}(Q) \tag{3}$$

where $\sigma_{\bar{P}D}$ is the measured cross-section for reaction (2), $\vec{k}$, $\vec{q}$ the $\bar{P}$ and spectator momenta in the laboratory system respectively, the solid angle (Ω) is defined with the z-axis along q, $F_D(q)$ is the Fourier transform of the deuteron wave function, and $\sigma_{\bar{P}N}(Q)$ the $\bar{P}$ annihilation cross-section on bound neutrons. For low spectator momenta this cross-section should approach the cross-section on free neutrons and if it behaves like $1/v$ then

$$|\vec{k} + \vec{q}| \cdot \sigma_{\bar{P}N}(Q) = \text{Const} \tag{4}$$

In our experiment $|\vec{k} + \vec{q}|$ is on the average, relatively large, and in the region where the $1/P_L$ behavior of cross-sections is approximately valid.

From the experimental data we shall extract using (3) the behavior of $|\vec{k} + \vec{q}|\,\sigma_{\bar{P}N}$ as a function of $\sqrt{Q}$ for two regions of spectator momenta by integrating over all beam and spectator momenta for given Q. It is interesting however, first to present (Fig. 12) the distributions in k, q, and $\hat{k}$, $\hat{q}$. We also present the same distributions in parallel for the quassi-elastic reactions

$$\bar{P} + D \rightarrow P_{vis} + \bar{P} + N_S \tag{5}$$

which will contaminate the direct ($\bar{P}N$) annihilations, if in addition to scattering the $\bar{P}$ annihilates on the spectator neutron, and they will produce effects which are reflections of the character of $\bar{P}$ +"P" $\rightarrow \bar{P}$ + P. It may be useful here to quote that 49% of all scatterings do not show a recoil (proton $\leq$ 90 MeV/c) and 20% of those with a visible recoil are due to coherent ($\bar{P}D \rightarrow \bar{P}D$) scattering. Coherent scattering has been identified and has been excluded from these distributions.

By comparing the results of two scans the overall spectator efficiency is found to vary smoothly from 0.95, at 100 MeV/c, to 0.98, at 150 MeV/c and beyond. In our studies we use only events with stopping spectators (see Fig. 12b) and applying appropriate geometrical corrections which turn out to be smooth and small. (These geometrical corrections have been tested in $\bar{P}$ captures at rest and the angular distributions become isotropic). Stopping protons are used only because in this case alone the resolution, $\delta Q_{\bar{P}N}$, is exceptionally small: $\sim$ ± 3 MeV. Furthermore, we treat

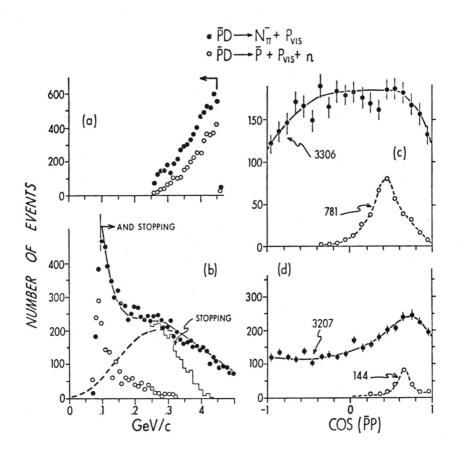

Fig. 12

(a) Distribution of events as a function of the antiproton momentum at the vertex. (b) Momentum distributions of the visible protons in ($\bar{P}N$) annihilations and "elastic". Fit is deuteron wave function plus $p^2 \, Exp(-\omega/T)$. (c) and (d) The $\bar{P}$ "spectator" antiproton angular distributions for low spectator momenta (.1 - .2) GeV/c and higher ones (> .2 GeV/c and stoppings). Curves are hand drawn.

separately the events with spectator momenta < 200 MeV/c and those
with higher ones. The lower momentum group seems to be part of the
spectator distribution while the higher one is more complex.

The angular distributions with respect to the beam are dramatically different (Fig. 12c, d); while the lower momentum group is
almost isotropic the higher one has a strong forward enhancement.
We note that the dependence of Q on this angle is most sensitive
for the higher momenta while at lower momenta it is equally sensitive to all three variables. Therefore, an enhancement in Q is
expected to reflect itself more significantly on the angular distribution at higher momenta and vice versa. Fig. 13 shows that this
angular anisotropy is reflected to a resonance like structure at
Q ~ 20 MeV and this peak can not be interpreted as a reflection of
$\bar{P}$"P" scattering with subsequent $\bar{P}N_s$ absorption. For one thing the
elastics produce narrower angular distributions but more importantly
there is no evidence of this in Fig. 12c. From the size of the
observed forward enhancemnt (~ 800 events) and the ratio of
"elastics" in these momentum ranges (781/144) ~ 4300 "elastic" events
should contribute to Fig. 12c with a sharp angular distribution if
the angular enhancement (Fig. 12d) were due to "elastics".

In Fig. 14 the $\bar{P}N$ annihilation cross-sections are presented,
apart from overall normalization constants for the low and the high
spectator momenta groups.

The following are their main features and our conclusions:

(i)   Copious production of the 20 MeV structure. It is parti-
      cularly associated with high spectator momenta or
      equivalently with high relative momentum which may suggest
      high spin. Note the ($\bar{P}N$) mass distribution for $\bar{P}$"P"
      elastics which rule out $\bar{P}$"P" elastic with subsequent re-
      absorption is the observed effect. Simple minded fits
      (Fig. 14) give for this structure (Q, Γ) = (20.0 ± .9,
      22.6 ± 1.0) MeV which are consistent with the observed
      rise in $\bar{P}D$ cross-sections below .35 GeV/c.

(ii) A signal at ~ 55 MeV (Fig. 14a) is consistent with the
      observations of Carrol et al.[1]. Our fits give
      (Q, Γ) = $(56.7 ^{+\ 2.5}_{-\ 1.5}, 4.4 ^{+}_{-} 3.6)$ MeV in good agreement with
      the results $(55.5 ± 2, 9 ^{+4}_{-3})$ MeV of Carrol et al.

(iii) Sharp increases at large negative values. We do not know
      their nature but they depend on the ($\bar{P}N$) relative momentum.
      Analysis of the events with (100 - 150) MeV/c spectator
      moment has also been done. The increase (Fig. 14a) for
      Q < -5 MeV disappears except for the point at - 7.5 MeV
      which shows ~ 5 σ deviations from its neighbors. We note
      that this point is in the region where an enhancement for
      $\bar{P}D \rightarrow \pi^-\pi^\circ P_s$ at rest has been observed[9].

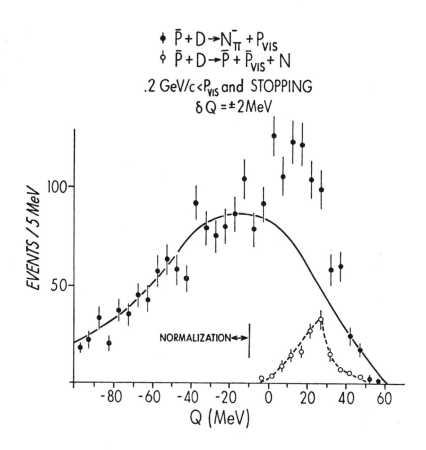

Fig. 13

Invariant mass − $2m_p$ distribution of $(\bar{P}N)$ annihilations with a proton of $>$ .2 GeV/c and Stopping. Curve is the expected distribution based on isotropic $\bar{P} - P_{vis}$ distribution.

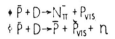

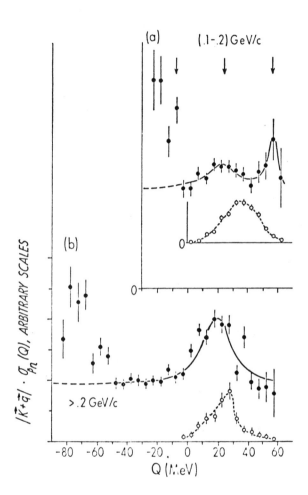

Fig. 14

See text for explanation. Distributions for "elastics" are the
measured Q distributions. (a) Fit to two Breit-Wigners and uniform
background. Results $(Q, \Gamma) = (22.7 \pm 2.4, 19.4 {}^{+2.4}_{-1.4})$,
$(56.8 {}^{+2.5}_{-1.5}, 7.6 {}^{+2.5}_{-1.3})$ MeV. (b) Fit to Breit-Wigner plus univorm
background. $(Q, \Gamma) = (19.4 \pm 1.0, 23.5 \pm 1.2)$ MeV.

In conclusion, two I = 1 $\bar{N}N$ inelastic and relatively narrow structures are firmly established at low energies.

C. $\bar{P}P \rightarrow 2\pi^0$

We have observed this interesting reaction at low momenta[13,10]. Note that because of parity and statistics this reaction does not proceed from odd $\bar{N}N$ orbitals. In fact it proceeds from a small fraction (1/4) of all possible $\bar{N}N$ states: I = 0, and $^3P_0$, $^3P_2$, $^3F_2$, $^3F_4$, $^3H_4$, $^3H_6$, etc.

Since it is forbidden from S-states by assuming "normal" low energy behavior this cross-section should go as $P_L{}^3$. Contrary to this prediction   it increases (Fig. 15) faster than $1/P_L$ at low momenta and faster than the annihilation cross-section. (See also Binnie's talk to this conference on anomalous threshold behaviors).

The angular distributions (Fig. 16) remain unchanged through the observed energy range. We fitted them in terms of initial states (L_J) using

$$p^2\frac{d\sigma}{d\Omega} = |\sum_J(A_J\sqrt{\frac{J+1}{2(2J-1)}}\ Y_J^1 + B_J\sqrt{\frac{J}{2(2J+3)}}\ Y_J^1)|^2$$

$$+ |\sum_J(A_J\sqrt{\frac{J}{2J-1}}\ Y_J^0 - B_J\sqrt{\frac{J+1}{2J+3}}\ Y_J^0)|^2$$

$$+ |\sum_J(A_J\sqrt{\frac{J+1}{2(2J-1)}}\ Y_J^{-1} + B_J\sqrt{\frac{J}{2(2J+3)}}\ Y_J^{-1})|^2$$

where $A_J = A_{L+1}$ and $B_J = B_{L-1}$ are complex numbers and have been optimized. The Results of the fits are compared with the data (Fig. 16) and the contributions of the various initial states are shown in Table II.

Table II. $2\pi^0$ Angular Distributions: Evidence for $\vec{S}\cdot\vec{L}$ Forces

$^{2S+1}L_J$	$<P_L> \sim 270$ MeV/c	$<P_L> \sim 520$ MeV/c	All P_L
$^3P_0$	73.0 ± 12.	79.4 ± 17.	75.6 ± 9.8
$^3P_2$	4.2 ± 0.6	5.5 ± 1.0	4.6 ± .5
$^3F_2$	17.1 ± 1.6	14.3 ± 2.0	15.8 ± 1.3
$^3F_4$	3.1 ± 0.3	0.66 ± 0.3	1.9 ± .2
$^3H_4$	2.6 ± 0.3	0.14 ± 0.5	2.6 ± .3

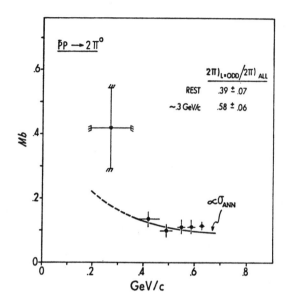

Fig. 15

$2\pi^o$ cross-sections at low momenta from the Columbia-Syracuse collaboration comparison with $\pi^+\pi^-$ data (see Ref. 10) the contribution to 2π of the odd $\overline{N}N$ orbitals is computed. No significant difference at rest and flight is apparent. The lowest point shows upper and lower limits. For this point great care was required to separate at rest from in-flight

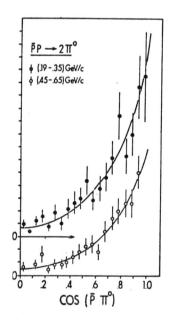

Fig. 16

Angular distributions. Fits are done to a series of all allowed initial $\overline{N}N$ states (see text, Table II).

Inspection of these results shows that high waves (at least L = 3) are important in all this low energy region. Furthermore, the L-1 states contribute much more than the L+1 (compare e.g. $^3P_0$, $^3P_2$), and that for the same J the L-1 is preferred more than the L+1 (compare e.g. $^3P_2$, $^3F_2$). These observations suggest strong initial state spin-orbit interactions which is an important ingredient of the predictions based on NN potentials[14] for $\overline{N}N$ resonant and bound states.

V. COMMENTS ON THE ABSENCE OF INTERACTIONS WITH THE SPECTATOR

We have stressed over the years[2,3] the remarkable observations[18,19] that the products of the annihilations at rest in deuterium do not see (interact) with the spectator. Based on the observed annihilation spectra, the known π-N differential cross-sections, and the observed spectator momentum distribution the characteristics for the secondary interactions $\pi+N_S \rightarrow \pi'+N$ can be predicted with certainty except for their frequency. It is found that in such collisions the scattered spectator will have mainly momenta greater than .2 GeV/c and the events with secondary interactions should exhibit Δ's in strong disagreemtn with our observations (Fig. 17). Another prediction for secondary interactions is that their fraction, in a given pion multiplicity, should be proportional to the multiplicity. There is no such increase of non-spectator ($\geq$ .2 GeV/c) events[11]. These observations can be summarized with the safe upper limit

$$0.1 \left\{ \sigma_{\pi N_S} \left\langle \frac{1}{4\pi r_D^2} \right\rangle \right\}$$

for the probability of pions undergoing secondary interactions with the spectator.

The expected amount of secondary interactions,

$$\sigma_{\pi N_S} \left\langle \frac{1}{4\pi r_D^2} \right\rangle ,$$

is strictly valid for spectator like interactions and annihilation times short enough ($\leq 10^{-23}$ sec) for the preservation of the spectator-annihilation region geometry which at t = 0 (the time of capture and the time that the spectator starts its free motion) is the deuteron geometry. Experimentally, it is observed that in about 70% of all annihilations the nucleon has a low momentum distribution which fits to the Fourier transform of the deuteron wave function. On this basis the above geometrical arguments should be valid except for the annihilation time. Using the relation of free motion between the spectator and the annihilation region the separation between source and target is given in terms of the deuteron wave function and the annihilation time. We then conclude, and in the absence of alternative quantitative suggestions, that $\tau_{ann} \geq 5 \times 10^{-22}$ sec $\approx 100 \times \lambda\pi/c$.

A similar conclusion is also reached from the absence of annihilations on the spectator neutron after "elastic" $\overline{P}$"P"

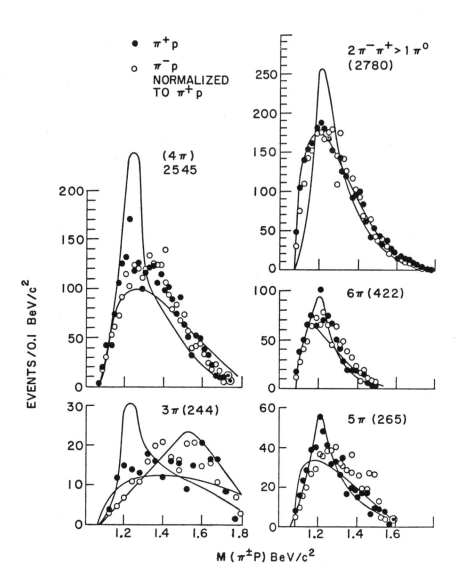

Fig. 17

M($\pi^{\pm}$P) distributions for at rest $\bar{P}$D annihilations with a proton
(> 250 MeV/c). Curves are based on calculating $\pi P_S \rightarrow \pi'p$ from
the known annihilation π-spectra, spectator distribution (Hamada-
Johnson) and the known πP phase shifts.

scattering. It is relevant to note here that the two step process
$K^-D \to \{(K^-P_s) \to \pi^-\Sigma^+\} + N_s \to \pi^- + (\Sigma^+N_s \to \Lambda P)$ for K^- interactions at
rest is dominant[20]. The angular distributions of the spectator
proton with the antiproton (Fig. 12c) is the most convenient for
searching for evidence for this two step process. If 100 events are
used as an upper limit for double scattering and taking into account
that ~ 10^4 events have undergone scattering with proton momentum
(0.1 - .2) GeV/c in the film in which these annihilations have been
found, we then deduce that < 1% of the scattered antiprotons
annihilate on the spectator and this is in agreement with the πN_s
results in annihilations at rest. These persistent and remarkable
absences of two step processes involving both nucleons of the
deuterium are true for at rest annihilations and low energy elastic
scattering and suggest that $\overline{N}N$ interactions are characterized with
unexpectedly long lifetimes.

On the basis of these results and the uncertainty principle,
$\tau\Gamma = h$, we get that $\Gamma \le 1.5$ MeV. If this is interpreted in terms of
s-channel resonances such resonances should dominate $\overline{N}N$ interactions
above and below threshold. If indeed this is the only interpretation
we will be faced with a new dimension in Boson Spectroscopy: a rich
micro-structure a la nuclear physics.

VI.   CONCLUSIONS

The present observations of antinucleon-nucleon interactions
safely warrants the following conclusions:

(i)   Cross-sections are parametrized well in terms of
       $A + B/P_L$. This parametrization defines sharply the high
       ($P_L > 1.9$ GeV/c) and the intermediate ($P_L \ge .5$ GeV/c)
       regimes.

(ii)   The energy dependent parts of cross-sections are I-spin
       dependent. The I = 0 terms are larger by a factor of two
       in the high and three in the intermediate regions than
       I = 1 terms.

(iii) The transition region is near the T and U bumps which
       have been observed in several experiments but whether
       these ~ 50 MeV wide structures represent resonances is
       still under debate.

(iv)   Recent high statistics and good resolution experiments in
       the resonance region (2m to ~ 2m + 70 MeV) establish the
       two narrow I = 1 structures in the annihilation channels
       with (Q, Γ)

$$(56.7 \, {}^{+ \, 2.5}_{- \, 1.5}, \, 4.4 \pm 3.6) \text{ MeV}$$

$$(20.0 \pm .9, \, 22.6 \pm 1.0) \text{ MeV}$$

(v) High $\bar{P}P$ angular momenta contribute to the $2\pi^\circ$ at low momenta and the cross-section does not behave as expected from centrifugal barrier arguments. Furthermore, the contributions of the various waves indicate strong spin-orbit interactions.

(vi) There are evidences for two $I^G = 1^+$ narrow (< 10 MeV) bound states with $Q = -$ ($0 - 10$) and $- 83$ MeV.

(vii) The persistent absence of interactions involving both the nucleons in spectator-like $\bar{P}D$ interactions strongly suggest that the $\bar{N}N$ interactions proceed via s-channel narrow (≤ 1 MeV) states.

REFERENCES

1. Carrol et al., Phys. Rev. Let. 32, 185 (1974).

2. Proceedings of the seminar on "Interactions of High Energy Particles with Nuclei and new Nuclear-like Systems", ITEP, 1973 (to be published).

3. CERN Report No. 72-10(73).

4. Abrams et al., Phys. Rev. D1, 1917 (1970).

5. Alspector et al., Phys. Rev. Let. 30, 511 (1973).

6. Cutts et al., preprint submitted to this conference. See also Phys. Rev. Let. 32, 950 (1973).

7. Eisenhandler et al., Physics Letters 47B, 531 (1973).

8. Eisenhandler et al., Physics Letters 47B, 536 (1973).

9. Gray et al., Phys. Rev. Let. 30, 1091 (1973).

10. Devons et al., Phys. Rev. Let. 27, 1614 (1971).

11. Gray et al., Phys. Rev. Let. 26, 1491 (1972).

12. Bizzarri et al., Let. N. C., 1, 749 (1969) and Kinsky et al., Bull. Am. Phys. Soc. 15, 638 (1970) and 16, 138 (1971).

13. Dris et al., Bull. Am. Phys. Soc. 18, 715 (1973).

14. Bogdanova et al. Phys. Rev. Let. 28, 1418 (1973). We acknowledge important communications on these matters with Prof. I. Shapiro.

15. Review of Particle Properties, Physics Letters 50B (1974).

16. Bizzarri et al., INFN/AE-73/8, Istituto Nazionale Fisica Nucleare report, 1973.

17. Tzanakos and Kalogeropoulos, Bull. Am. Phys. Soc. 18, 715 (1973) and Reference 2.

18. L. Gray, Ph. D. Dissertation, Syracuse University, 1969.

19. Bizzarri et al., Let. N. C. 2, 431 (1969).

20. Dahl et al., N. C. 27, 342 (1963).

MULTIHADRON PRODUCTION IN e+e- INTERACTIONS

W. Chinowsky
University of California, Berkeley

It should be recognized directly that e+e- annihilations are not the best source of meson resonant states. Why that is so can be understood already from values of total cross sections for hadron production. These are shown as a function of energy in Fig. 1. The higher energy data are mostly from the SLAC-LBL[1] collaboration at SPEAR. Included also are earlier results from CEA[2] and Frascati[3]. It must be emphasized that the quoted total cross sections are all to some extent dependent on the production model used to calculate detection efficiency. In the case of the SPEAR data it was assumed that the production proceeds according to the requirements of Lorentz invariant phase space only, with multiplicity distribution and average momentum of prongs adjusted to agree with the data. The average calculated detection efficiency as a function of energy is shown in Fig. 2. Normalization is provided by the collection of Bhabha scatters detected, with the assumption that this elastic cross section is given correctly by QED. Radiative corrections are approximately 1% and only mildly dependent on the assumed dependence of the cross section on S. It is estimated that less than 5% of the total cross section results from two photon exchange processes. That conclusion can be significantly modified only if the cross section for hadron production in photon-photon annihilations[4] is anomalously large and strongly increasing with energy. These hadron yields are in violent disagreement with expectations of a naive parton model of the structure of hadrons which requires that the cross section decrease as E^{-2}. Partons with structure can of course be conceptualized in such a way as to yield any desired energy dependence of the annihilation cross section. A parton form factor due to exchange of "gluons" yields an e+e- annihilation cross section depending on S as $1/S \times (1 + 2S/M_G^2)$ [5], where M_G is the gluon mass. A gluon mass of 10 GeV/c^2 cannot then account for the observed constant cross section. It is fair to say that there does not as yet exist any model leading to a satisfactory understanding of the behavior of the total cross section[6].

Another characteristic of these events important for the identification of meson states is the large multiplicity. Fig. 3 shows the observed charged-particle multiplicity corrected for detection efficiency as a function of energy. The average multiplicity is increasing slowly, if at all, from ∿3.5 to ∿4.0 in the energy range 3.0 GeV-5.0 GeV. The average total energy in charged particles is shown in Fig. 4. The fraction of total energy in charged particles decreases slowly, if at all, with increasing energy.

Meson production processes may be represented by the single photon exchange diagrams

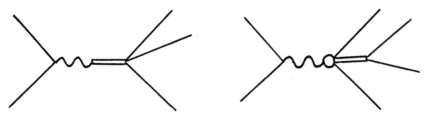

as well as the two photon exchange

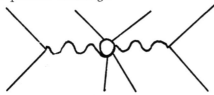

The last produces states even under charge conjugation, the others odd.

Many examples of vector meson production and decay according to the first diagram exist in the literature.[7] These are revealed as narrow maxima in the cross section for annihilation into a particular channel. Two results of this kind, from Frascati,[8] have been submitted to this conference. The next two figures show cross sections for $\pi^+\pi^-\pi^+\pi^-$ and $\pi^+\pi^-\pi^0\pi^0$. Both peaks may be interpreted as resonances. The first is already known as ρ'. The second is newly reported here. A resonance fit to the data below 1.5 BeV is shown in Fig. 7. It may be noted that such a bump at 1250 MeV has been observed in the spectrum of meson masses recoiling from the proton in the reaction $\gamma p \to p\pi^+\pi^-$ + neutrals.[9] Those data are consistent with production of a resonance that may be either $B^0 (J^P = 1+)$ or a new $\rho'^0 (J^P = 1^-)$.

The results from SPEAR for the $\pi^+\pi^-\pi^+\pi^-$ channel are shown in Fig. 8 together with earlier Frascati results. This shows characteristic threshold behavior of the cross section for exclusive channel particle production. Clearly, a resonance interpretation is not definitively established by the peak in cross section.

Similar behavior is seen in the cross section for annihilation into six charged pions, shown in Fig. 9. Again there is a steep rise from threshold, the cross section reaching a maximum $\sim$1 BeV above threshold. This is followed by a rather less precipitous decrease at higher energy. Again it would appear hardly necessary to interpret this peak in the $6\pi^{\pm}$ cross section as indicating a resonant state.

The situation with the two photon exchange processes is still quite primitive. At SPEAR, no multihadron events have been identified as resulting from two photon exchange. A Frascati group[10] has reported to this conference two examples of the process $e^+e^- \to e^+e^-\pi^+\pi^-$ with $\pi^+\pi^-$ effective mass in the $\varepsilon(660)$ region. With these they obtain a $\gamma\gamma$ width, $\Gamma\varepsilon\gamma\gamma = 9.6^{+13.3}_{-8.0}$ KeV and an asymptotic ratio $\sigma(e^+e^- \to \text{hadrons})/\sigma(e^+e^- \to \mu^+\mu^-) = 5.8^{+3.2}_{-3.5}$ Also, two multihadron events produced by two photon exchange have been detected.[11] In one event, the multihadron effective mass was 1400 MeV/c^2 and in the other 800 ± 90 MeV/c^2. If the latter is interpreted as a decay of an η' meson, it follows that the $\gamma\gamma$ partial width is $\Gamma_{\eta' \to \gamma\gamma} = 11^{+15}_{-8}$ KeV. If the event is not a resonance decay, it follows that an upper limit is $\Gamma_{\eta' \to \gamma\gamma} < 33$ KeV with 95% confidence. Both these results were obtained with an experimental arrangement [12] permitting detection of a final state electron or positron together with the hadrons.

Since both momenta and time of flight of the produced charged hadrons are determined with the SPEAR magnetic detector, masses are measured. With the limitations of time and momentum resolution imposed by the detector characteristics, π^s and k^s are distinguished if their momenta are less than $\sim$600 MeV/c; protons are identified if their momenta are less than $\sim$800 MeV/c. Fractional yields of negative particles at 4.8 Bev are shown in Fig. 10. Positive particle yields are consistent with these but the proton sample particularly is more contaminated with background and so is not shown here. Qualitatively similar results are found at lower energy.

A search for production of mesons ρ^0, ϕ^0 and $K^*(890)$ was made among the sample of events with identified charged particles. Representative effective mass distributions are shown in Figs. 11, 12 and 13. There is slight evidence for production, but both the data and analysis are still in too crude a state to permit sensible estimates of relative yields.

In summing up, these results indicate the validity of the first statement. The e^+e^- annihilations are rather poor producers of meson resonant states and not likely to contribute much new knowledge of their properties. The interest in the study of the process is rather in elucidation of the fundamental problem of the structure of the elementary particles than in the pursuit of details of the hadron final state interactions.

REFERENCES

1. The participants in the SLAC-LBL experiment at SPEAR are:
   J.-E. Augustin, A. M. Boyarski, M. Breidenbach, F. Bulos,
   J. T. Dakin, G. J. Feldman, G. E. Fischer, D. Fryberger,
   G. Hanson, B. Jean-Marie, R. R. Larsen, H. L. Lynch, D. Lyon,
   C. C. Morehouse, J. M. Paterson, M. Perl, T. Pun, B. Richter,
   and R. F. Schwitters (SLAC)
   G. S. Abrams, D. Briggs, W. Chinowsky, C. E. Friedberg,
   G. Goldhaber, R. J. Hollebeek, J. A. Kadyk, G. H. Trilling,
   J. S. Whitaker, and J. E. Zipse (LBL).
2. A. Litke et al., Phys. Rev. Lett. 30, 507 (1973)
   G. Tarnopolsky et al., Phys. Rev. Lett. 32, 432 (1974).
3. F. Ceradini et al., Phys. Lett. 47B, 80 (1973)
   C. Bacci et al., Phys. Lett. 44B, J33 (1973).

4. S. Brodsky, SLAC-PUB-1322 (Oct. 1973) (Invited talk presented to the International Colloquium on Photon-Photon Collisions in Electron-Positron Storage Rings, College de France, Paris, Sept. 1973).
5. M. Chanowitz and S. Drell, Phys. Rev. Lett. $\underline{30}$, 807 (1973).
6. J. D. Bjorken, SLAC-PUB-1318 (Oct. 1973) (Invited paper presented at the 1973 International Symposium on Electron and Photon Interactions at High Energies, Bonn, Germany, Aug. 1973).
7. G. Cosme et al., Phys. Lett. $\underline{48B}$, 155 (1974)
    G. Cosme et al., Phys. Lett. $\underline{48B}$, 159 (1974)
    M. Bernardini et al., Phys. Lett. $\underline{46B}$, 261 (1973).
    Earlier references may be found in the review paper presented by C. Mencuccini to the Adriatic Summer Meeting on Particle Physics, Rovinj, (Sept.-Oct. 1973).
8. M. Conversi, L. Paoluzi, F. Ceradini, M. L. Ferrer, R. Santonico, M. Grilli, P. Spillantini and V. Valente.
9. J. Ballam et al., SLAC-PUB-1364 and LBL-2474 (1973).
10. S. Orito, M. L. Ferrer, L. Paoluzi and R. Santonico (Submitted to Phys. Lett. B).
11. G. Barbiellini, F. Ceradini, M. L. Ferrer, S. Orito, L. Paoluzi, R. Santonico and T. Tsuru, LNF-74/10(P) (1974).
12. G. Barbiellini and S. Orito, LNF-71/17 (1971) and Proc. First Conf. on Meson Resonances and Related Electromagnetic Phenomena, Bologna (1971).

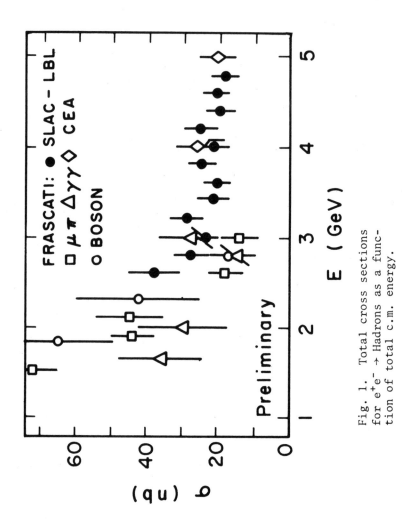

Fig. 1. Total cross sections
for e⁺e⁻ → Hadrons as a function of total c.m. energy.

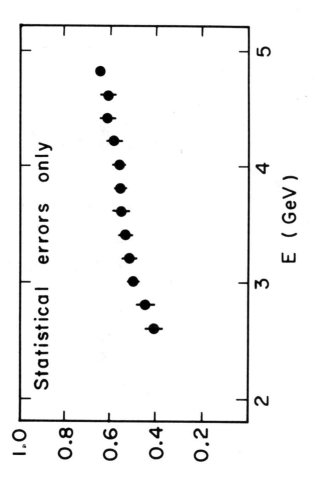

Fig. 2. Average detection
efficiency of SPEAR magnetic
detector as a function of
energy.

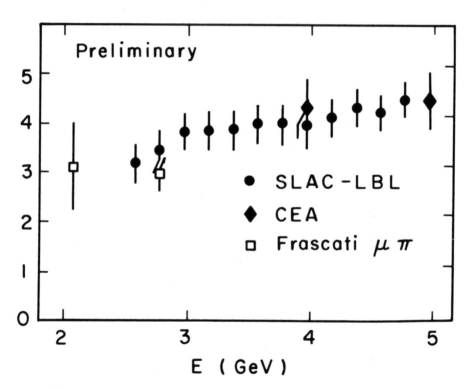

Fig. 3. Average charged-
hadron multiplicity as a
function of energy.

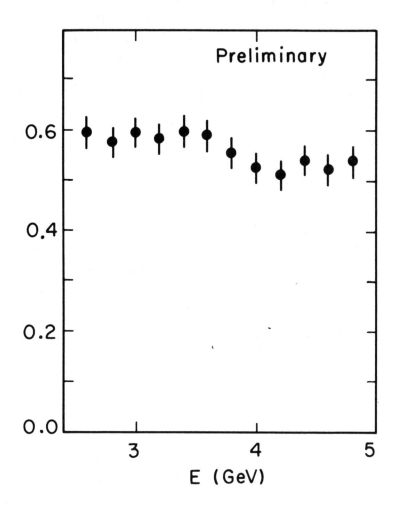

Fig. 4. Average fraction of
total energy in charged
hadrons as a function of
energy.

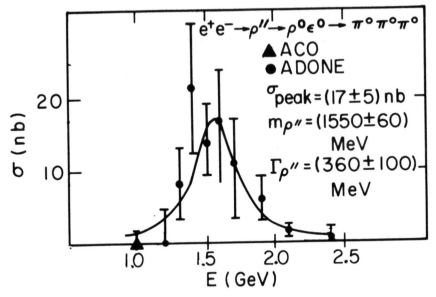

Fig. 5. Energy dependence of the cross section
for $e^+e^- \to \pi^+\pi^-\pi^+\pi^-$. Data from ADONE and ACO.

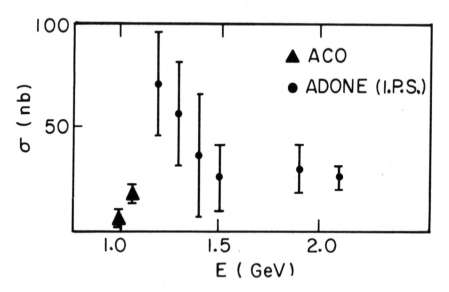

Fig. 6. Energy dependence
of the cross section for
$e^+e^- \to \pi^+\pi^-\pi^0\pi^0$.

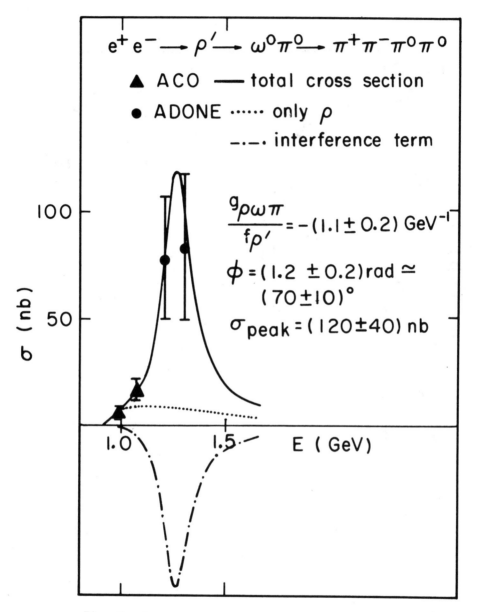

Fig. 7. Best-fit theoretical curve for the assumed process $e^+e^- \to \rho'(1250) \to \omega^0\pi^0 \to \pi^+\pi^-\pi^0\pi^0$, including contributions from $\rho(760) \to \omega^0\pi^0 \to \pi^+\pi^-\pi^0\pi^0$.

134

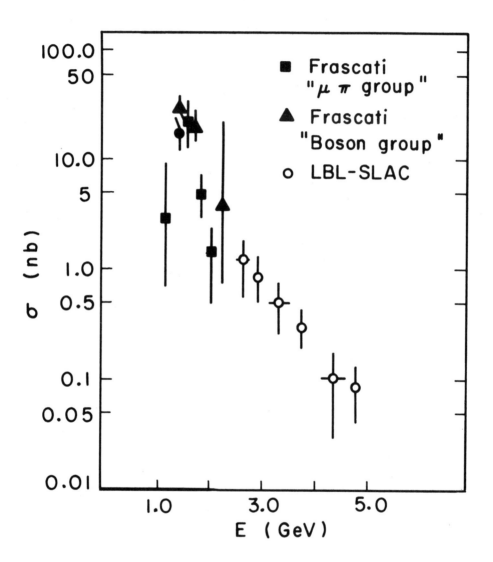

Fig. 8. Cross section for
$e^+e^- \to \pi^+\pi^-\pi^+\pi^-$. Data from
SPEAR and ADONE.

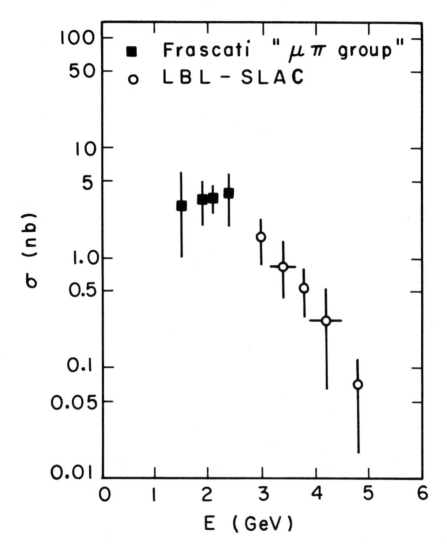

Fig. 9. Cross section for
$e^+e^- \rightarrow \pi^+\pi^-\pi^+\pi^-\pi^+\pi^-$. Data from
SPEAR and ADONE.

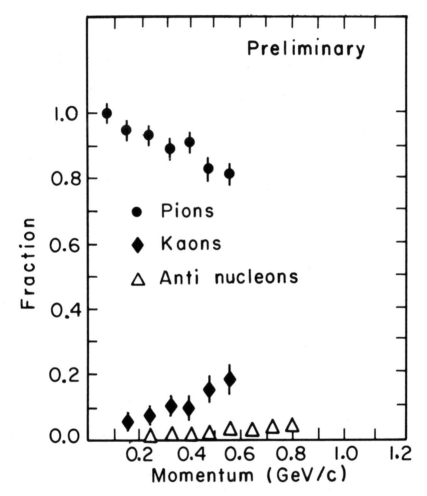

Fig. 10. Fractional yields of
π^-, K^- and antiprotons at
4.8 GeV e^+e^- c.m. energy as a
function of momentum.

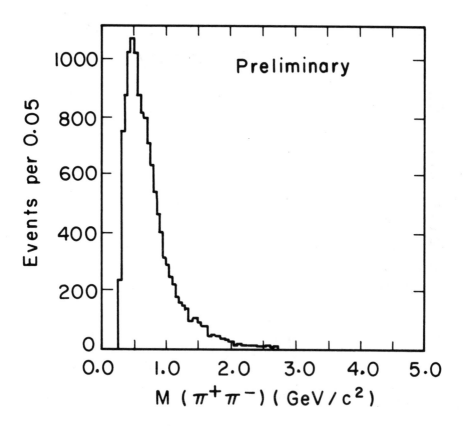

Fig. 11. Distribution in $\pi^+\pi^-$
effective mass at 4.8 GeV
e^+e^- c.m. energy.

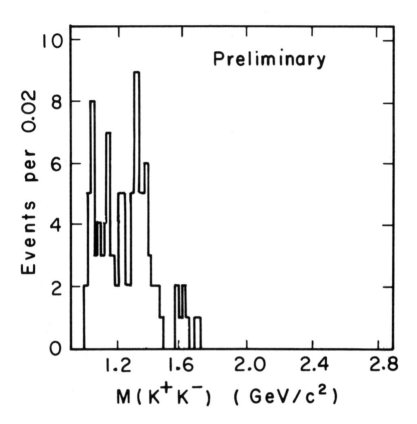

Fig. 12. Distribution in K+K-
effective mass at 4.8 GeV e+e-
c.m. energy.

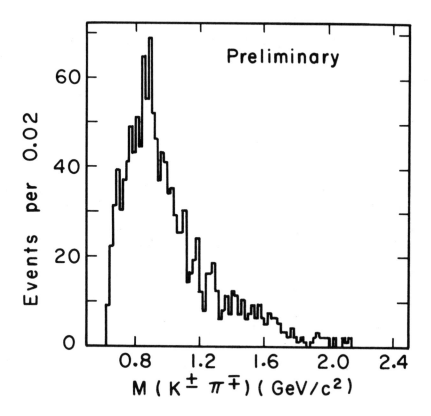

Fig. 13. Distribution in $K^{\pm}\pi^{\mp}$
effective mass at 4.8 GeV e⁺e⁻
c.m. energy.

REVIEW OF STRANGE MESONS[*]

R.L. Eisner

Brookhaven National Laboratory, Upton, New York 11973

ABSTRACT

A review of strange meson spectroscopy is presented.
The problem areas are still the status of the strange members of the two 1^+ nonets as well as the situation in the
high-mass K^* region.

I. INTRODUCTION

In the short time I have here today I will try to present a general review and point out some of the problems in strange meson
spectroscopy. In Fig. 1 is summarized what is essentially our present
knowledge of the status of bump-hunting. Here, at the intersection
of the grid of well-known nonets are indicated the resonances (or
enhancements) which are reasonably well established with the given
two-particle decay. Except for a few missing states (e.g. D meson)
most all known resonances are found with at least one of the possible
listed decay modes. The symbol N at an intersection spot means that
nothing of significance has been observed with the designated decay.
We will primarily be concerned with in:

Sect II: What we know well: 0^-, 1^-, 2^+ nonet
Sect III: Problem Areas: The Low-Mass Enhancements (LME)
Sect IV: High-Mass K^*'s
Sect V: Summary and Conclusions.

II. WHAT WE KNOW WELL

A. 0^- nonet strange member - K(495): no comment.

B. 1^- nonet strange member - K(890): no new results. The data on
K(890) properties is rather good. The mass difference between the
neutral and charged state has been measured[1] $(\Delta M(K^{*0}-K^{*-}) = 5.7 \pm 1.7$
MeV) and shown to be in the same direction as that of the ground
state K(495).

C. 2^+ nonet strange member - K(1420):
  1. J^P: Aguilar-Benitez, et al.,[1] have performed a J^P analysis
of the K(1420) produced in the reaction $K^-p \rightarrow K^-(1420)p$ for combined
data at 3.9 and 4.6 GeV/c. The analysis was done on a sample of
non-peripherally produced events [away from the π-exchange region so
that S-D interference effects are minimized]. The authors assume
that the K(1420) does not interfere with the background and perform
a simple background subtraction on the Jackson $\cos\theta$ and ϕ angular

---

*   Work performed under the auspices of the U.S. Atomic Energy
    Commission.

distributions. The assignment $J^P=1^-$ is ruled out; however, 2^+ and 3^- cannot be distinguished. If some dynamical assumptions are employed and if all $\rho_{mm'}$ with $m,m' > 1$ are set equal to zero, then the $\cos\theta$ distribution is found not to be fitted by the 3^- hypothesis. Thus, 2^+ is strongly favored.

2. <u>Branching Ratios: $K\pi\pi/K\pi$</u>: There have been some discrepancies in the $K\pi\pi/K\pi$ branching ratio of the K(1420). It has been shown[1] that this number is best obtained when the K(1420) is produced in charge-exchange reactions rather than in reactions [i.e. $Kp \rightarrow (K\pi\pi)p$] which have strong background problems [i.e. Q production]. A new result submitted to the conference by the Brussels/CERN/Mons/Munchen Collaboration[2] [reaction $K^+n \rightarrow K(1420)p$ at 4.6 GeV/c] gives

$$\frac{K(1420) \rightarrow K\pi\pi}{K(1420) \rightarrow K\pi} = 0.37 \pm 0.10$$

which is in excellent agreement with previous determinations from the reaction $K^-p \rightarrow K(1420)n$:

$$
\begin{aligned}
3.9 \text{ GeV/c [BNL]}^1&: \quad .40 \pm .10 \\
4.6 \text{ GeV/c [BNL]}^1&: \quad .46 \pm .10 \\
4.1, 5.5 \text{ GeV/c [ANL]}^3&: \quad .49 \pm .10
\end{aligned}
$$

III. PROBLEM AREAS

A. <u>Q Region</u>

The Q region has been extensively studied, though not with the statistics which presently make up the A_1 data sample. As such, the status of the Q is even more uncertain than is that of the A_1. SU(3) and the quark model have an interest in the Q region: a brother of the A_1 [$J^{PC} = 1^{++}$] and the B meson [$J^{PC} = 1^{+-}$] are both expected to be present somewhere in the Q region [i.e. $M(K\pi\pi) < 1.5$ GeV].

The present experimental evidence argues against a simple single [Breit-Wigner-type] resonance in the Q region. [This is not to say that the data are in agreement with a multi-resonance hypothesis.] Let us summarize:

<u>Arguments Against a Single Resonance in Q Region:</u>
(1) A compilation of the data[4] on the reaction $K^+p \rightarrow (K\pi\pi)^+p$ for momenta between 7.3 and 13.0 GeV/c reveals a non-Breit-Wigner shape of the Q enhancement. As shown in Fig. 2, a sharp rise near threshold and a rapid drop at $M(K\pi\pi) \sim 1.28$ GeV are observed.
(2) The slope of $d\sigma/dt$ rapidly varies across the $K\pi\pi$ mass in the Q region for Q produced in the reactions $K^+p \rightarrow Q^+p$ [5] [Fig. 3] and $K^0(\bar{K}^0)p \rightarrow Q^0(\bar{Q}^0)p$ [Ref. 6].

(3) Assuming a 1^+ hypothesis, Bingham, et al.,[5] have fitted the Dalitz plot in the reaction $K^+p \rightarrow K^+\pi^+\pi^-p$ as a function of $K\pi\pi$ effect ive mass with a matrix element consisting of interfering ρ and K^* amplitudes. The results shown in the lower portion of Fig. 3 for α, the ratio of the $K\rho$ to $K^*\pi$, indicate a large contribution from $K\rho$ at low $M(K\pi\pi)$ [$\lesssim 1.30$ GeV] for all the data studied [i.e. $2.5 \leq P_{LAB} \leq 12.7$ GeV]. This ratio is found to decrease rapidly with increasing $K\pi\pi$ mass.[7] Such an effect is not consistent with a single resonance interpretation.

On the other hand, there is very little evidence for two distinc resonances (i.e. K_A and K_B). In fact, let us suppose that a bona fid K_B was present in the Q enhancement. From generalized C-parity argu- ments and from the fact that P is an SU(3) singlet, we would expect the $g_{KK_BP} = 0$ so that $K\rho \xrightarrow{P} K_BP$. This would lead to the expectatio that

(a) The portion of the Q peak corresponding to the K_B [only Regg exchange] would decrease faster with energy than that of the K_A [whic could couple to P]. This would lead to a distinct difference in the energy dependence of the mass spectra. However, as shown[5] in Fig. [top], there is no significant difference between the energy dependen of any of the mass segments of the Q. All portions are observed to f with $n \simeq 0.6$ [using $\sigma \sim p^{-n}$]. Such a behavior also argues against an significant difference in the structure of the Q at different inciden momenta.

(b) The K_B should not be strongly produced in the reaction $K^+d \rightarrow Q^+d$. As such, a difference between the Q structure in the reactions $K^+p \rightarrow Q^+p$ and $K^+d \rightarrow Q^+d$ might result. The comparison of these two ma spectra by Firestone, et al.,[9] [at 12 GeV/c, see Fig. 4 -bottom] r veals, apart from the influence of deuterium effects, no significant changes in the shape of the $K\pi\pi$ low-mass enhancement.

However, remarks (a) and (b) above are tempered by some theoretical possibilities regarding the 1^{++} and 1^{+-} nonets. In par- ticular, as shown by Kane,[10] it is possible [theoretical] that the nonet states $|K_A\rangle$ and $|K_B\rangle$ mix to produce the physically observed $|Q_A\rangle$ and $|Q_B\rangle$ states; viz.

$$|Q_A\rangle = |K_A\rangle \cos\phi + |K_B\rangle \sin\phi$$

$$|Q_B\rangle = -|K_A\rangle \sin\phi + |K_B\rangle \cos\phi$$

With such mixing, any statement regarding couplings of $|Q_A\rangle$ and $|Q_B\rangle$ t P would depend critically on the mixing angle, ϕ. If, for example, $\phi \simeq 45°$, no simple statement such as points (a) and (b) above could b made.

B. The Rest of LME (for now?)

We show, for completeness, other strange LME which have so far been observed [Fig. 5]:

$$Kp \rightarrow Lp^{(11)}; \; L \rightarrow K(1420)\pi \text{ [to be discussed below]}$$
$$Kp \rightarrow (K\omega)p^{(12)}$$
$$Kp \rightarrow (K\phi)p^{(12)}$$
$$Kp \rightarrow (Kf)p^{(13)}$$

C. Additional Complications to the Resonance Interpretation of LME

1. Galtieri, et al.,[12] [12 GeV/c] have shown that to obtain a LME in the reaction

$$Kp \rightarrow (X\pi)p$$

it is not necessary that X be in the resonance region [i.e. be a K(890) or K(1420)]. This is seen in Fig. 6 where the $X\pi$ spectrum is shown for different X selections. The $X\pi$ LME traveling peaks are evident.

2. Cohen, et al.,[14] have compared the reactions

$$\pi^+ p \rightarrow (\pi^+ \rho^0)p$$

and

$$\pi^- n \rightarrow (\pi^- \rho^-)p$$

at 7 GeV/c. With similar selections they observe LME in $\rho^0 \pi^+$ [A_1] and in $\rho^- \pi^-$ [exotic]. These spectra are shown in Fig. 6 in which the solid curves represent the predictions of a double Regge-pole model [DRPM]. They argue that if one accepts the duality notion that agreement with experiment and DRPM implies a resonant interpretation for LME, one is forced to have an exotic $\rho^- \pi^-$ resonance.

D. Search for Non-Diffractive K_A and K_B

Given the problems discussed above, it appears best to search for the two K^*'s [i.e. K_A and K_B] in channels not dominated by the seemingly nonresonant LME. If the broad structures corresponding to the Q, A_1, A_3, etc., are indeed composed of resonant states, then there is no reason for their not being produced in nondiffractive channels. The following examples of searches for nondiffractive states are not exhaustive, but serve to give some feeling as to the present state of the art.

a. $KN \rightarrow (K^*\pi)^0 N'$

Werner, et al.,[15] have made a compilation of all available data on $K^*\pi$ states [$\sim$ 100 events/μb] produced in charge-exchange reactions. The appropriate effective mass spectra are shown in Fig. 7. Whereas some amount of a broad Q enhancement [e.g. cross-hatched area in Fig. 7a] can be accommodated by the data there is no evidence, apart from the K(1420), in the $K\pi\pi$ $I_z = 1/2$ spectra for any narrow enhancements.

b. $\pi^- p \to \Lambda(K\pi\pi)$

Figure 8 shows the $K\pi$ and $K\pi\pi$ effective mass spectra at 4.5 and 6.0 GeV/c incident π^- momenta of Crennell, et al.[16] In addition to the K(1420) there is evidence at both momenta for production of a narrow ($\sim$ 60 MeV wide) $K\pi\pi$ state at a mass $\simeq$ 1300 MeV. The lower portion of the figure shows the $\pi\pi$ and $K\pi$ mass projections as a function of $K\pi\pi$ mass for the 4.5-GeV/c data. In the K(1300) region there is evidence for a surplus of ρ^- events ($\sim 3\sigma$) with little evidence for an accumulation of $K^*(890)$ events above background. This leads the author to suggest that K(1300) decays dominantly into Kρ.

c. In $\bar{p}p$: The C Meson

Evidence has been presented by Astier, et al.,[17] for production of a $K\pi\pi$ enhancement, deemed the C meson, at a mass of $\simeq$ 1250 MeV, produced in $\bar{p}p$ annihilations at rest. The $K\pi\pi$ effective mass spectrum is shown in Fig. 9 for all reactions in which the C meson is claimed to be produced. A Zemach analysis gives $J^P(C)=1^+$ with,

$$M(C) = 1242 \, {}^{+\,9}_{-10} \text{ MeV}$$

$$\Gamma(C) = 127 \, {}^{+\,7}_{-25} \text{ MeV} \, .$$

A large C $\to$ Kρ branching fraction is claimed.

Problems:

(1) S-wave capture is assumed and used to prove both the existence as well as the J^P of the C. The new results[18] that the reaction $\bar{p}p \to \pi^0\pi^0$ is observed at rest (not allowed by S-wave capture) cast some doubt on this assumption.

(2) The $K\pi\pi$ enhancements [Fig. 9] are all at the peak of the phase space.

(3) As shown in Fig. 9, the C meson is not produced in $\bar{p}p$ annihilations at 0.7[19] or 1.2[20] GeV/c.

E. Conclusions on Non-Diffractive Q Search

Where do we stand with non-diffractive Q production? I feel that the data are not of such statistical quality that the statement c be easily answered; however, one can make a decent hand-waving argumen for production of a $J^{PC} = 1^{++}$ nonet member [K_A] with mass between 1.2 and 1.35 GeV. This follows if we accept and associate with the same object:

(a) The K(1300) observed in $\pi^- p \to \Lambda K(1300)$; large K$\rho$ decay mo

(b) The C meson; large Kρ decay mode.

(c) The lower portion of the diffractive Q [i.e. with M($K\pi\pi$) $\leq$ 1.3 GeV which shows a Kρ decay mode].

These observations tie together in that (a)-(c) all give an enhancement with a dominant Kρ decay mode. I deduce the enhancement to belong to the C=+1 nonet from the observation that the Kρ content of the Q does not change with energy [see Fig. 3] which implies dominant P coupling. The outcome of all this is that there appears to be no strong evidence for production of the K_B in all the data investigated. Clearly, more data on nondiffractive K$\pi\pi$ production is needed.

IV. HIGH-MASS K*'S

A. Non-Charge-Exchange Channels and the Region Between 1.6 and 2.0 GeV: the L Meson

Problem: The experimental situation is best described as being confused. The problem centers about the M, Γ and branching ratios of the so-called L meson. In particular, the question is whether or not the K$\pi\pi$ enhancement observed in reactions such as

$$K^{\pm}p \rightarrow (K\pi\pi)^{\pm}p$$

is uniquely a K(1420)π threshold enhancement [i.e. the sister of the $2^-S(f\pi)$ possibly non-resonant state].

The experimental situation is here summarized through a compilation of most of the available data on L production[21-26]:

1. M,Γ of the L: [See Fig. 10]

Comment: The quoted values for M(L) are all consistent. The Berkeley experiment[21] quotes a large width [$\Gamma \approx 250$-300 MeV, but no fit]. The quoted widths of the other experiments are not consistent with this value. In particular, the K$^-$p experiments at 4.6,[23], 10.0[24] and 10.9-15.9[26] GeV/c all quote widths about half as wide as that of the 12.0-GeV/c experiment. As is clear, mass and width determinations crucially depend on background parametrizations. Therefore, the errors on Γ, in particular, are subject to systematic uncertainties which could be much larger than the quoted statistical errors.

2. Branching Ratios: Or is the L entirely a K(1420)π threshold effect? [See Fig. 11 for a compilation of the different experimental techniques which try to answer this question.]

Comments

a. $K^+p \rightarrow K^+\pi^+\pi^-p$ at 12 GeV/c[21]: Figure 11 shows the "K(1420)π" [i.e. after a K(1420) mass slice]. A broad enhancement is observed. The authors claim that the L signal [Fig. 10] is undiminished by this cut [i.e. L $\rightarrow$ K(1420)π is consistent with 100%]. Needless to say such a statement relies heavily on background uncertainties, etc. For instance, in the K(890)π spectrum the authors claim no L $\rightarrow$ K(890)π; however, a background could be drawn in Fig. 11 which could produce a sizable enhancement in the L region.

b. $K^+p \rightarrow K^+\pi^+\pi^-p$ at 10 GeV/c[22]: In an attempt to determine the K(890), K(1420) and ρ content as a function of K$\pi\pi$ effective mass, the authors have sliced the $K^+\pi^+\pi^-$ mass spectrum and have fitted the corresponding Dalitz plots. The results are shown in Fig. 11 [note th 100-MeV bin size]. I have indicated [dotted line] the bins correspond ing to the L peak in the total K$\pi\pi$ mass spectrum of Fig. 10. The authors claim evidence for a significant increase in the amount of bot K(890)π and K(1420)π in the L region. It does appear that the observe peak in the total K$\pi\pi$ mass spectrum is contributed to by many differen states and not entirely by K(1420)π.

c. $K^-p \rightarrow K^-\pi^-\pi^+p$ at 10 GeV/c[24]: A different tack: Try to show that [L $\rightarrow$ K(1420)π]/[L $\rightarrow$ K$\pi\pi$] is not 100%. Figure 11 shows the K$\pi\pi$ effective mass spectrum for events not in the K(1420) $\rightarrow$ $K^-\pi^+$ region [i.e. 1.32-1.52 GeV]. Clearly, an excess of events is apparent in the L region. The number of L events in this mass distribution, as determined by a Breit-Wigner fit, is 166 $\pm$ 31 compared to 321 $\pm$ 60 events [see Fig. 10] in total. The tails of the K(1420) are not sufficient to explain the difference between these numbers and that expected for a pure L $\rightarrow$ K(1420)π decay mode.

d. $K^-p \rightarrow K^-\pi^-\pi^+p$ at 4.6 GeV/c[24]: Figure 11 shows the Kπ mass spectrum corresponding to a K$\pi\pi$ slice covering the L enhancement. Only a small amount of K(1420) is observed which is not sufficient to account for all the L. The authors quote the branching ratio

$$\frac{L \rightarrow K(1420)\pi}{L \rightarrow Total} = (20 \pm 20)\% \ .$$

e. $K^-d \rightarrow K^-\pi^+\pi^-d$ at 12.6 GeV/c[25]: Figure 11 shows the K^*_{890} and $K^*_{1420}\pi$ effective mass distributions. Even with the small statisti it appears that the number of L events in the latter spectrum cannot possibly account for the entire observed L peak in Fig. 10.

3. Summary of Properties of the L

Reaction	M(quoted)	Γ(quoted)	Branching Ratio + other comments (quoted)
12 GeV/c $K^+p\rightarrow(K\pi\pi)^+p$	1780	250-300	L consistent with K(1420)π, but no numbers
10 GeV/c $K^+p\rightarrow(K\pi\pi)^+p$	1761±10	91±12	L not all K(1420)π
4.6 GeV/c $K^-p\rightarrow(K\pi\pi)^-p$	1745±20	100±50	L not all K(1420)π
10 GeV/c $K^-p\rightarrow(K\pi\pi)^-p$	1780±15	138±40	L not all K(1420)π [Also, see PWA in talk of D. Hansen, this conference[27]]
12.6 GeV/c $K^-d\rightarrow(K\pi\pi)^-d$	≈ 1740	≈ 130	L not all K(1420)π
10.9-15.9 GeV/c $K^-p\rightarrow L^-p$	1767±6	100±26	No information

4. Best Bet on L
   a. Certainly some K(1420)π threshold effect.
   b. Which cannot explain entire "L" enhancement.

Figure 12a,b shows the results of a partial-wave analysis of combined data at 10.0 and 16.0 GeV/c on the reaction $K^-p \rightarrow K^-\pi^+\pi^-p$.[27] Of interest is the distribution of the difference between the total mass spectrum and that corresponding to the intensity of the partial wave of the LME [i.e. for the L → K(1420)π, the 2⁻ S wave]. As observed in Fig. 12b, a significant non-K(1420)π signal is present in the L region; the authors claim that
   c. The L region is very complicated with many partial waves (not necessarily resonant) contributing.

5. Other Enhancements (Decay Modes) in the L Region
   a. Colley, et al.,[22] [10 GeV/c, K^+p] have presented a claim for a $K^*\rho$ enhancement in the L region. Their results are suggestive, but more data are needed for a stronger demonstration of the K^*-ρ correlation.
   b. Previously published claims for L → Kω [i.e. with M(Kω) = M(L) ≈ 1760-1780 MeV] are not verified.

Two new results which show evidence for a narrowish $K\omega$ enhancement with mass $\sim$ 1700 MeV have been submitted to the conference. Chung, et al.,[28] [$K^-p \to K^-\omega p$ at 7.3 GeV/c] present the background-subtracted $K\omega$ spectrum shown in Fig. 13a in which a clear > 4σ bump is observed. A fit with an S-wave Breit-Wigner over a linear background yields the following resonance parameters:

$$M = 1710 \pm 15 \text{ MeV}$$
$$\Gamma = 110 \pm 50 \text{ MeV.}$$

The $K\omega$ mass spectrum of the ABCLV collaboration[29] [$K^-p \to K^-\omega p$ at 10 + 16 GeV/c] is shown in Fig. 13b in which an enhancement is also observed with a mass near 1700 MeV. It would be nice to have more statistics on this effect so that a definitive J^P test could be performed and better branching ratios obtained.

B. **High-Mass K^*'s: Natural Parity States Observed in Charge-Exchange Channels**

There have been three reports of structure in the $K\pi$ system in the mass region above the K(1420). They have come from reactions of the type $KN \to (K\pi)N'$ where $|\Delta Q(N \to N')| = 1$.

The Experiments:

1. 9.0 GeV/c [Purdue][30]: $K^+n \to p + K^{*o}$

2. 12.0 GeV/c [Berkeley][31]: $K^+n \to p + K^{*o}$

3. 7.3 GeV/c [BNL][32]: $K^-p \to n + K^{*o}$

There are inconsistencies in the reported properties of the claimed K^* states which are now summarized:

1. Purdue [Carmony, et al.[30]]: $K^+n \to p + X^o$ at 9 GeV/c

a. $X^o \to K^+\pi^-$: The $K^+\pi^-$ effective mass spectrum [4581 events] is shown in Fig. 14b. A broad structure is observed in the region between $1.6 \le M(K^+\pi^-) \le 2.2$ GeV; there is some indication of a narrow enhancement above background at $M(K\pi) \sim 1760$ MeV.

b. $X^o \to K^o\pi^+\pi^-$ [Note: K^o = seen + unseen]: Figure 15a shows the $K^o\pi^+\pi^-$ effective mass spectra. A very significant enhancement is observed at a $M(K^o\pi^+\pi^-) \sim 1760$ MeV with some indication for additional structure at $M(K^o\pi^+\pi^-) \sim 2100$ MeV [2-4σ].

The result of a simultaneous fit to the $K\pi$ and $K\pi\pi$ spectra [assuming both enhancements are decay modes of the same object] gives:

(1) K(1760): M = (1769 $\pm$ 12) MeV

$$\to \begin{matrix} K\pi \\ K\pi\pi \end{matrix} \qquad \Gamma = (130 \pm 50) \text{ MeV}$$

(2) Extremely weak evidence for a state with M = 2115 $\pm$ 45 MeV, Γ = 300 $\pm$ 100 MeV decaying into both Kπ and K$\pi\pi$. I do not find this claim to be statistically compelling.

2. Firestone, et al.[31] [Berkeley] have studied the same reactions, but at 12.0 GeV/c.

a. $X^o \to K^+\pi^-$: The $K^+\pi^-$ effective mass spectrum after a $|t'|$ < 0.2 GeV^2 selection is shown in Fig. 14c. A big, broad $\sim$ 400-MeV-wide object is observed. There is no evidence for a narrow K(1760).

b. $X^o \to K^o\pi^+\pi^-$: The $K^o\pi^+\pi^-$ spectrum from the reaction $K^+n \to K^o\pi^+\pi^-p$ is shown in Fig. 15d. No compelling signal at either 1760 or 2100 MeV is observed.

3. Aguilar-Benitez, et al.[32] [BNL] have studied the reaction $K^-p \to K^-\pi^+n$ at 7.3 GeV/c. If, as expected, π exchange dominates the reaction, then at the same energy the properties of $K^-p \to K^*n$ and $K^+n \to K^{*o}p$ should be similar.

a. $X^o \to K^-\pi^+$: The $K^-\pi^+$ effective mass spectrum [5330 events] is shown in Fig. 14a. Here, a narrow 4σ enhancement is observed at M(Kπ) = 1760 MeV.

b. $X^o \to \bar{K}^o\pi^+\pi^-$: The $\bar{K}^o\pi^+\pi^-$ spectrum as shown in Fig. 15c shows little if any evidence for activity at either 1760 or 2100 MeV.

Additional data to those in Ref. 27 have been presented to this conference by the BNL group. The $K^-\pi^+$ [$\bar{K}^o\pi^+\pi^-$] effective mass spectra are shown in Fig. 16a [16b]. The K(1760) $\to K^-\pi^+$ signal is, indeed, very strong [$\sim$ 5σ]. No evidence is seen for a K$\pi\pi$ decay mode.

Summary

Experiment	Kπ	K$\pi\pi$	Observations
9.0 GeV/c K^+n	Broad Enhancement with possible narrow peak at 1760 MeV	2 Enhancements 1760: Good 2100: Poor-fair [2-4σ]	Narrow K$\pi\pi$ Forces Kπ(2100). Total fit gives Γ(2100) = 300 $\pm$ 100 MeV (?)
12.0 GeV/c K^+n	Broad enhancement No narrow structure	Nothing	Appears all peaks have disappeared.
7.3 GeV/c K^-p	Narrow structure at 1760 MeV	Little evidence	Kπ(1760) is consistent with 9.0 GeV/c K^+n result;

Additional Remarks

1. A closer examination of the data of Carmony, et al.,[30b] yields an inconsistency in the results supporting the claimed $K\pi\pi$ enhancements. This is apparent from an inspection of the two insets of Fig. 15. Here, the effective mass spectra from the reactions

$$K^+n \rightarrow K^0\pi^+\pi^-p \quad \text{(a)}$$

$$\text{and} \quad K^+n \rightarrow K^+\pi^-\pi^0p \quad \text{(b)}$$

are compared. Carmony, et al., quote $\sim 50\%$ $K^*\pi$ and $\sim 50\%$ $K\rho$ branching fractions for K(1760) which leads to the expectation of a ~ 120-event K(1760) signal in reaction (b) [based on the number of events in reaction (a)]. Such an effect is clearly not present which may necessitate a re-evaluation of the significance of the claim for the K(1760) $\rightarrow K\pi\pi$ decay mode in these data.

2. Some new observations in the reaction $K^-p \rightarrow K^-\pi^+n$ at 7.3 GeV/c: There have been suggestions[33] that a good place to look for resonance states with weak production amplitudes is in their interference with the much stronger diffraction dissociation (D.D.) amplitude, i.e.

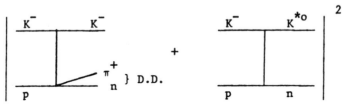

Figure 17a shows the π^+n effective mass spectra in which the low-mass diffractively produced component is evident. In order to look at the D.D.-$K\pi$ overlap, only those events with $M(\pi^+n) < 1.8$ GeV and $|t_{K-K}| < 0.5$ GeV^2 are selected [corresponding to the shaded area in Fig. 17a]. The resultant $K\pi$ effective mass spectrum is shown in Fig. 17b. The K(1760) signal is now greater than a 5σ effect. In addition, there is evidence for possible structures in the $K\pi$ mass region above 2.0 GeV. Clearly more statistics are needed in order to conclusively demonstrate additional resonance structure, but the idea of pulling out small resonant signals through their interference with the much stronger D.D. amplitude is extremely attractive and warrants further searches.

Conclusion:

1. Narrow K(1760) → Kπ seems good, but J^P is not established.
2. Claim for Kππ of Carmony, et al.,[30] needs re-examination in light of discrepancy between K(1760) → $K^0\pi^+\pi^-$ and K(1760) → $K^+\pi^-\pi^0$ branching fractions.
3. Some evidence for additional Kπ states above 1800 MeV in $K^-p \to K^-\pi^+n$ at 7.3 GeV/c. More statistics needed.

V. SUMMARY

1. Better understanding of LME is needed.
2. Are we going to have filled A_1 and B nonets? Are LME to be included in them? (Remember, the Q and A_1 are not isolated examples!)
3. Will increased statistics yield higher-mass states? Will rarer types of decay modes be found [N in Fig. 1 filled?]?
4. There appears to be good evidence for 1^-, 2^+, 3^- ππ and Kπ decay modes of resonance states. Will the trend continue [i.e. 4^+, 5^-, ...]?
5. In particular:
    More data on nondiffractive channels could prove fruitful:

$$K^+p \to \Delta^{++} + X^0$$

$$K^-p \to \Lambda + X^0$$

$$K^-p \to \Sigma^+ + X^-$$

$$K^-p \to n + X^0, \text{ etc.}$$

ACKNOWLEDGEMENTS

The support and advice of Dr. N.P. Samios are gratefully acknowledged.

REFERENCES AND FOOTNOTES

1. M. Aguilar-Benitez, et al., Phys. Rev. D4, 2583 (1971).
2. Munich/Brussels/Mons/CERN Collaboration, paper No. 42.
3. Schweingruber, et al., Phys. Rev. 166, 1317 (1968).
4. A. Firestone, in Experimental Meson Spectroscopy, ed. by C. Baltay and A. Rosenfeld, p. 229 (1970).
5. H.H. Bingham, et al., Nucl. Phys. B48, 589 (1972).
6. G.W. Brandenburg, et al., Nucl. Phys. B45, 397 (1972).
7. A word of caution about the significance of the $K\rho/K^*\pi$ branching fraction: The analysis quoted has been performed on data of the K^+p world data collaboration. This data essentially contains, from the desired $K^+p \to K^+\pi^+\pi^-p$ reaction, only the highest probability hypothesis. Therefore, the $K^+-\pi^+$ ambiguity [which is quite large in the

Q region] could result in an apparent but false ρ signal in the $\pi^+\pi^-$ effective mass spectrum. This follows because true $K^*(890) \to K^+\pi^-$ events, if incorrectly identified as "π^+"π^-, tend to peak at $\sim$ 700 MeV and can, therefore, mimic a true ρ signal.

8. A question has been raised [S. Flatté] as to why the $K^*(1420)$ [$\to K^+\pi^+\pi^-$] production, which is dropping with energy, is not apparent in the 1.35- to 1.45-GeV region of Fig. 4. The answer comes from a study of the energy dependence of the reaction $K^+p \to K^+(1420)p \to (K\pi)^+p$ and from knowledge of the $K(1420) \to K\pi\pi/ K(1420) \to K\pi$ branching fraction. It shows, for $p_{LAB} > 4$ GeV/c, that the $K(1420) \to K^+\pi^+\pi^-$ production only accounts for < 10% of all the cross-section in the region 1.35-1.45 GeV. Hence, the fall with energy of the $K(1420)$ would not be apparent in Fig. 4. Clearly, if the production cross-section of the $K_B \to K^+\pi^+\pi^-$ were at the same level as the $K(1420)$, then it also could not be observed through the energy dependence exhibited in Fig. 4.

9. A Firestone, et al., Phys. Rev. D5, 505 (1972).
10. G. Kane, Phys. Rev. 156, 1738 (1967).
11. P.J. Davis, et al., Nucl. Phys. B44, 344 (1972).
12. A. Barbaro-Galtieri, et al., Phys. Rev. Lett. 22, 1207 (1969).
13. R. Barloutaud, et al., DPLPE 73-01 (Saclay preprint).
14. D. Cohen, et al., Phys. Rev. Lett. 28, 1601 (1972).
15. B. Werner, et al., Phys. Rev. D7, 1275 (1973).
16. D.J. Crennell, et al., Phys. Rev. D6, 1220 (1972).
17. A. Astier, et al., Nucl. Phys. B10, 65 (1969).
18. S. Devons, et al., Phys. Rev. Lett. 27, 1614 (1971).
19. M. Aguilar-Benitez, et al., Nucl. Phys. B14, 195 (1969).
20. J. Duboc, et al., Nucl. Phys. B46, 429 (1972).
21. 12-GeV/c K^+p: A. Barbaro-Galtieri, et al., Phys. Rev. Lett. 22, 1207 (1969).
22. 10-GeV/c K^+p: D.C. Colley, et al., Nucl. Phys. B26, 71 (1971).
23. 4.6-GeV/c K^-p: M. Aguilar-Benitez, et al., Phys. Rev. Lett. 25, 54 (1970).
24. 10-GeV/c K^-p: J. Bartsch, et al., Phys. Lett. 33B, 186 (1970).
25. 12.6-GeV/c K^-d: D. Denegri, et al., Nucl. Phys. B28, 13 (1971).
26. 10.9-15.9 GeV/c K^-p: H.R. Blieden, et al., Phys. Lett. 39B, 668 (1971).
27. Private communication from G. Jones, and talk by D. Hansen, this conference. The data is that of the ABCLV collaboration.
28. S.U. Chung, et al., BNL preprint NG-276.
29. ABCLV collaboration, paper No. 82.
30. a. D.D. Carmony, et al., Phys. Rev. Lett. 27, 1160 (1971).
    b. D.D. Carmony, Invited Talk at Berkeley Conference (1973).
31. A. Firestone, Phys. Lett. 36B, 513 (1971).
32. M. Aguilar-Benitez, et al., Phys. Rev. Lett. 30, 672 (1973).
33. D.H. Miller, Invited Talk at Erice Summer School, 1973.

	2+ NONET				0− NONET			
	f⁰	f'	**K(1420)**	A₂	η'	η	K	π
π	A₃	N	L	g(?)	N	δ, A₂	K(890) K(1420) K(1760)	ρ S* f⁰ g
K	LME	N	N	N	N	N	φ, A₂ f' S*	K(890) K(1420) K(1760)
η	N	N	N	N	N	N	N	δ, A₂
η'	N	N	N	N	N	N	N	N
ρ	N	N	N	N	N	N	Q, K(1420)	A₁ A₂ ω(1675)
φ	N	N	N	N	N	N	E, F₁	Q K(890)
ε	N	N	N	N	N	N	LME	B, g

(Row groups at right: π, K, η, η' — 0⁻ NONET; ρ, φ, ε — 1⁻ NONET)

Fig. 1. Grid representing decay modes of well-known resonances.

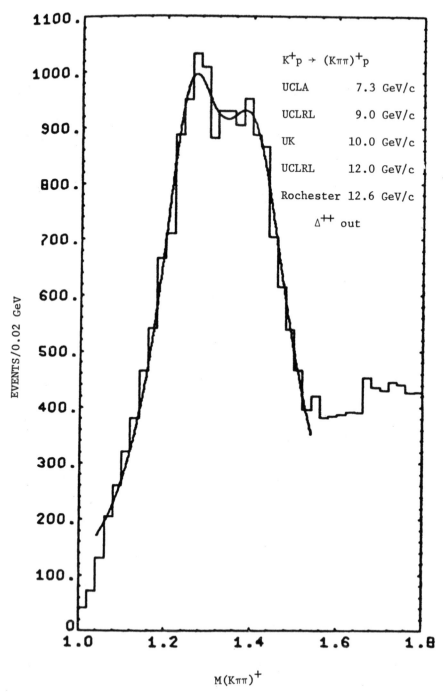

Fig. 2. Compilation of the $(K\pi\pi)^+$ spectrum from the reaction $K^+p \to$ $(K\pi\pi)^+p$ (Firestone).

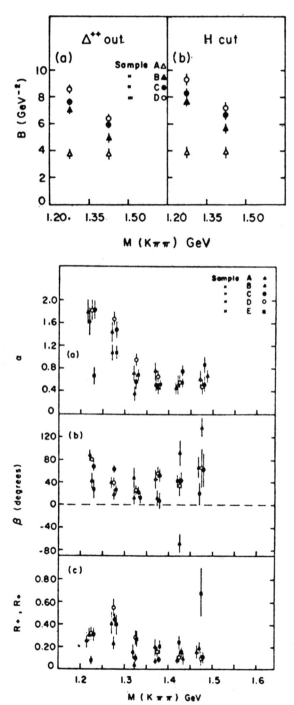

Fig. 3. TOP: Slope of dσ/dt as a function of (Kππ)⁺ mass for the
reaction K⁺p → (Kππ)⁺p for a variety of incident K⁺ momenta.
BOTTOM: Results of the Kππ Dalitz plot fit of Bingham, et al.

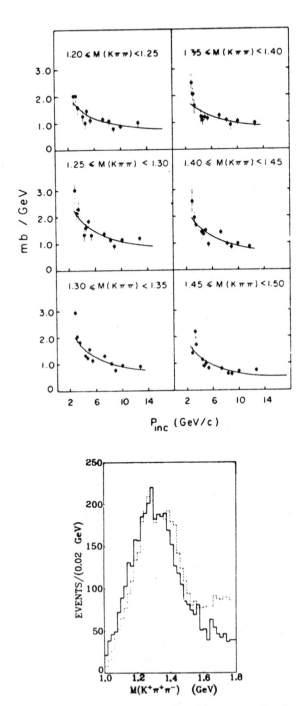

Fig. 4.   TOP: σ vs. P_inc for various (Kππ)+ intervals from the reaction
K+p → K+π+π−p.
BOTTOM: Comparison of the shape of the Q+ produced in associa-
tion with proton [dashed] or deuteron [solid].

157

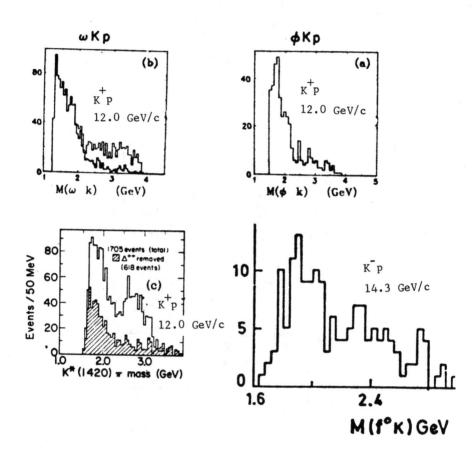

Fig. 5. Compilation of data on LME.

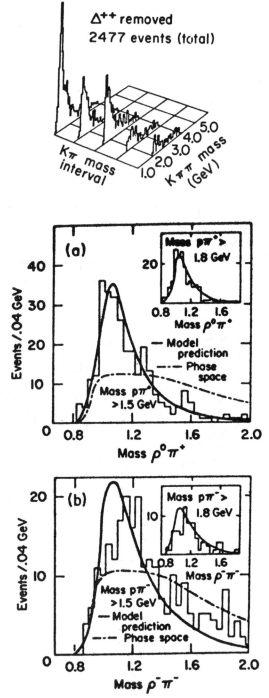

Fig. 6. Complications to a fundamental interpretation for LME:
TOP: Traveling $(K\pi\pi)^+$ peaks for $(K\pi)^0$ slices which are not
necessarily in the resonance region.
BOTTOM: Comparison of $\rho^0\pi^+$ and $\rho^-\pi^-$ effective mass spectrum by
Cohen, et al.

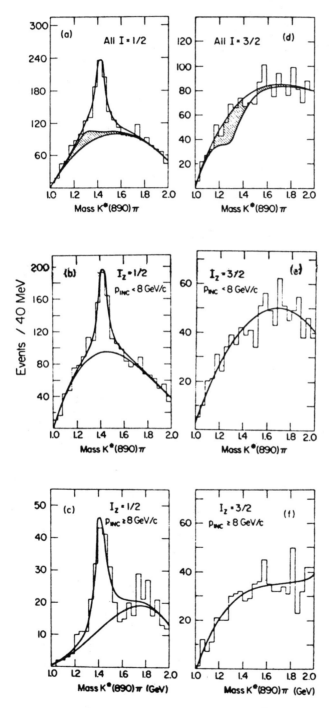

Fig. 7. a-c. Compilation of data on neutral K*π production from charge-exchange reactions.

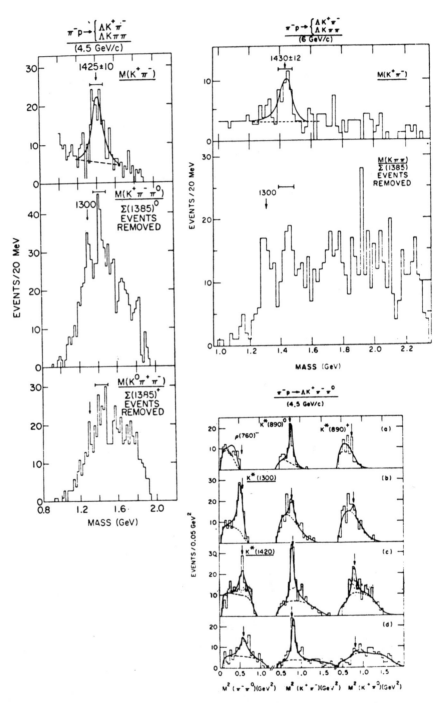

Fig. 8. Evidence for the existence of K(1300) produced in the reaction
$\pi^- p \to \Lambda K\pi\pi$ at 4.5 and 6.0 GeV/c.

C Meson

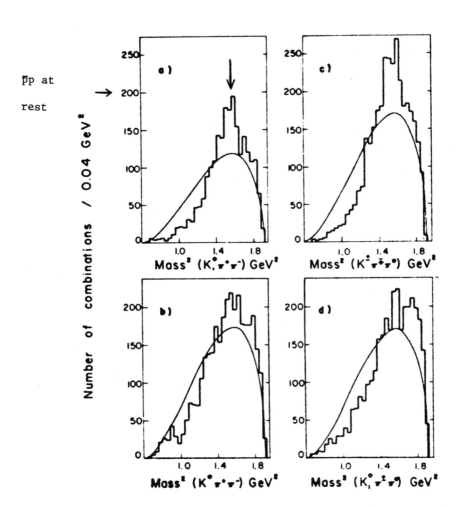

Fig. 9. a. Evidence for the existence of the C meson [p̄p at rest].

<u>N o</u> <u>C</u> <u>M e s o n</u>

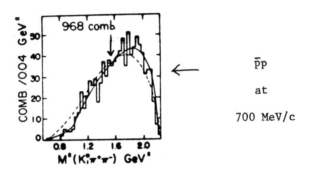

$\bar{p}p$

at

700 MeV/c

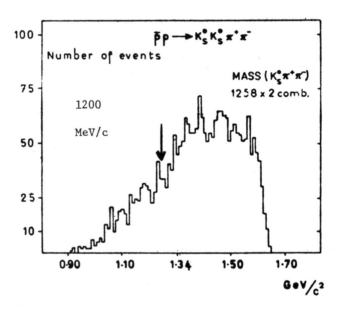

Fig.9. b. Evidence for no signal for C—meson production [$\bar{p}p$ at 700 and 1200 MeV/c].

L M e s o n

K^+ interactions

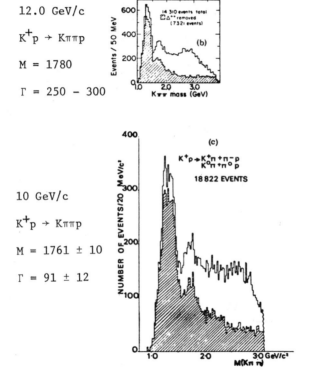

12.0 GeV/c

$K^+p \rightarrow K\pi\pi p$

M = 1780

Γ = 250 - 300

10 GeV/c

$K^+p \rightarrow K\pi\pi p$

M = 1761 ± 10

Γ = 91 ± 12

Fig. 10. Compilation of data on the mass and width of the L meson.
a. K^+ interactions

L Meson

K⁻ interactions

10 GeV/c

K⁻p → Kππp

M = 1780 ± 15; Γ = 138 ± 40

4.6 GeV/c

K⁻p → Kππp

M = 1745 ± 20

Γ = 100 ± 50

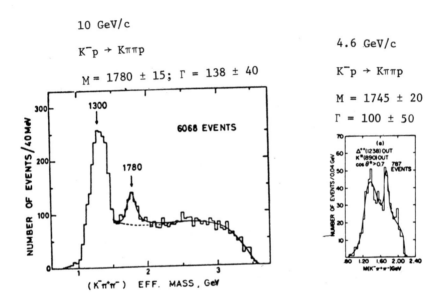

K⁻p → K*⁻p

M = 1767 ± 6; Γ = 100 ± 26

K⁻d → Kππd

M ≈ 1740

Γ ≈ 130

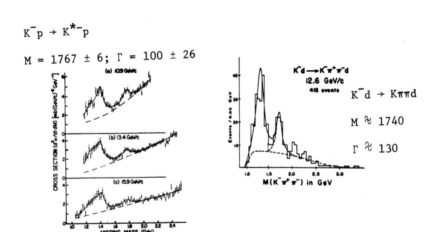

Fig.10. b. K⁻ interactions

Decay Modes of L

K^+ interactions

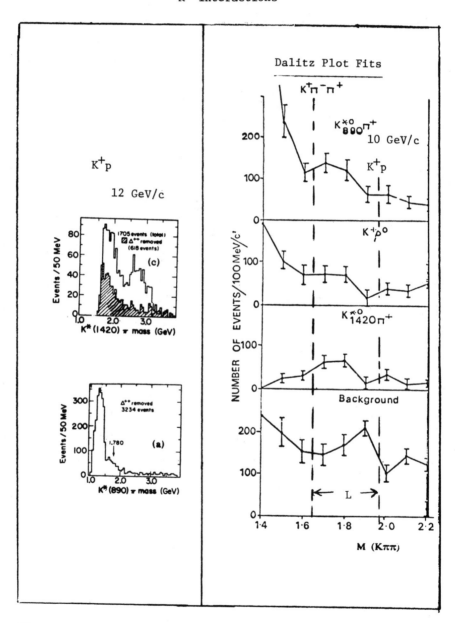

Fig. 11. Compilation of data on the decay modes of the L meson.
a. K^+ interactions

Decay Modes of L

K⁻ interactions

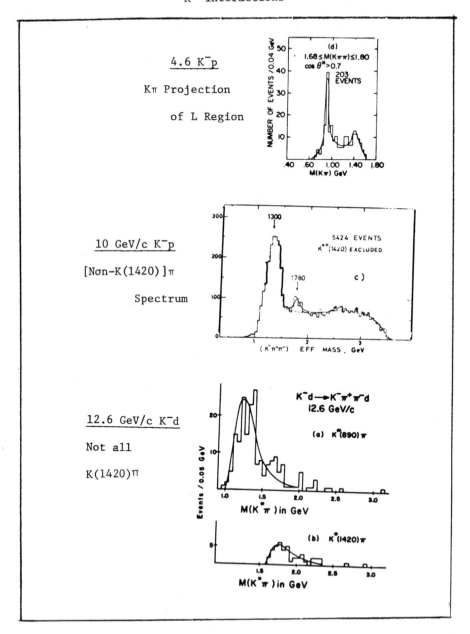

4.6 K⁻p

Kπ Projection

of L Region

10 GeV/c K⁻p

[Non-K(1420)]π

Spectrum

12.6 GeV/c K⁻d

Not all

K(1420)π

Fig.11. b. K⁻ interactions

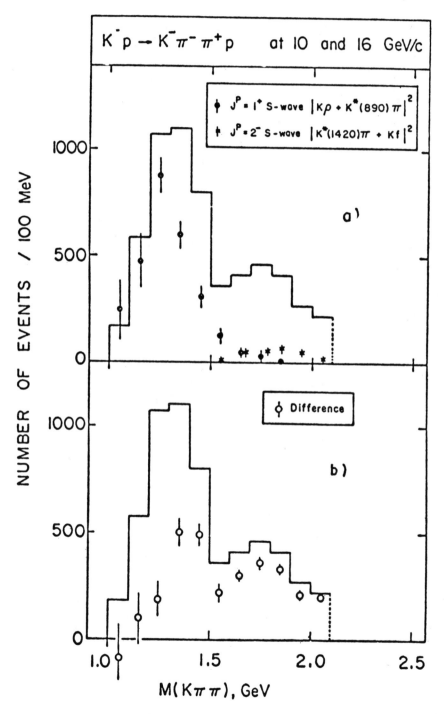

Fig. 12.   PWA results on the (Kππ)⁻ system.

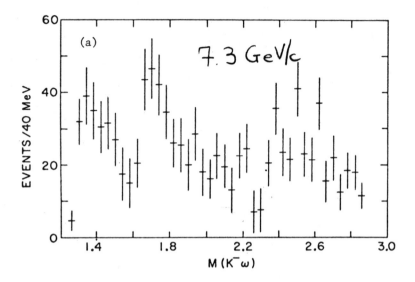

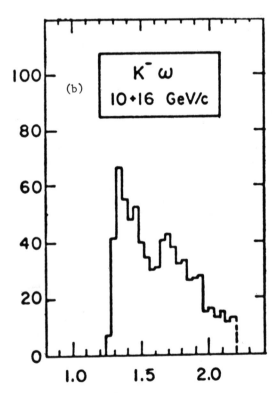

Fig. 13. New results supporting a Kω resonance at mass ∿ 1700 MeV:
a. BNL: 7.3 GeV/c K⁻p → K⁻ωp.
b. ABCLV collaboration: K⁻p → K⁻ωp.

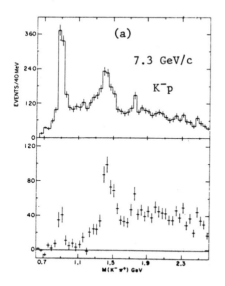

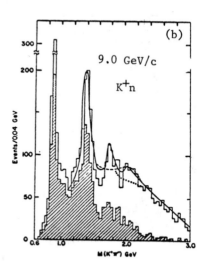

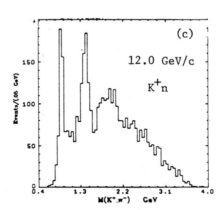

Fig. 14.   Compilation of Kπ spectra from charge-exchange reactions.

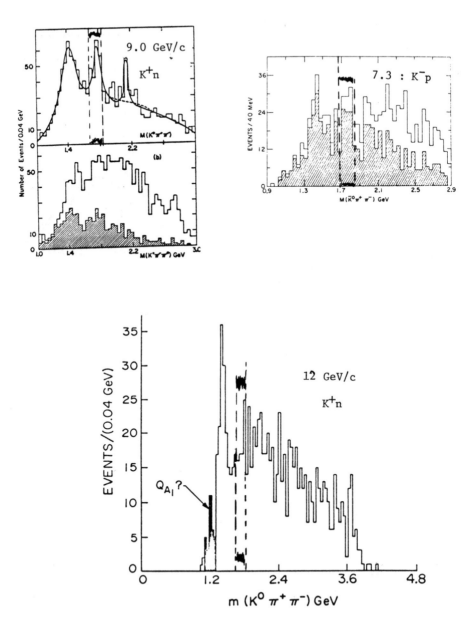

Fig.15. Compilation of Kππ spectra from charge-exchange reactions.

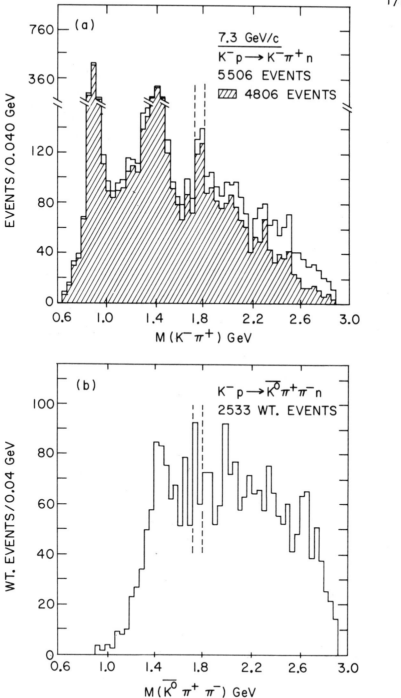

Fig.16. a. $K^-\pi^+$ effective mass spectrum from the reaction $K^-p \rightarrow K^-\pi^+n$
at 7.3 GeV/c.
b. $\bar{K}^0\pi^+\pi^-$ effective mass spectrum from the reaction
$K^-p \rightarrow \bar{K}^0\pi^+\pi^-n$ at 7.3 GeV/c.

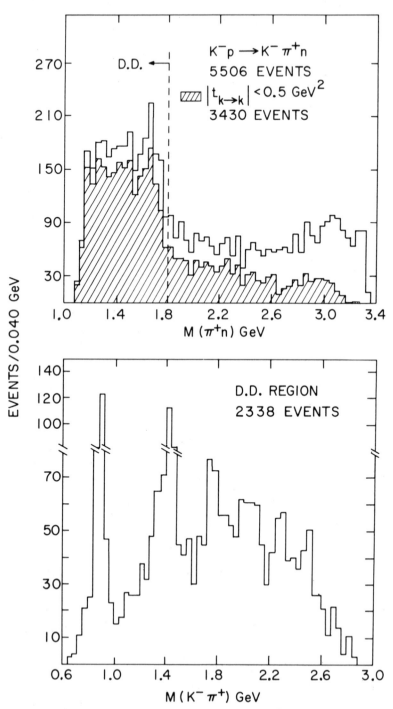

Fig.17. a. π^+n effective mass spectra from the reaction $K^-p \rightarrow K^-\pi^+n$
              at 7.3 GeV/c.
         b. $K^-\pi^+$ effective mass spectra for events with $M(\pi^+n) < 1.8$
              GeV and $|t_{K-K}| \leqslant 0.5$ GeV^2 [i.e. in D.D. region].

NEW RESULTS ON Kππ

J. D. Hansen

The Niels Bohr Institute, Copenhagen, Denmark

ABSTRACT

New results on the reaction $K^- p \to K^- \pi^+ \pi^+ p$ at incident momenta from 10 to 40 GeV/c and on the reaction $K^- p \to \bar{K}^0 \pi^+ \pi^- n$ at incident momenta from 8 to 16 GeV/c are presented. Both reactions show spin parity structures similar to the reaction $\pi p \to \pi \pi^- \pi^+ p$. Details of the Q-region are discussed.

INTRODUCTION

I will talk about new results on the reaction

$$K^- p \to K^- \pi^- \pi^+ p \tag{1}$$

and the reaction

$$K^- p \to \bar{K}^0 \pi^+ \pi^- n \tag{2}$$

Contributions pertinent to reaction 1 were submitted by the Aachen-Berlin-CERN-London-Vienna Collaboration at 10 and 16 GeV/c [1], by the Rutherford-Saclay-Ecole Polytechnique Collaboration at 14.3 GeV/c [2] and by the Urbana-CERN-IHEP Collaboration at 40 GeV/c [3].

The contribution pertinents to reaction 2 was a joint paper by the Aachen-Berlin-CERN-London-Vienna and Athens-Democritus-Liverpool-Vienna Collaborations at 8, 10 and 16 GeV/c [4]. In the talk reaction 1 will be compared to the reaction

$$\pi p \to \pi \pi^- \pi^+ p \tag{3}$$

at similar energies. Several contributions to reaction 3 were presented to this conference. I will use reference 5 as an example. The analysis technique used in all of the contributions is the partial wave analysis technique developed first for reaction 3 at Urbana, Illinois, hereafter referred to as PWA, as discussed in reference 6. The plan for the talk is: First to discuss general features of reaction 1, a reaction where $\Delta Q = 0$. Secondly to discuss reaction 2. This reaction has $\Delta Q = 1$. Thirdly I will return to reaction 1 and discuss detailed results obtained in the Q-region. Finally I will summarize the results.

METHOD OF ANALYSIS

The only analysis method to be discussed here is the PWA as developed at Urbana for reaction 3 and later extended to all 4 body reactions at CERN. A detailed description of this can be found in

reference 6. We are here only interested in the part of reaction 1 and 2 where the $K\pi\pi$ effective mass $m_{K\pi\pi}$ is small and where the fourmomentum transfer t is small. A 4 body final state is described by 9 variables. However, as the proton is unpolarized and the polarization of the outgoing proton not measured the matrix element cannot depend on the azimulthal angle around the beam. This reduces the number of variables to 8. These eight variables naturally divides into two groups, one of 3 and one of 5 variables. The first group consists of the "production" variables s, total C.M.S. energy squared, t and $m_{K\pi\pi}$ and the second of the "decay" variables α, β, γ, s_1 and s_2 where α, β and γ are Euler angles, and s_1 and s_2 are Dalitz plot variables. The matrix element is then expanded as an infinite sum of amplitudes $|J^P M \eta \ell j n >$ which is referred to as states or waves. The dependence on the proton helicity is ignored in this talk. J is the spin of the $K\pi\pi$ system, P its parity, M the J_z component and η the eigenvalue of the reflection operator in the production plane. For M = 0 η = +(-) correspond to natural (unnatural) spin parity exchange. For M $\neq$ 0 the same holds to order 1/s. The final state is first combined to a system n with spin j then combined with the lone-meson with an orbital angular momentum ℓ. The dependence on the Euler angles and the other s_m, n $\neq$ m is given without any assumptions. However, to do fits several assumptions have to be made:
- J and j finite. This assumption makes the sum finite.
- The overall amplitude is a sum of two finite sums. This because both K^* and ρ signals are observed in the data.
- The dependence on s_n is assumed to be the one observed in $K\pi$ and $\pi\pi$ phaseshift analysis including momentum barrier factors. This is often referred to as the isobar model.
- States with same $J^P M \eta$ are normally assumed to be coherent i.e. the complex ratio of the proton spin flip amplitude to the nonflip amplitude is the same for such states. This assumption plays a role in the detailed discussion of the Q-region.

In reference 7, where the 3π system is studied the additional assumption is made that <u>all</u> states are coherent.

Fits with this explicit dependence on the "decay" variables are done in interval of the production variables s, t, $m_{K\pi\pi}$. The parameters are the density matrix elements and some complex branching ratios.

$$\Delta Q = 0$$

Reaction 1 is studied in 4 experiments from 10 to 40 GeV/c. In all of the experiments the K^- in the final state is identified by kinematic fitting. When the energy change upon interchange of the mass of the two negative particles is small (5 MeV at 10, 14.3 and 16 GeV/c and 3 MeV at 40 GeV/c) the event is considered ambiguous and the event is excluded from the analysis. In the lower energy experiments also events in the $\Delta^{++}(1236)$ region are excluded. The analysis takes these cuts into account and corrects the results for the cuts. This is only possible when the description as in the PWA is complete.

The four experiments give similar results.

Figure 1 shows the effective $K\pi\pi$ mass at 40 GeV/c. A clear Q-peak is observed. A shoulder is observed in the L-region. Shown shaded are the events used in the PWA-analysis. It is apparent that the cut is most severe in the Q-region.

Figure 2 shows the $J^P M\eta$ composition as function of $m_{K\pi\pi}$ in the 40 GeV/c experiment. Here there is summed over all decay modes and their interferences for a given $J^P M\eta$ state. The same is shown for the 10 and 16 GeV/c experiments combined in figure 3. As previously stated very good agreement between the experiments is observed. In the Q-region the $J^P M\eta = 1^+0^+$ state dominates, but other states are present as well especially the 0^- state. In the L-region no single state dominates. All spins from zero to three are present. Note that the 0^- state peaks at a higher mass value than the 1^+ state does. Thus the Q- and L-regions are complex. Further one notices a dominance of production of unnatural spin parity states (with M = 0) and natural spin parity exchange. These two points are demonstrated in figure 4, where the appropriate contributions are added for the 10 and 16 experiments.

Before leaving the L-region we note that the $2^-s K^*(1420)\pi$ state shows the clearest peak in that region. As the spin structure is very complex in every mass bin it might be useful to plot the average spin as function of the mass. This is done for some 3π [8] and $K\pi\pi$ [1] experiments in figure 5.

The line is a fit of the form $<J> = a(M - m_0)$, with $m_0 = m_\pi$ and m_K. In both fits a value of $a = 1.1 \pm 0.1$ is found. The experience gained in doing the PWA fits taught us that J_{max} grows roughly as M^2 thus the system becomes very complicated when the mass exceeds 2 GeV.

A claim often made is that the Q is 1^+s and the L is 2^-s. To study this the appropriate sums are made in figure 6. As can be seen these states do not account for the observed mass plot especially in the L-region. This is in accordance with figures 2 and 3.

Next we compare $K\pi\pi$ with 3π in table I. The values for the 3π system are taken from reference 5. Within the errors everything agrees.

$$\Delta Q = 1$$

In reaction 2 no problem with the identification of the $\bar{K}^0$ is encountered as the $\bar{K}^0$ is observed to decay in the bubble chamber. Unfortunately only 1/3 of the $\bar{K}^0$ have a visible decay. This leads to very low statistics.

In figure 7 is shown the effective $\bar{K}^0\pi^+\pi^-$ mass. A clear $K^*(1420)$ peak is observed over a rising background. The cross section of the $K^*(1420)$ is only half of the cross section of the "background". The results of the PWA are shown in table II for a fit between 1.04 and 1.56 GeV. The new result is that the dominant state is 1^+ followed by 2^+ and 0^- and that the ratio of 0^- to 1^+ is roughly the same as for reaction 1 which have $\Delta Q = 0$. The best fit is obtained with a rising mass dependence of the 1^+ and 0^- state. This is in good agreement with the mass plot but very different from the structure found in the $\Delta Q = 0$ reaction 1 as it was observed in figures 2 and 3.

For $\Delta Q = 0$ reactions states which obey the rule $\Delta P = (-1)^{\Delta J}$ dominate. This is the Gribov-Morrison rule. It has been thought to be a characteristic of diffraction, but states obeying the same rule are now also found to dominate $\Delta Q = 1$ reactions. The implications of the rule will therefore have to be reconsidered. Another characteristic of diffractive reactions is that the cross section is roughly independent of energy. When the cross section as function of laboratory momentum is written as $\sigma \sim p^{-n}$ values of n $\sim$ .5 are found. For the $\Delta Q = 1$ reaction a value of n = 1.5 ± .1 is found when all events in the region 1.04 to 1.56 GeV are used.

The different mass and energy dependence are hard to explain in a resonance model but are in good agreement with what one would expect from a simple Deck model.

THE Q-REGION

In this section we will discuss the main contribution to the Q-region which is the 1^+ state. In figure 8 is shown some of the contributions to the 1^+ system as observed in the 40 GeV/c experiment. The interference terms are left out. In figure 9 is shown the contribution from $K^*\pi$ and $K\rho$ in the 10 and 16 GeV/c experiments. In both figures dominance of $K^*\pi$ is observed with all other contributions being similar in magnitude. It has been claimed in the past [9-11] that the ratio, R of $K\rho$ to $K^*\pi$ is highest near the $K\rho$ threshold. However, in none of these experiments ambiguous events were treated properly. If a K^* is misidentified the apparent $\pi^+\pi^-$ mass will peak in the ρ-region when the $K\pi\pi$ mass is not too high. Therefore wrong identification of events would produce ρ events even if there were none. Consequently previous claims [9-11] can only be taken as a guidance. In figure 10 is shown the ratio R for various fits. Except in one fit the $1^+S\,K^*\pi$ and $1^+S\,K\rho$ states are assumed to be coherently produced. Within errors all results agree for $K\pi\pi$ masses above 1.3 GeV. In the interval 1.2 to 1.3 some discrepancies are found. However, as was observed at 14.3 GeV/c the results depends very strongly on the exact set of states used, and still the fits are equally probable.

The interference between the $1^+S\,K\rho$ state and the $1^+P\,K\pi$ state is very large and negative. The difference of the 10 and 19 amplitude fits is that in the 10 amplitude fit the contributions from $1^+S\,K\rho$, $1^+P\,K\pi$ and interference are all small while all are big in the 19 amplitude fit, however the sum of the contributions is essentially constant. That the amount of ρ cannot be determined near threshold because of its near complete overlap with the $\kappa(\epsilon)$ in the other system has been observed before [12]. At 14.3 GeV/c fits were performed with some 1^+ states not fully coherent with other 1^+ states using the 10 amplitudes from the previous fit. In the fit shown, the $1^+S\,K^*\pi$ and $1^+P\,K\pi$ is treated as one system, and the $1^+S\,K\rho$ and $1^+P\,K\epsilon$ as another. The ratio R goes then up and is constant around 0.3. That the ratio goes up might be expected from the fact that the $\overset{*}{K}\pi$ and $K\rho$ interference is positive, figure 9. This interference is decreased as the coherence is found to be only $\sim.2\pm.25$, see figure 11a. Finally the solid curve is phasespace times 3/4 which is what is expected from a decay of a single SU(3) system with $C = \pm 1$. The data does not support the assumption of a single state. This because the coherence must then be one which it is not and if fits are performed with the coherence constraints, then the fits do not follow the expected curve. Note that the coherence between states of different spin parity, figure 11b, is close to .8 thus not too inconsistent with the additional assumption made in reference 7, where reaction 3 is studied. Finally some phases of 1^+S $\overset{*}{K}\pi$ relative to other states were presented, some of these are shown in figure 12. No phase variation is observed.

SUMMARY

The $\Delta Q = 0$ reaction $K^-p \to K^-\pi^+\pi^-p$ is studied in the laboratory momentum range from 10 to 40 GeV/c and compared to the $\Delta Q = 0$ reaction $\pi^-p \to \pi^-\pi^-\pi^+p$ and to the $\Delta Q = 1$ reaction $K^-p \to \bar{K}^0\pi^+\pi^-n$ in the range from 8 to 16 GeV/c. In the $\Delta Q = 0$ reaction it is found that:
- Produced unnatural spin parity states dominate.
- Natural spin parity exchange dominates.
- States with $M = 0$ dominate.
- Many states contribute to the Q- and L-regions in such a way that the average spin increases linear with the mass.
- Both $J^P = 0^-$ and 1^+ states peak in the Q-region. The dominant 1^+ state peaks at a lower mass than the 0^- state does.
- The 2^-S $\overset{*}{K}(1420)\pi$ state shows clearest peak in the L-region.
- The amount of the 1^+S $K\rho$ state near threshold is not well known. One reason is that ambiguous events are excluded both from the data and the theoretical distribution. This leads to a loss of information especially in the region of $\pi\pi$ masses close to the ρ mass when the $K\pi\pi$ mass is low. Another reason is the large overlap of the 1^+S $K\rho$ and 1^+P $\kappa\pi$ states.
- No phase variation is observed.
- The coherence is large except possible between the 1^+S $\overset{*}{K}\pi$ and $1^+S K\rho$ states which have a coherence of only $\sim .2\pm.25$. Note that these states are normally assumed to be fully coherent. Other states than the $1^+S K\rho$ state seem not to be affected by the removal of this assumption.
- The $K^-p \to K^-\pi^-\pi^+p$ reaction is very similar to the $\pi^-p \to \pi^-\pi^-\pi^+p$ reaction.

Further it is found that the $\Delta Q = 1$ reaction shows the same characteristics as the $\Delta Q = 0$ reaction does, except that:
- More of the $J^P = 2^+$ state is produced, but still only about 1/3 of the events in the Q-region belongs to this state. The rest consists of $J^P = 0^-$ and 1^+ events.
- The mass distributions of the unnatural spin parity states 0^- and 1^+ are increasing with mass i.e. do not show the Q-peak which is observed in the $\Delta Q = 0$ reaction.
- The cross section for the $\Delta Q = 1$ falls faster with incident momentum than the corresponding cross section for $\Delta Q = 0$ does.

Finally we note that the Gribov-Morrison rule which is a characteristic of $\Delta Q = 0$ reactions is also valid for $\Delta Q = 1$ reactions.

REFERENCES

1. Spin parity structure of the Q and L enhancements in
   $K^-p \rightarrow K^-\pi^-\pi^+p$ at 10 and 16 GeV. CERN/D.PH.II/PHYS,
   M. Deutschmann, G. Otter, G. Rudolf, H. Seyfert, H. Wieczorek,
   H. Böttcher, W.D. Nowak, S. Nowak, V.T. Cocconi, M.J. Counihan,
   J.D. Hansen, G.T. Jones, G. Kellner, W. Kittel, A. Kotanski,
   D.R.O. Morrison, D. Sotiriou, T.C. Bason, P.J. Dornan, P.R.
   Thornton, Ph. Katz, D. Kisielewska, M. Markytan and J. Strauss.
   Aachen-Berlin-CERN-London-Vienna Collaboration.
2. A study of the Q enhancement in the reaction $K^-p \rightarrow K^-\pi^-\pi^+p$ at
   14.3 GeV/c, RL 74-073.
   S.N. Tovey, J.D. Hansen, K. Paler, T.P. Shah, J.J. Phelan, R.J.
   Miller, S. Borenstein, B. Chaurand, B. Drevillon, G. Labrosse,
   A. Borg, D. Denegri, Y. Pons and M. Spiro.
   Rutherford-Saclay-Ecole Polytechnique Collaboration.
3. Analysis of the reaction $K^-p \rightarrow K^-\pi^-\pi^+p$ at 40 GeV/c.
   COO-1195-
   R. Klanner, G. Ascoli, R. Sard etc..
   Urbana-CERN-IHEP Collaboration.
4. Evidence for unnatural spin-parity states of $(\bar{K}^0\pi^+\pi^-)$ in the
   charge exchange reaction $K^-p \rightarrow \bar{K}^0\pi^+\pi^-n$.
   CERN/D.PH.II/PHYS 74-10
   Aachen-Berlin-CERN-London-Vienna Collaboration.
   Athens-Democritus-Liverpool-Vienna Collaboration.
5. Partial wave analysis of the 3π system produced in the reaction
   $\pi^+p \rightarrow \pi^+\pi^+\pi^-p$ at 8, 16 and 23 GeV/c.
   CERN/D.PH.II/PHYS 74
   Aachen-Berlin-Bonn-CERN-Heidelberg Collaboration.
6. Formalism and assumptions involved in partial wave analysis of
   three-meson systems.
   CERN/D.PH.II/PHYS 73-34
   J.D. Hansen, G.T. Jones, G. Otter and G. Rudolph.
7. Amplitude analysis of $(3\pi)^+$ production at 7 GeV/c.
   M. Tabak, E.E. Ronat, A.H. Rosenfeld, T.A. Lasinski and
   R. Cashmore.
   LBL and SLAC
8. G. Ascoli et al., Phys.Rev. D7 669 (1973).
9. K.W. Barnham et al., Nucl.Phys. B25 49 (1970).
10. H.H. Bingham et al., Nucl.Phys. B48 589 (1972).
11. R. Barloutaud et al., Nucl.Phys. B59 374 (1973).
12. B. Weinstein et al., Phys.Rev. D8 2904 (1973).

Table I

QUANTUM	$\pi^+ p \to (\pi^+\pi^+\pi^-)p$ at 8,16 & 23 GeV/c	$K^- p \to (K^-\pi^+\pi^-)p$ at 10 and 16 GeV/c
Unnatural parity states	92%	89%
Natural parity exchange	100%	95%
$(J^P=0^-)/(J^P=1^+)$	0.26	0.33
$(J^P=0^-)/(\text{unnatural state})$	0.15	0.19
$J^P=1^+$	53%	57%
$J^P=2^+$	8%	11%
$J^P=1^+$ in A_1 (or Q) region	79%	70%

Table I. Comparison of $K\pi\pi$ and 3π. Errors are of about 5%.

Table II

$K^- p \to \bar{K}^0\pi^+\pi^- n$ at 8, 10 and 16 GeV/c

$1.04 < M(\bar{K}^0\pi^+\pi^-) < 1.56$ GeV

J^P	No. of events
0^-	91 ± 17
1^+	160 ± 28
2^+	134 ± 28

Table II. Number of events in J^P states.

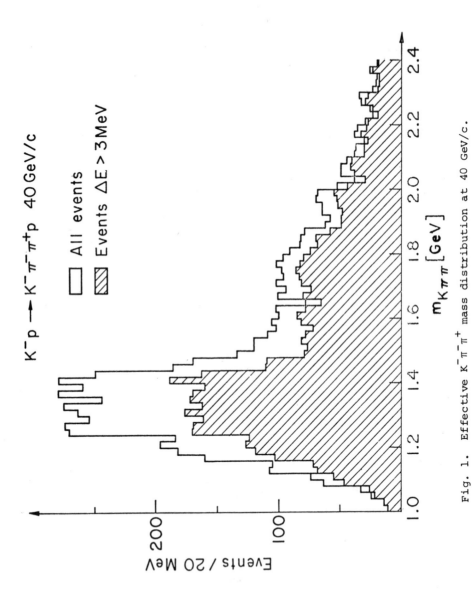

Fig. 1. Effective $K^-\pi^-\pi^+$ mass distribution at 40 GeV/c.

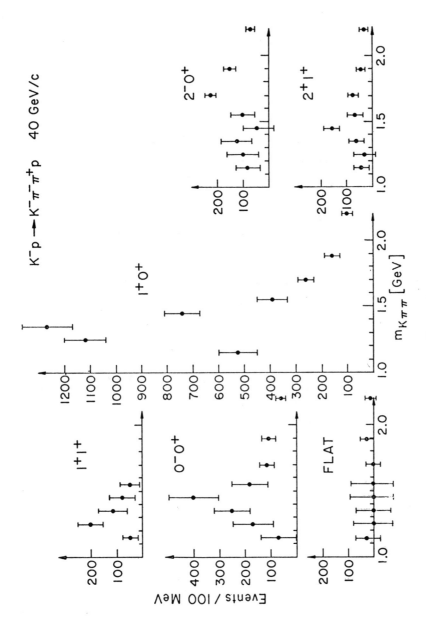

Fig. 2. Number of events in $J^P M\eta$ state vs. mass at 40 GeV/c.

182

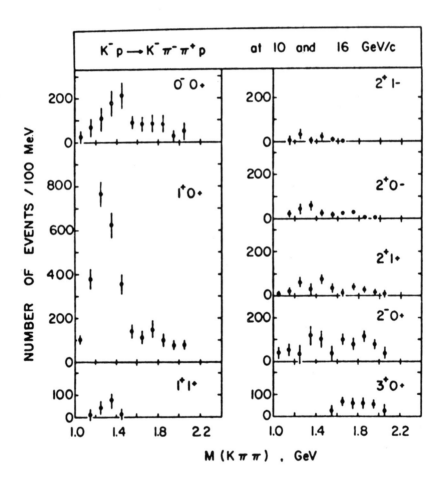

Fig. 3. Number of events in $J^P M\eta$ state mass at 10 and 16 GeV/c.

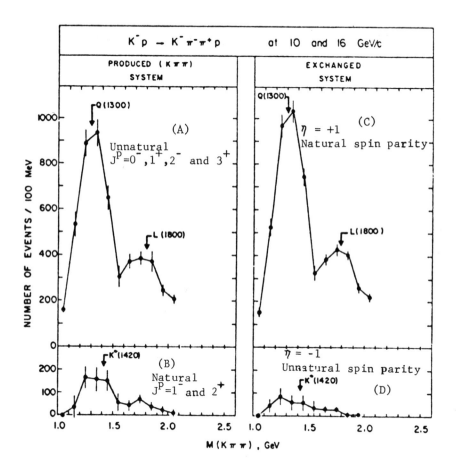

Fig. 4. A) Number of events from unnatural spin parity states
           vs. mass.
         B) Number of events from natural spin parity states
           vs. mass.
         C) Number of events produced by natural spin parity
           exchange.
         D) Number of events produced by unnatural spin parity
           exchange.

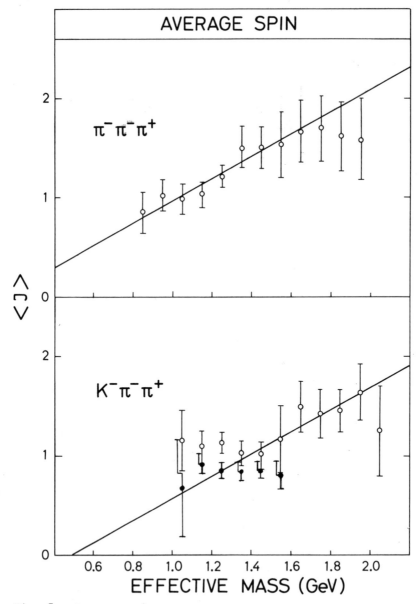

Fig. 5. Average spin vs. mass
a) $\pi^- p \to \pi^- \pi^- \pi^+ p$ 11 to 25 GeV/c
b) $K^- p \to K^- \pi^- \pi^+$ Φ 10 and 16 GeV/c
● 14.3 GeV/c

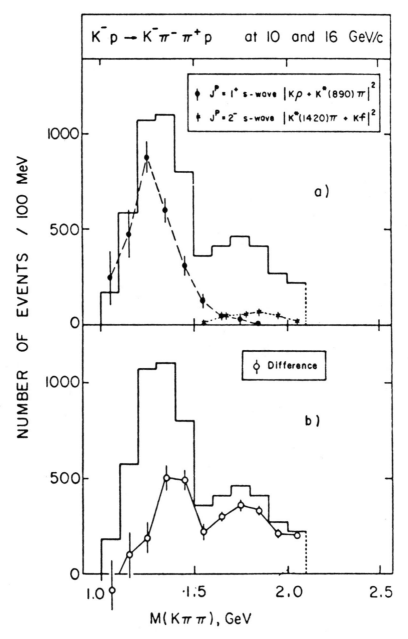

Fig. 6. Number of events corresponding to $J^P L = 1^+ s$ and $2^- s$ vs. mass.

186

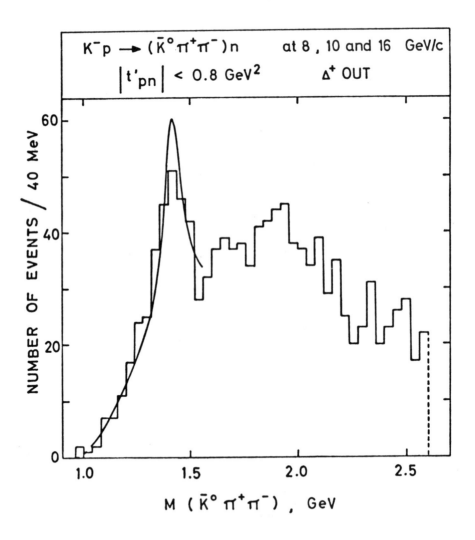

Fig. 7. Effective $\bar{K}^0\pi^+\pi^-$ mass distribution at 8, 10 and
16 GeV/c.

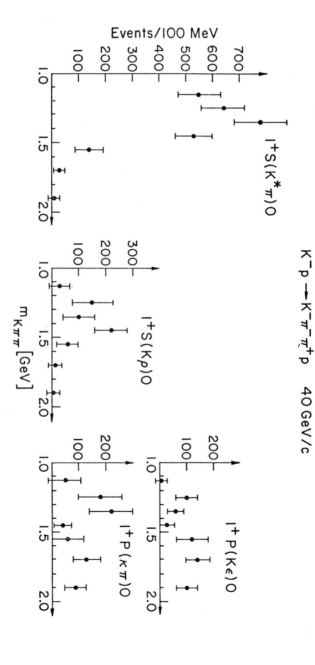

Fig. 8. Breakdown of 1⁺ state at 40 GeV/c.

188

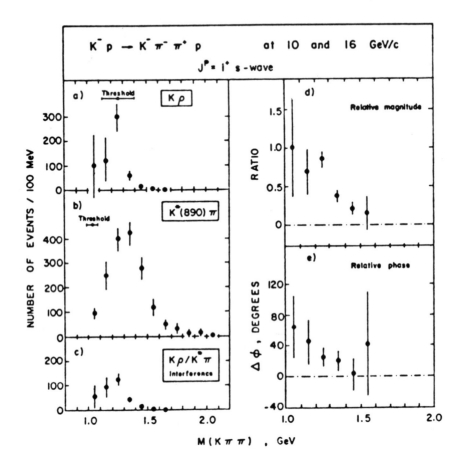

Fig. 9. a), b), and c) breakdown of 1^+ state at 10 and 16 GeV/c.
d) Magnitude of K amplitude relative to K (890) π amplitude.
e) Relative Kρ/K (890) phase.

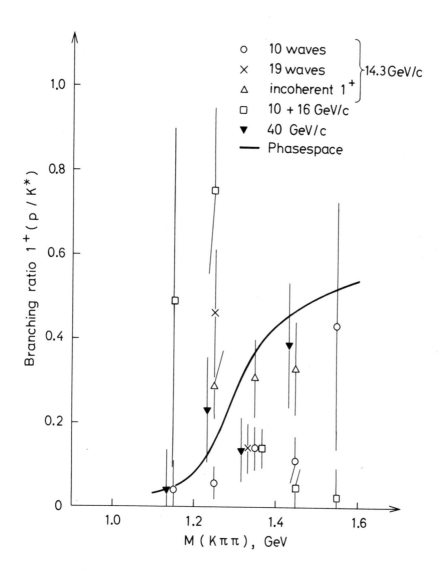

Fig. 10. Ratio of number of Kρ to K(890)π events of 1⁺
state for various fits.

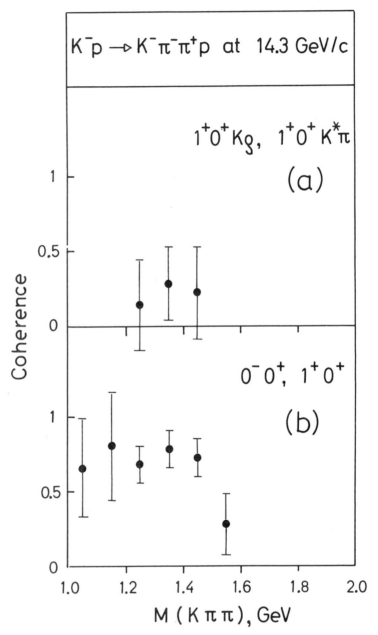

Fig. 11. Coherence at 14.3 GeV/c.
a) 1^+0^+ K vs. 1^+0^+ K (890)
b) 0^-0^+ vs. 1^+0^+

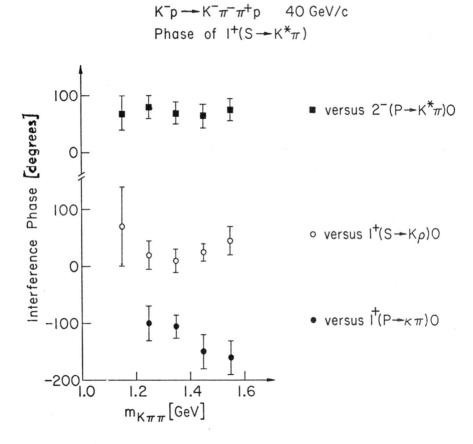

Fig. 12. Phase of 1⁺s K (890) at 40 GeV/c.

SURVEY OF THEORETICAL ANALYSES OF
ππ AND πK INTERACTIONS

Pran Nath
Department of Physics, Northeastern University
Boston, Massachusetts 02115

ABSTRACT

We examine the present status of ππ and πK theory and
its relationship to experiment. The restrictions on
scattering amplitudes from axiomatic principles, their
consequences for partial waves and applications for ππ
phenomenological analyses are given. Field theory models,
Pade approximants and calculations using super-propagator
methods are discussed. Finally soft pion theorems and
hard meson current algebra analyses and unitarization
schemes are presented. Comparison of theory to existing
data is given.

TABLE OF CONTENTS

I.   INTRODUCTION

We present here a survey of the theoretical analyses on ππ and
πK scattering and give a comparison of these analyses with experiment.
Determinations of ππ and πK scattering up to 2 GeV are now available
as a result of the recent high energy single pion production experiments in reactions involving πp and Kp. (For a review see references
1-4). Additional information in the low energy domain comes from a
number of other processes. The colliding beam experiment $e^+e^- \to$
$\pi^+\pi^-$[5] gives us the ρ-parameters and the pion form-factor. The K_{e4}
decay $(K^{\pm} \to \pi^+\pi^-e^{\pm}\nu)$ determines $\delta_0^0 - \delta_1$ [1] in the threshold range
$M_{\pi\pi} < m_K$ [6]. The $K^0 \to 2\pi$ decay contains the information on
$\delta_0^0(m_K^2) - \delta_0^2(m_K^2)$ [7].

$\eta \to 3\pi$ and $K \to 3\pi$ decays also are a possible source of information on low energy $\pi\pi$ amplitudes. However, there are some theoretical problems to contend with before one can extract useful information from these processes [8]. The experiment on K_{e3} decay ($K^\pm \to \pi^0 e^\pm \nu$) gives information on $<\pi|S^\mu|K>$ [9] (S^μ is the strangeness changing current) which enters in the low energy πK interactions. There also exist data on the process $\pi\pi \to \overline{KK}$ [1-3] which is inter-related to both the $\pi\pi$ and the πK process.

First we review the general properties of pion-pion amplitudes arising from the axiomatic principles of a quantum field theory. As a consequence of these general properties, the partial waves must obey a number of physical and unphysical region constraints. The unphysical region constraints are the Martin inequalities [10] and the Roskies sum rules [11-12] which involve partial wave amplitudes in the region $0 \leqslant s \leqslant 4m_\pi^2$ (see fig. 1). Another type of constraints for partial wave amplitudes have been derived by Roy [13] and take the form of a closed system of partial wave integral equations valid for $-4m_\pi^2 \leqslant s \leqslant 60m_\pi^2$. Recently a new set of equations valid over a somewhat larger domain has been obtained by Mahoux, Roy and Wanders [14]. We discuss the applications of these equations and also discuss calculations based on other constraints for phase shift analyses. One interesting observation resulting from these analyses is that analyticity, crossing and unitarity do not uniquely determine the low energy pion-pion amplitudes [15-16], and one must supplement the analyses with additional experimental constraints. Another observation which emerges from these analyses concerns the use of the unphysical region constraints. The problems of extrapolation from the unphysical into the physical domain for phenomenological amplitudes which are necessarily approximate tends to diminish the usefulness of the unphysical region constraints as tools for determining physical region amplitudes (see the discussion in sec. III).

Recently, attempts have been made to calculate pion-pion amplitudes from field theory models. Here, one considers phenomenological Lagrangians which satisfy current algebra and are in addition renormalizable. The perturbation expansion for amplitudes calculated from such Lagrangians when summed through the technique of Pade approximants (the convergence of the approximants is assumed) produces an amplitude which is unitary at any level of the (diagonal) Pade approximation. The above program has been implemented by Basdevant and Lee [18] for the case of the renormalized Gell-Mann-Levy σ-model [17]. They compute the amplitudes in perturbation up to second order (see fig. 9) and unitarize using the [1,1] Pade approximant. This technique allows the generation of dynamical resonances obtained through summation of a perturbation series. Thus the calculation of Ref. 18 generates a ρ and a f^0 (not originally present in the Lagrangian) although the mass of f^0 is rather low and the width of ρ rather small. An apparent weakness of this approach is in the requirement of renormalizability for the underlying phenomenological model.

In the last two years a new class of calculations using the non-linear chiral Lagrangians and super-propagator techniques have appeared. Since the underlying theory is non-renormalizable in the conventional sense, one loop contributions to the tree amplitudes involve subtraction constants which cannot be removed by the usual renormalization constraints. The super-propagator method allows one to determine these constants by the consideration of super-propagator diagrams (see fig. 10). The method was first applied to the pion-pion problem by Lehmann [19]. However, he used a perturbation expansion which lacked point transformation covariance i.e. independence in the definition of the pion field co-ordinates allowed within the chiral SU(2)xSU(2). This defect has been corrected subsequently by Ecker and Honnerkamp [20] who used the super-propagator method to calculate $\pi\pi$ phase-shifts up to one GeV [21]. Analogous calculations for physical mass pions involving massive super-propagators [22] have also been done [23]. A discussion of the results and comparison with experiment is given in sec. IV.

The pion-pion scattering is ideally suited for the application of current algebra principles because the external particles as well as the spectrum of intermediate states appearing in the scattering process are intimately related to the vector and the axial vector currents, their divergences and commutators (similar statements also hold for πK scattering). Thus for example, the pion may be interpolated by the divergence of the axial current, the ρ by the vector current and the ε-mesons by the sigma-commutator. The pion-pion system in the region of elastic scattering may then be constrained by the following set of current algebra conditions (i) the currents obey the SU(2)xSU(2) equal-time commutations relations of Gell-Mann [24] (ii) the vector currents are conserved (CVC) and (iii) the axial vector currents are partially conserved (PCAC). Weinberg [25] used these current algebra conditions along with the assumption of an iso-scalar σ-commutator and a gentleness hypothesis (as one limits the pion momentum q^μ to zero) to obtain the low energy pion-pion parameters. The present status of Weinberg predictions and experiment including the K_{e4} data is discussed in sec. V. In the hard pion calculations which allow one to extend the predictions of current algebra far beyong threshold the gentleness hypothesis of soft pion physics is replaced by the assumption of single meson saturation of intermediate sums and smoothness hypothesis at the meson vertices (for review see Ref. 26 and 27). At threshold the hard meson results give well defined corrections to Weinberg scattering lengths and these corrections obey the Olsson sum rule [28]. In going beyond the threshold region one must use a unitarization procedure and here a number of schemes exist as discussed in Sec. V. With a proper unitarization and with the inclusion of the intermediate resonances that appear to exist experimentally the domain of hard meson phenomenology may be extended considerably beyond threshold. Phase shift analyses based on hard meson current algebra up to 2 GeV are presented in sec. V.

πK scattering is governed by the chiral SU(3)xSU(3) current algebra. The algebraic symmetry possessed by currents may in general allow for arbitrary symmetry breakdowns for the phenomenological Lagrangians that describe the particle interactions. Sometimes ago Glashow and Weinberg [29], Gell-Mann, Oakes and Renner [30] proposed a simple model for the chiral symmetry breakdown in that the symmetry breaking interaction belongs to the $(3,3^*) + (3^*,3)$ representation of $SU(3) \times SU(3)$. One of the consequences of this hypothesis is that it gives a small value for the ξ parameter in the K_{e3} decay [9]. The value of ξ has not been settled experimentally, though the most recent data favors a small ξ [31]. It is interesting that the $(3,3^*)$ symmetry breaking scheme follows from the hard meson analysis. It has been shown by Arnowitt, Friedman, Nath and Suitor [32] that one may deduce the $(3,3^*) + (3^*,3)$ breaking from the assumptions of pole dominance and chiral $SU(3) \times SU(3)$ algebra. The application of hard meson $SU(3) \times SU(3)$ current algebra to πK scattering produces stringent constraints and after the masses and widths of resonance states are specified, the amplitudes depend only on the ratio of the leptonic decay constants for K and π, i.e., on F_K/F_π. In sec. VI we present the phase-shift analysis of current algebra πK amplitudes and compare the results with experiment. A brief summary of other attempts on πK scattering is also given.

Finally we comment on the structure of the resonance spectrum and related topics. The S^* resonance which is near the $K\bar{K}$ production threshold is of necessity a two channel phenomenon [33]. We discuss another possible component to its structure. In the vicinity of the $K\bar{K}$ threshold one has in fact two resonance structures, the S^* and the S. One may conceive of these states as arising from the collision of two overlapping resonances which can produce both a broad and an arbitrarily narrow resonance [34].

In a final section we conclude by giving a comparison of the relative advantages and disadvantages of the various theoretical techniques and an outlook on the future theory. Those readers not interested in the details can skip immediately to this section for a summary of the results of this survey.

II. GENERAL PROPERTIES OF PION-PION AMPLITUDES AND RIGOROUS CONSTRAINTS ON PARTIAL-WAVES

The amplitude $A^I(s,t,u)$ [35] describing the scattering process of Fig. 1 (where I signifies the iso-spin) satisfies the following properties. These properties can be shown to hold on the basis of general axioms of field theory and $\pi\pi$ mass spectrum.
(i) $A^I(s,t,u)$ satisfies crossing symmetry [36].
(ii) $A^I(s,t,u)$ satisfies fixed-t dispersion relations for t in the range $-28m_\pi^2$ to $4m_\pi^2$ [37].
(iii) The partial wave expansion

$$A^I(s,t,u) = \sum_{\ell=0}^{\infty} (2\ell+1) \frac{1+(-1)^{\ell+I}}{2} A_\ell^I(s) P_\ell(\cos\theta) \qquad (2.1)$$

converges in the Lehman-Martin ellipse where $t = -2k^2(1-\cos\theta)$ and $k^2 = (s-4m_\pi^2)^{\frac{1}{2}}/2m_\pi$.

Due to the normalization of $A^I(s,t,u)$ [35] and Eq. (2.1), $A_\ell^I(s)$ satisfies in the elastic region the unitarity condition

$$\text{Im}A_\ell^I(s) = ks^{-1/2}|A_\ell^I(s)|^2 \qquad (2.2)$$

(a) Partial Wave Constraints on the Unphysical Interval o $\leqslant s \leqslant 4m_\pi^2$.

The study of partial wave constraints in this region are motivated by the hope that such constraints would help in a construction of the low energy pion-pion interaction. The constraints studied here have generally been of two types. The first type are those arising from requirements of crossing symmetry alone. These types of constraints have been discussed by Balachandran and Nuyts [11] and by Roskies [12]. The constraints take the form of moment sum rules and form a closed set of relations when one retains a finite set of partial waves [38]. The second type of constraints are those arising from crossing symmetry and positivity of absorptive parts. These constraints take the form of inequalities and have been studied by Martin and others [10]. All the above inequalities are local in that they involve values of partial waves or their derivatives in the interval s=o and s=$4m_\pi^2$ [39]. In addition, there are a variety of other schemes that exploit the positivity of the absorptive parts to obtain constraints on the amplitudes in the interval s=o to s=$4m_\pi^2$[40]We shall return to the possible usefulness of these schemes for the construction of low energy pion amplitudes in sec. III.

(b) Crossing and Analyticity Constraints in the Physical Region.

The partial wave constraints discussed in (a) require a knowledge of the amplitudes in unphysical domain o $\leqslant s \leqslant 4m_\pi^2$. Sometimes ago, Roy [13] derived a system of exact integral equations for pion-pion scattering on the basis of crossing symmetry and fixed-t dispersion relations. These equations express the real part of a partial wave in terms of scattering lengths and a sum of integrals over the physical values of the absorptive parts for all partial waves. However, the equations hold only in the limited domain $-4m_\pi^2 \leqslant s < 60m_\pi^2$. The equations have the general form

$$A_\ell^I(s) = s_\ell^I + \sum_{I'} \int_0^1 dx P_\ell(x) \int_{4m_\pi^2}^\infty ds' \sum_{\ell'=0}^\infty (2\ell'+1) K_{\ell,\ell'}^{II'}(s,s',x) \text{Im} A_{\ell'}^{I'}(s')$$

(2.3)

Where s_ℓ^I is a polynomial in s determined in terms of I=o,2 s-wave scattering lengths and $K_\ell^{II'}$ is a complicated kernel defined in ref. 13. It may be noted that eqs. (2.3) are derived from the principles of axiomatic field theory and do not assume for instance, the existence of a Mandelstam representation. We also note that eqs. (2.3) involve all partial waves and require a considerable amount of input information to be of practical value.

Recently, Mahoux, Roy and Wanders [14] have obtained a set of equations valid for $-28m_\pi^2 \leqslant s \leqslant 125.31m_\pi^2$. The new equations are derived from axiomatic analyticity and three channel crossing symmetry. However, one does not obtain a unique set of equations but a whole family of such sets with each set holding only over a part of the accessible values of s. At least two values of parameter labelling of the family are needed to cover the entire interval $[4m_\pi^2, 125.31m_\pi^2]$.

III. PHENOMENOLOGY BASED ON RIGOROUS CONSTRAINTS

We now proceed to discuss the phenomenology of low and inter-mediate energy pion-pion scattering. We first discuss those analyses which are built directly rising the properties of amplitudes given in sec. II. At the end we shall discuss briefly other analyses which make additional physical assumptions. We order our discussion according to models based on the following assumptions:

a. Forward Dispersion Relations
b. Roy Equations
c. Unphysical Region Constraints
d. Analyses Using Additional Assumptions.

a. Forward Dispersion Relations. These analyses are well known and we shall discuss them only briefly here. The forward dispersion relations for the amplitude A_t^I with t-channel i-spin I [41], in the unsubtracted form are given by

$$A_t^I(\omega) = \frac{2}{\pi} \int_1^\infty \frac{\omega' d\omega' \operatorname{Im} A_t^I(\omega')}{(\omega'^2 - \omega^2)} \tag{3.1}$$

where $\omega = (s-u)/4m_\pi^2$. Using the $I = 1$ component of eq. (3.1) [42], Olsson [28] has derived sum rules for $(2a_o^o - 5a_o^2)$ and a_1^1 (where a_ℓ^I denotes the scattering length for the ℓ-th partial wave with iso-spin I). The sum rules are

$$(2a_o^o - 5a_o^2) m_\pi^{-1} = \frac{3}{4\pi^2} \int_1^\infty \frac{d\omega}{(\omega^2-1)^{1/2}} \Delta_{\pi\pi} \tag{3.2}$$

with

$$\Delta_{\pi\pi} = \sigma_{tot}(\pi^-\pi^+) - \sigma_{tot}(\pi^+\pi^+)$$

and

$$18a_1^1 m_\pi = \frac{3}{4\pi^2} \int_1^\infty \frac{d\omega}{(\omega^2-1)^{1/2}} \frac{\omega-1}{\omega+1} \Delta_{\pi\pi} + \frac{6}{\pi^2} \int_1^\infty \frac{\omega d\omega}{(\omega^2-1)^{3/2}} \sigma_1 \tag{3.3}$$

where σ_1 represents the $I_s = 1$ cross-section. These sum rules have been examined by Olsson and Tryon. Olsson determines a value of $(0.040 \pm 0.005)m_\pi^{-3}$ for a_1^1. For $(2a_o^o - 5a_o^2)$ the value of $(0.55 \pm 0.05)m_\pi^{-1}$ was found [27]. Using eqs (3.2) and (3.3) a sum rule for the difference $c = (2a_o^o - 5a_o^2)m_\pi - 18a_1^1 m_\pi^3$ can be obtained which is even more convergent than either of the original sum rules. Saturation of this sum rule gives a value of $c = -(0.07 \pm 0.02)$ [27]. We note here that the values of the para-meters determined above have generally been found to be rather insensitive to variations in the assumed values of the s-wave amplitudes used in the saturation of the sum rules.

Morgan and Shaw [15] have used forward dispersion relations and unitarity to constrain amplitudes in the threshold region using known information in the resonance region. The method they employ involves dividing eq. (3.1) into three domains consisting of a low energy region $o \leqslant \sqrt{s} \leqslant E_1$ where one intends to determine the amplitudes, an intermediate region $E_1 \leqslant \sqrt{s} \leqslant E_2$ where the amplitudes are assumed known (this includes the ρ-mass region) and a high energy region $E_2 \leq \sqrt{s} \leq \infty$. This last region is approximated by a polynomial in $\sqrt{s}$ for $\sqrt{s} \leq m_\rho$. The subtraction constants appearing in the polynomial are then determined through a set of prescribed consistency conditions [44].

The s-wave threshold parameters determined from the preceding analysis are not unique but rather cover a range of values corresponding to the range of input s-wave amplitudes used in the ρ-region. Plotted in the $(a_0^{\,0}, a_0^{\,2})$ plane, the solutions show a remarkable property in that the computed $a_0^{\,0}$ and $a_0^{\,2}$ fall in a very narrow band (called the universal curve) for the great variety of s-wave inputs in the ρ-region.

One may understand the above effect in part in terms of the Olsson sum rule eq. (3.2). As mentioned above, saturation of this sum rule gives a value of $(2a_0^{\,0} - 5a_0^{\,2}) = (0.55 \pm 0.05)m_\pi^{-1}$ which is found to be fairly insenstivie to the variations in the s-wave input. The constancy of the difference $2a_0^{\,0} - 5a_0^{\,2}$ leads to a linear relationship between $a_0^{\,0}$ and $a_0^{\,2}$ which is seen to hold over part of the universal curve.

A unique prediction for $a_0^{\,0}$ and $a_0^{\,2}$ results if we supplement the above analysis by experimental determinations of $a_0^{\,0}/a_0^{\,2}$. This quantity has been obtained by Gutay et al. [45] from the study of forward-backward asymmetry in $\pi^+\pi^-$ production. Cline et al. [46] also have measured this quantity from the change branching ratios $\sigma(\pi^0\pi^0)/\sigma(\pi^+\pi^+)$ and $\sigma(\pi^0\pi^0)/\sigma(\pi^+\pi^-)$. The experimental value [47]

$$a_0^{\,0}/a_0^{\,2} = -3.2 \pm 1.0 \tag{3.4}$$

yields then $a_0^{\,0} = (0.16\pm0.04)m_\pi^{-1}$ and $a_0^{\,2} = (-0.05\pm0.01)m_\pi^{-1}$. For $a_1^{\,1}$ one obtains $a_1^{\,1} = (0.035\pm0.0027)m_\pi^{-3}$. These determinations are consistent with Weinberg soft pion results [25].

In searching for favored solutions among the input amplitudes additional constraints such as Eq. (3.4) and prediction of the correct charge branching ratios must be used. In the analysis of Ref. 15 such criterion produce the favored solution - Between - Down (BDI, BDII). This may be compared with experimental data from the Berkeley (Protopopescu et al. [48]) and CERN experiments [49] on $\pi^+\pi^-$ production. A number of different analyses using the latter data has been made. These are labeled A^{50}, B^{51}, C^{52}, D^{53} and E^{54}. The analyses from Berkeley and CERN experiments agree qualitatively for the $I = 0$ s-wave phase-shift, and are also compatible with the recent data on $\pi^0\pi^0$ production [55]. The comparison of $\delta_0^{\,0}$ with experiment is given in Fig. 2. In Fig. 3 we plot $\delta_0^{\,2}$ against experiment taken from Baton et al. [56], Colton et al. [57], Sander et al., Baker et al., and Baubillier et al. [58], Cohen et al. [59], Walker et al.[60] and Hoogland et al. [61]. A comparison also of the P-wave is given in Fig. 4 with experiment taken from Ref. 51. The overall agreement for all the phase-shifts is quite good, though the $I = 2$ phases show an upward trend at the high end not indicated by experiment.

b. Roy Equations. We next discuss another type of calculations (16,62-64) which are based on the Roy equations [13] eq. (2.3). As discussed in sec. II, these equations are a set of partial wave integral relations which determine $\text{Re}A_\ell(s)$ (for $-4m_\pi^2 \leq s \leq 60m_\pi^2$)

provided $\text{Im}A_\ell(s)$ is specified in the entire physical region $o \leqslant s \leqslant \infty$ and for all partial waves. In the region of elastic unitarity one may use eq. (2.3) along with the unitarity relation to provide a set of integral equations for $\text{Im}A_\ell$ valid in the region of elastic unitarity. To implement this program, one obtains a set of pheno-menological equations from eq. (2.3) by separating out explicitly the partial waves of interest (s and p) over an energy region up to a maximum value of Λ. Eq. (2.3) becomes [65]

$$A_\ell^I(s) = s_\ell^I(s) + B_\ell^I(s) + \underset{I'\ell'=0}{\Sigma\Sigma} \int_{4m_\pi^2}^{\Lambda^2} ds' K_{\ell\ell'}^{II'}(s,s') \text{Im}A_{\ell'}^{I'}(s') \quad (3.5)$$

where $B_\ell^I(s)$ represents the so-called driving term and is the difference of the last terms on the right hand sides of eqs (2.3) and (3.5). Thus far eq. (3.5) is exact and also empty unless something further is said about B_ℓ^I.

Now $B_\ell^I(s)$ is determined if one has knowledge of (i) $\text{Im}A_{\ell'}^I(s)$ for $\ell' \geqslant 2$ and $4m_\pi^2 \leqslant s \leqslant \Lambda^2$ and (ii) $\text{Im}A_\ell^I(s)$ for all ℓ and $\Lambda^2 \leqslant s \leqslant \infty$. This is a request for a considerable amount of input to be obtained from extraneous sources (not necessarily governed by the axiomatic principles of a quantum field theory) and no doubt tends to tarnish some of the initial rigour that was used to obtain eq. (3.5).

For the practical applications of eq. (3.5) one constructs a model for the driving term B_ℓ^I. Phenomenologically a value of Λ is chosen above which the amplitude is to be represented by a Regge pole and a pomeron exchange. The authors chose Λ^2 to be $110m_\pi^2$. This specifies one part of the driving term (part (ii)) valid for $\Lambda^2 \leqslant s \leqslant \infty$. The other part (part (i)) of the driving term requires a knowledge of $\text{Im}A_{\ell'}^I(s)$ for $\ell' \geqslant 2$ in the range $4m_\pi^2 \leqslant s \leqslant \Lambda^2$. Here the approximation used is more drastic in that the resonance contribution alone is retained and all other low energy contributions ignored [67].

Having determined the driving term, eq. (3.5) may be used along with the unitarity relation [67] to obtain a set of non-linear integral equations for $\text{Im}A_\ell^I(s)$ in the range $4m_\pi^2 \leqslant s \leqslant \Lambda^2$. The program has been implemented in detail by Basdevant, Froggart and Petersen [16] who discovered a multiplicity of solutions, all satisfying the integral equation constraints. The solutions obtained depend on five [68] independent parameters which may be chosen to be a_0^0, a_0^2, m_ρ, Γ_ρ and m_ϵ (defined by $\delta_0^0(m_\epsilon) = 90^\circ$). The arbitrariness of the solutions demonstrates that the constraints of analyticity, crossing and unitarity do not fully determine the low energy amplitudes.

One may then further restrict the scattering in the threshold domain by requiring agreement of the solutions obtained with all the known data up to the resonance domain. In the analysis of ref. 16 s-wave data from three phase-shift analyses (Saclay [56], Berkeley [48] and CM-EMI [52] in the range $500 < \sqrt{s} < 900$ Mev were used in addition to p-wave resonance parameters.

The analysis shows that for any given set of experimental data (Saclay, Berkeley or CM-EMI) s-wave scattering lengths a_0^0 and a_0^2

are not uniquely determined but are rather related by a universal curve (see fig. 5), as in Ref. 15. The Saclay [56] and Berkeley [48] data restricts $a_o{}^o$ in the range $-0.05 < a_o{}^o < 0.6$ whereas the phases of Estabrook et. al. [52] (CM-EMI) give $a_o{}^o > 0.15$. However, as in Ref. 15 the p-wave scattering length is stable and close to the soft pion value.

We give a comparison of the analysis with all the data in figs (6) - (8) where we have chosen $a_o{}^o = 0.17m_\pi{}^{-1}$ and fits corresponding to CM-EMI s-wave inputs are shown. For I = 1 phase-shift $\delta_1{}^1$ one finds a significant deviation between theory and experiment near threshold. In the $(k^3/\sqrt{s})\cot\delta_1{}^1$ plot, the data near threshold lies lower than the predictions of theory (the same also holds for current algebra predictions). Alternately, the trend of the data appears to produce a larger scattering length (by almost a factor of 2!) than predicted by all present theories. This represents perhaps the most significant conflict between theory and experiment in the region $\sqrt{s} \lesssim 1$ GeV.

There is more data in the threshold region from a new CERN experiment [69] but the situation as regards to theory and experiment mentioned above remains unchanged.

At the present time experimental data on the $\pi\pi$ phase-shift up to approximately 2 GeV is available and it is now of interest to extend the theoretical analyses in this domain. In this context the new set of partial-wave equations derived by Mahoux, Roy and Wanders [14] with their extended domain in s $(-28m_\pi{}^2 \lesssim s \lesssim 125.31m_\pi{}^2)$ provide one possibility.

Recently, Bonnier and Johanneson [70] have attempted to study the influence of the 1 GeV - 2 GeV data on low energy $\pi\pi$ amplitudes by use of supplementary conditions expected to enhance the role of 1 GeV - 2 GeV data at low energy and the usual Roy equations. They favor a rather large s-wave scattering length, with $a_o{}^o$ in the range 0.5 and 1.1.

Finally we comment on the possible role that the parametrizations may play in the calculations discussed above. Here we illustrate a source of ambiguity that occurs in the commonly used parametrization of the amplitude using the K-matrix. One writes

$$A_\ell{}^I(s) = [K_\ell{}^I(s)^{-1} + C_\pi(s)]^{-1} \tag{3.6}$$

where the specific form of $K_\ell{}^I$ need not concern us here. The usual choice for $C_\pi(s)$ is the function [71]

$$C_\pi(s) = \frac{1}{\pi}\left(\frac{s-4m_\pi{}^2}{s}\right)^{1/2} \log\frac{\sqrt{4m_\pi{}^2-s}+\sqrt{-s}}{2} \tag{3.7}$$

The function $C_\pi(s)$ is quite well behaved in the region $o \lesssim S \lesssim 4m_\pi{}^2$. Above threshold, however, $C_\pi(s)$ has a real part which grows as ℓns and produces a significant correction to the amplitude even at 1 GeV. Now the choice of the form eq. (3.7) for C_π is quite arbitrary restricted only by the unitarity requirement that its right hand cut discontinuity be $-ik/\sqrt{s}$. We view models of C_π such as eq. (3.7) as suspect unless otherwise theoretically justified especially in an energy region where they could produce large modifications of the amplitudes (e.g. above 1 GeV) [72].

c. Unphysical Region Constraints. Analyses of pion-pion amplitudes have also been attempted through a direct application of the partial wave constraints in the unphysical domain $0 \le s \le 4m_\pi^2$. As discussed in sec. II above, these constraints consist of Roskies relations [12] and certain inequalities obtained by Martin and others [10]. One such calculation is that of Kang and Lee [73] who construct a model of s and p-wave amplitudes using partial wave dispersion relations and unitarity [74]. The left hand cut is approximated by a few poles and the pole structure is then determined by the requirement that the solutions to the dispersion relations satisfy Roskies sum rules and in addition the current algebra constraints at threshold. The final solutions are checked against the Martin inequalities [76].

The analysis produces a solution where the ϵ and ρ mesons are generated as dynamical bound states of the two pion system. At threshold the amplitudes given by the analysis are close to the current algebra result but deviations occur as we go beyond threshold. The I = 0 s-wave resonates at about 700 MeV, a somewhat low value in view of the current experimental data. The position of the P-wave resonance is in agreement with experiment but the P-wave phase shifts are a rather poor fit to the data. A comparison of theory and experiment is given in Figs. (2)-(4).

In view of the large deviation of the theory from experiment in the P-wave, one may view the present calculation as an illustration of the weakness of the unphysical region constraints for the determination of physical region amplitudes [77].

Roskies relations and Martin inequalities have sometimes been used as a test of the crossing symmetry and positivity properties of amplitudes generated through various theoretical models. As a rule there is a loss of crossing symmetry when an amplitude is arranged to satisfy unitarity and one may wish to test this loss through the satisfaction of the unphysical region constraints [80]. However, as we have already remarked, the usefulness of this procedure is severely limited by the fact that the physical region amplitudes must be extrapolated into the unphysical regions before the constraints may be imposed.

d. Analyses Using Additional Assumptions. Next we briefly consider models which are based on additional physical principles such as duality [82]. A specific realization of this principle is in the Veneziano model [83] which has been used by Lovelace [84] for a phenomenological description of the $\pi\pi$ (and πK) scattering. The partial waves given by the Veneziano model are real and must be unitarized before comparison with experiment and the unitarization scheme adopted by Lovelace is similar to the one given by Eqs. (3.6)-(3.7) [85,86] but involves both the $\pi\pi$ and the $K\bar{K}$ channels.

Aside from the technical problems concerning how best to unitarize the Veneziano model, the waning interest in this model is due primarily to the lack of experimental support for many of the Veneziano predictions. Thus the Lovelace-Veneziano prediction of the ϵ(700 MeV) meson has disappeared from data (due to the choice of the down solution in the ρ-region) and the ρ' prediction

(m_ρ' = 1288 MeV, $\Gamma_{\rho'\to\pi\pi}$ = 105 MeV) has yet to be seen. The predicted f → 2π partial width of 98 MeV is also rather small (the experimental value is $\Gamma_{f\to2\pi}$ = 141 ± 25 MeV). On the credit side the model predicted on s* with an m_{s*} = 912 and Γ_{s*} = 15 MeV. Though the predicted mass is a bit low (most present experiments give m_{s*} ≃ 1 GeV the predicted width is within the spectrum of observed values [87]. However, the coupling to the $K\bar{K}$ channel is too weak. In sum, despite some successes, the unitarized Veneziano formula does not predict the correct meson spectrum (coupled to the ππ system) as presently observed.

Finally, we comment on the past ππ phase-shift analyses based on assumptions of partial wave dispersion relations [89]. If one assumes the asymptotic form of the Veneziano model, the partial wave amplitudes may involve essential singularities at infinity and not obey dispersion relations [90]. However, it has been argued by Tryon [91] that in a limited energy domain even amplitudes such as those given by the Veneziano model may be approximated by functions which do obey dispersion relations. The analyses such as those of ref. 89, though of questionable rigorous validity, assume then an approximate numerical validity.

IV. PION-PION DYNAMICS FROM FIELD THEORY MODELS, PADE APPROXIMANTS AND SUPER-PROPAGATORS

The purpose here is to construct unitary amplitudes starting from some presumably fundamental chiral Lagrangian. Now Lagrangian field theory has to cope with two essential difficulties. The first one is that given a renormalizable Lagrangian the usual perturbation expansion of the s-matrix in the coupling constant will diverge. The second difficulty is that the Lagrangian at hand (such as the nonlinear chiral Lagrangian) may not be renormalizable in the conventional sense. To overcome the first problem one needs an algorithm for summation of divergent series and here the Pade approximant technique has been suggested [92]. Regarding the second problem, certain classes of Lagrangians such as the nonlinear σ model, considered non-renormalizable in the conventional sense, may now be treated by the so-called super-propagator techniques.

We begin our discussion by considering the chiral symmetric phenomenological pion Lagrangian which provides the basic ingredients of the field theory models. We shall examine the various topics discussed above in the following order:
a. Phenomenological Pion Lagrangians
b. Pade Approximant Algorithm
c. Super-Propagator Method

a. Phenomenological Pion Lagrangians. Pion interactions may be constructed by considerations of chiral symmetry and chiral SU(2) x SU(2) algebra of charges proposed by Gell-Mann [93]. The algebra involves the vector charges Q_a and the axial-vector charges Q_a^5 (a = 1, 2, 3) which produce two SU(2) algebras for the chiral combinations $Q_a^\pm = 2^{-1}(Q_a \pm Q_a^5)$:

$$[Q^+_a, Q^-_b] = 0$$

$$[Q^{\pm}_a, Q^{\pm}_b] = i\ \epsilon_{abc}\ Q^{\pm}_c \tag{4.1}$$

The charge algebra given by Eqs. (4.1) is chirally symmetric. However, the phenomenological Lagrangian which satisfies the algebra is not invariant under the SU(2) x SU(2) group of transformations. This arises due to the non-conservation of the axial charge Q_a^5 (and the existence of the PCAC condition).

A specific example of such a phenomenological Lagrangian, which we shall refer to in the following, is the well-known Gell-Mann-Levy model which satisfies both the constraints of current algebra and the PCAC condition. This Lagrangian is given by

$$\mathcal{L} = \tfrac{1}{2}((\partial_\mu \sigma)^2 + \partial_\mu \pi_a \partial^\mu \pi_a) - \frac{\mu^2}{2}(\sigma^2 + \pi_a \cdot \pi_a)$$

$$- \frac{\lambda}{4}(\sigma^2 + \pi_a \cdot \pi_a)^2 + C\sigma \tag{4.2}$$

where π_a represents the pion field and σ an iso-spin scalar field. Nonlinear realization of the σ-model where the only independent field appearing in the Lagrangian is the pion field may be simply obtained by setting $\sigma^2 + \pi_a^2 = F_\pi^2$ and substituting the value of σ obtained from here in Eq. (4.2). The above Lagrangians when used in the tree approximation give scattering amplitudes which correctly yields the soft-pion Weinberg result for the scattering lengths [25].

We next proceed to consider methods which have been used to treat the Lagrangians considered above beyond the tree graph approximation with a view to obtaining unitary amplitudes. As mentioned above, the Pade approximants and the super-propagator method have been used recently in this context by a number of authors. We now turn to a discussion of the pion-pion scattering based on these techniques.

b. Pade Approximant Algorithm. The scattering amplitudes given by phenomenological Lagrangians considered above are real in the tree graph approximation. Beyond threshold, however, such an amplitude violates the requirement of unitarity. We examine now the efforts to satisfy both the current algebra and unitarity conditions by considering higher order corrections from the renormalizable chiral Lagrangian and "summing" the resulting divergent expansion through the Pade approximant algorithm [94].

Consider then the amplitude $A(s,t)^I$ which we assume has been computed in a perturbation expansion in the coupling constant λ

$$A(s,t)^I = \sum_{I}^{\infty} \lambda^n A^{(n)}(s,t)^I . \tag{4.3}$$

Using Eq. (4.3), one may obtain the partial s-matrix $S_\ell^I(\lambda)$ also in a power series in λ. The [N,M] Pade approximant to $S_\ell^I(\lambda)$ is then the ratio of two polynomials $P_N(\lambda)/Q_M(\lambda)$ $(= S_\ell^{I[N,M]})$ with order N and M in powers of λ satisfying the condition

$$S_\ell^I(\lambda) = P_N(\lambda)/Q_M(\lambda) + 0(\lambda^{N+M+1}) . \qquad (4.4)$$

For $N = M$ one obtains the diagonal Pade approximants $S_\ell^{I}[N,N]$. It can be shown that the Pade approximants are both unique and unitarity[95]. The usefulness of the method, of course, resides in the hope that the Pade approximants are convergent.

Basdevant and Lee [18] have applied the Pade program on the σ-model given by Eq. (4.2). This Lagrangian is renormalizable [103], and in every order of perturbation theory satisfies constraints of current algebra [94]. The theory given by Eq. (4.2) has three parameters μ, λ and c but after the (renormalized) pion mass m_π and the pion decay constant F_π are specified the theory depends only on one renormalized coupling constant λ_γ (m_σ is not independent but determined in terms of the other constants).

The calculation of Ref. 18 was carried out up to second order perturbation theory (involving only single closed loops) and unitarized using the [1,1] Pade approximant. The unitarized amplitudes were checked for the satisfaction of crossing symmetry and found to satisfy it to a good approximation. However, the satisfaction of current algebra is no longer guaranteed. The Pade unitarization generates dynamically both the ρ and the f^o not present in the original Lagrangian. However, an exotic $I = 2$ resonance also appears in the D-wave. The ρ width given by the model is small. The s-wave threshold parameters are $a_o^o = 0.09 \ m_\pi^{-1}$ and $a_o^2 = -0.025 \ m_\pi^{-1}$. A comparison of the phase-shifts obtained in this model with experiment is given in Figs. (11) and (12) [96]. The sign and the general magnitude of the $I = 2$ phase-shift is correct for $\sqrt{s} < 1$ GeV: For $I = 0$, however, there are significant deviations near the high energy end of the computed values.

A great weakness of the program discussed above is in its severe limitations that the basic Lagrangian be renormalizable. In Sec. V, we shall consider other programs which do not suffer from this limitation. We may, however, mention that the Pade approximant technique has also been used on certain non-renormalizable Lagrangians in the one-loop approximation [97]. One such calculation is due to Basdevant and Zinn-Justin who have studied the model of pion interaction with Yang-Mills field. The other is due to Jhung and Willey who use the nonlinear sigma model (see Figs. 11-14).

c. Super-Propagator Method. A number of authors have recently calculated pion-pion scattering based on the super-propagator technique [21,23]. Since this development is of a technical nature we shall describe the method only in physical terms and then discuss the results.

In the calculation of higher order corrections to the tree amplitudes in the nonlinear pion Lagrangian, one is faced with undetermined subtraction constants in the scattering amplitude. This arises due to the fact that the nonlinear Lagrangian is not a renormalizable one in the usual sense. The super-propagator formalism allows one to determine these constants. The method consists in summation of graphs shown in Fig. 10 through a Sommerfeld-Watson representation. The renormalized one loop approximation to the amplitude is then obtained by determination of the coefficient of the term F_π^{-4} in the representation.

The above technique was first used by Lehmann [19] whose perturbation expansion, however, lacked point transformation covariance with respect to the chiral group SU(2) x SU(2). This was subsequently corrected by Ecker and Honnerkamp [20,21].

A comparison of the results obtained from the above procedure with experiment is given in Figs. 11-13 where the calculations also include the nucleon one loop contributions [21]. The I = 2 phase-shift gives a reasonable fit to the data up to about 1 GeV. The P-wave is sensitive to the value of the axial vector coupling constant g_A and a good fit results if one uses the Goldberger-Treiman value for this constant rather than its experimental determination. The I = 0 s-wave is a reasonable fit at the low end of the data but deviates sharply from experiment at 1 GeV. Inclusion of the entire baryon octet loop rather than just the nucleon loop enhances the calculated phase-shift by no more than an additional $\sim 10^\circ$ at 1 GeV [23].

V. SU(2) × SU(2) CURRENT ALGEBRA AND PION-PION SCATTERING

In the preceding section we discussed theories which attempt to compute unitarity amplitudes starting from some basic or fundamental Lagrangians. The difficulties that plague such a program are many as already discussed. First, the Lagrangian may not be a renormalizable one and second, assuming it is, its perturbation expansion would diverge and one requires special techniques (not yet on a secure foundation) to sum such expansions. The criterion for what constitutes a fundamental Lagrangian is also obscure. An alternative technique to construct scattering amplitudes satisfying the full content of current algebra restriction at all energies is given by the hard meson method [26]. No problems regarding what is, or what is not fundamental concern us here since one takes all possible interaction structures relevant in any given energy domain. For the purpose of perspective and relative comparison we shall first give a brief description of the soft pion results for pion-pion amplitudes and then discuss the hard pion current algebra analyses and results. Thus we divide our discussion as follows:

a. Soft Pion Analysis

b. Hard Pion Current Algebra

a. Soft Pion Analysis. The soft pion current algebra analysis is based on the following four postulates:

(i)   SU(2) x SU(2) current commutation relations of Gell-Mann [24].

(ii)   Conservation of the vector current (CVC).

(iii) Partial conservation of the axial vector current (PCAC).

(iv)   Gentleness hypothesis in the region $0 \le q^\mu \le m_\pi$.

The gentleness hypothesis (iv) implies that the scattering amplitudes do not change appreciably as we go from threshold to the soft pion limit $q^\mu \to 0$. Weinberg [25] deduced the pion-pion scattering lengths on the basis of conditions (i)-(iv) and also the assumption of an iso-scalar σ-commutator, i.e., that

$$\delta(x^o - y^o)[A_a{}^o(x), \partial_\mu A^\mu{}_b(y)] \sim \delta^4(x - y)\sigma(x)\delta_{ab} \qquad (5.1)$$

where $\sigma(x)$ is an isoscalar field. The expressions obtained were [98]

$$a_o{}^o = 7(32\pi)^{-1}m_\pi F_\pi{}^{-2}, \quad a_o{}^2 = -2(32\pi)^{-1}m_\pi F_\pi{}^{-2} \text{ and } a_1{}^1 = (24\pi m_\pi F_\pi{}^2)^{-1}.$$
$$(5.2)$$

Using the experimental value of $F_\pi = 94$ MeV, one obtains the result

$$a_o{}^o = 0.16 \, m_\pi{}^{-1}, \quad a_o{}^2 = -0.045 \, m_\pi{}^{-1} \text{ and } a_1{}^1 = 0.033 \, m_\pi{}^{-3}. \qquad (5.3)$$

s-wave effective ranges were also obtained.

　　b. Hard Pion Current Algebra. The results given by the soft pion method can only hold in the vicinity of threshold. In order to exploit the full content of current algebra on scattering amplitudes beyond threshold, one must replace the gentleness hypothesis (iv) of soft pion physics by the assumption of single particle saturation of intermediate sums and a postulate of smoothness in particle momenta at the meson vertices. The hard meson method has been developed in a number of essentially equivalent forms. These consist of the Ward identity method of Gerstein, Schnitzer and Weinberg [99,100], effective Lagrangian approach of Arnowitt, Friedman, Nath and Suitor [101-103], dispersion method [104] and related approaches [105].

　　Application of hard pion current algebra to pion-pion scattering was made sometimes ago by Arnowitt, Friedman, Nath and Suitor [106]. The calculation was based on the single particle saturation of both the currents as well as the σ-commutator as defined in Eq. (5.1). The pion-pion amplitudes which satisfy rigorously all the constrains of current algebra include beside the usual pole diagrams, a set of sea-gull terms as shown in Fig. 15. Next we discuss three regions, threshold, near threshold and beyond threshold, and compare the predictions of current algebra with experiment.

1. Threshold Parameters. The s- and P-wave scattering lengths given by the hard pion method are close to the Weinberg values. They are

$$32\pi m_\pi a_o{}^o = 7m_\pi{}^2 F_\pi{}^{-2}[1 + \tfrac{1}{7}\varepsilon_\sigma(12\lambda^2 + 12\lambda\lambda_2 + 5\lambda_2{}^2)] ,$$

$$32\pi m_\pi a_o{}^2 = -2m_\pi{}^2 F_\pi{}^{-2}[1 - \varepsilon_\sigma\lambda_2{}^2] , \qquad (5.4)$$

$$24\pi m_\pi a_1{}^1 = F_\pi{}^{-2}[1 + \varepsilon_\sigma\lambda\lambda_2 + 6\varepsilon_\rho(1 - \tfrac{1}{4}\lambda_A)^2] ,$$

where λ, λ_A govern the width of the ε and ρ mesons [107] and λ_2 is an anomalous $\pi\pi\varepsilon$ coupling [106]. In the presence of more than one ε-mesons one sums over the correction terms for each ε meson in Eq. (5.4). Now $m_\pi{}^2/m_\rho{}^2 \sim 1/30$ and $m_\pi{}^2/m_\varepsilon{}^2 \sim 1/35$ for the s*. Thus unless λ_A, λ, and λ_2 are abnormally large, the hard pion corrections to Eq. (5.3) are only few percent. Thus essentially the hard pion results for $a_o{}^o$ and $a_o{}^2$ are those given by Eq. (5.3). We note that the smallness of the total hard pion correction is due to the smallness of the individual contributions. This is in contrast to the current algebra pion-nucleon scattering lengths [108,109] where the hard pion corrections due to the nucleon, N^* (3-3 resonance), and the ϵ are individually large (up to 40% of the soft pion values) and the validity of the Weinberg results is only accidental due to mutual cancellation among the correction terms.

We now proceed to compare the current algebra scattering length results with experiment. A determination of δ_0^o in the threshold region from the K_{e4} decay has been obtained by many different experimental groups; Beier et al. [110], Zylbersztejn et al. [111], Schweinberger et al. [112], and Ely et al. [113]. Analysis of the K_{e4} data has been done recently by Tryon [114]. Männer [69] at this conference has discussed the determination of threshold parameters obtained from a new CERN pion production experiment. Other determinations are due to Jones, Allison and Saxon [115] and Bunyatov [115] (who uses soft pion theory on $\pi^\pm p \rightarrow \pi^\pm \pi^+ n$). We summarize these results below

	a_o^o-Experiment	a_o^o-Current Algebra
Männer [69]	$(0.44 \pm 0.1)m_\pi^{-1}$	
K_{e4} data [114]	$(0.26 \pm 0.08)m_\pi^{-1}$	$0.16\ m_\pi^{-1}$ (5.5)
Jones et al. [115]	$-0.06\ m_\pi^{-1} < a_o^o < 0.03\ m_\pi^{-1}$	
Bunyatov [115]	$(0.18 \pm 0.02)m_\pi^{-1}$	

We note that current algebra prediction lies right about the middle of the experimental determinations. However, a more definitive determination to settle the scattering length question is desirable.

The a_o^o/a_o^2 experimental determination from the forward-backward asymmetry in $\pi^+\pi^-$ production and charge branching ratios is given by Eq. (3.4) and is consistent with the current algebra prediction of $\simeq -3.5$. However, for $I = 1$ phase-shift δ_1^1, the trend of the data near threshold in the $(k^3/\sqrt{s})\cot\delta_1^1$ plot appears to produce a larger scattering length than given by current algebra (see the discussion following Eq. 3.5) or by any present theory.

In sec. III we stated that the difference $C = (2a_o^o - 5a_o^2)m_\pi - 18a_1^1 m_\pi^3$ satisfies a rather convergent sum rule and saturation of this gives a value of $C = -(0.07 \pm 0.02)$. The sum rule is rather sensitive and tests a theory to beyond the ten percent level. Thus for the Weinberg prediction Eq. (5.2), $C = 0$. For the hard meson current algebra the corrections to the soft pion Weinberg satisfy the Olsson sum rule. We have

$$C_{HM} = 18a_1^1 m_\pi^3 (-6\epsilon_\rho (1 - \tfrac{1}{4}\lambda_A)^2 + \epsilon_\sigma \lambda^2) . \qquad (5.6)$$

Numerically $C_{HM} = -(0.073 \pm 0.004)$ [27] in excellent agreement with the sum rules determinations given above.

2. Low Energy Region. In the vicinity of threshold, one may expand the scattering amplitude $k^{2\ell}(A_\ell^I)^{-1}$ in powers of momenta. For the

s-wave, the region of validity of this expansion is very small as a low energy zero appears in the amplitude. The expansion of A_o^I itself, however

$$A_o^I = 2m_\pi [a_o^I - \frac{1}{2}(a_o^I)^2 r_o^I k^2 + \ldots] \qquad (5.7)$$

gives a rather convergent result in the low energy domain. Using current algebra values [116] for a_o^I and r_o^I, one obtains at the K-mass $\delta = \delta_o^0 - \delta_o^2 \simeq 40^\circ$. This result may vary by a few percent as the ε-coupling constants vary over a range of physically reasonable values. From the $K \to 2\pi$ experiment one has $\delta = - (39 \, ^{+13}_{-18})^\circ$ [117].

3. Resonance Region. Beyond the threshold region, current algebra amplitudes must be unitarized before one compares the results with experiment. The question of how to do this is subject to the same problems and criticisms mentioned in the analyses based on the Roy equations (see Sec. IIIb). A particularly simple unitarization is obtained by setting the current algebra amplitudes equal to the K-matrix so that $(\sqrt{s}/k^{2\ell+1})\cot\delta_\ell^I = (A_\ell^I)_{C.A.}$. Results of the current algebra analysis of Arnowitt, Friedman, Nath and Suitor [106,34] are presented in Figs. (16)-(19). The I = 2 s-wave current algebra prediction is in excellent agreement with experiment all the way up to 1.5 GeV (we note that the I = 2 channel is very elastic up to and even beyond 1 GeV due to a lack of $K\bar{K}$ production). The I = 1 phase shift δ_1^1 is also an excellent fit over the entire energy domain up to 1.8 GeV. [In the current algebra analysis of this phase shift, the ρ' (at 1600 MeV) was not included which explains the absence of the dip [118] in the current algebra results in this region. Actually, the presence of this dip is now in doubt (see Männer's review at this conference).] We add, however, that the essential element in the understanding of the phenomena in this energy domain, is in the smallness of the background - a feature which is correctly predicted by the current algebra analysis.

The I = 0, s-wave amplitude is essentially elastic up to the $K\bar{K}$ threshold beyond which a strong inelasticity sets in. In Fig. 16_2 we present the results of the current algebra analysis up to the $K\bar{K}$ threshold using elastic unitarity. The agreement over the entire domain is quite fair with some deviation occurring after the phase-shift crosses 90°. This presumably arises due to the influence of the $K\bar{K}$ production yet to be included in the analysis.

One rather interesting feature of the I = 0, s-wave amplitude is in the remarkably small width of the S^* which is an order of magnitude smaller than other widths associated with the $\pi\pi$ system. The S^* production and its associated parameters are well-known to be strongly influenced by the $K\bar{K}$ production [33,119]. We [34] draw attention here to another component in its structure. The collision of two broad overlapping states in the K-matrix can give rise to one broad and one arbitrarily sharp resonance in the S-matrix. If m_1 and m_2 represent the position of the poles in the K-matrix and if $2m\Gamma$ (where $m^2 = (m_1^2 + m_2^2)/2$) represents their common residue, the positions of the poles in the complex s-plane are given by (assuming $k/\sqrt{s} \simeq 1/2$ in the resonance region)

$$m^2 - im\Gamma \pm \sqrt{\Delta^4 - m^2\Gamma^2} \qquad (5.8)$$

where $\Delta^2 = (m_2^2 - m_1^2)/2$. In the overlapping mode ($m\Gamma > \Delta^2$), the imaginary part in (5.8) receives an additional contribution of $\pm i \sqrt{m^2\Gamma^2 - \Delta^4}$. In the limit $m\Gamma \gg \Delta^2$, one finds the generation of one broad and one arbitrarily narrow resonant state. The very narrow width of the S^* could very well arise at least in part from a collision of two ε-meson poles in the K-matrix giving rise to the production of S^* and S resonances in the amplitude.

The D-wave of the pion-pion system (not considered previously) has recently been analyzed by Arnowitt and Nath [34] within the general framework of current algebra. The analysis of this channel requires the inclusion of the iso-spin zero, $J^P = 2^+$, f^0 meson state in the calculation. The f^0 meson interactions may be included using formalisms developed recently satisfying constraints of current algebra and scale invariance [120]. The results of the phase-shift analysis for $I = 0$ are given in Fig. 19. An excellent agreement with experiment results up to 1.4 GeV. Beyond this energy the predicted results lie higher than experiment. The D-wave, however, is characterized by an appreciable $K\bar{K}$ production. Also in this energy domain the four pion production (in the $\rho\rho$, $\sigma\sigma$ or πA_1 mode) could have significant effects on phases determined from partial wave unitarity.

Finally, we mention that there exist a number of other procedures which are motivated by the same general desire to combine current algebra and unitarity. A unitarization scheme for hard pion amplitudes has been discussed by Schnitzer [121]. His approximations to the s- and P-wave phase-shifts are given in Figs. 20-22. Also shown are modifications of his calculation by Borges [122] to achieve improved crossing symmetry. Other attempts to unitarize the current algebra amplitudes are given in Refs. 123-124.

VI. SU(3) $\times$ SU(3) CURRENT ALGEBRA, $(3,3^*) + (3^*,3)$ BREAKING AND PION-KAON SCATTERING

πK scattering is very stringently constrained by the SU(3) x SU(3) current algebra conditions. Here one deals with octets (or nonets) of vector and axial vector currents which obey the following conditions which are the natural generalization of the SU(2) x SU(2).

(i)   SU(3) x SU(3) [or U(3) x U(3)] current commutation relations of Gell-Mann [24]

(ii)   Conservation of the vector currents (except the strangeness changing ones)

(iii) Partial conservation of all the axial vector currents. A natural additional assumption is the dominance of the divergence of the strangeness changing vector current S^μ by an $I = \frac{1}{2}$, $J^P = \frac{1}{2}^+$ kappa meson [125],

$$\partial_\mu S^\mu = F_\varkappa m_\varkappa^2 \varkappa . \qquad (6.1)$$

A key question is the form of the breakdown of chiral symmetry. The simplest possibility is the so-called $(3,3^*) + (3^*,3)$ break-ing [29,30], and we shall first discuss the relationship of this breaking to the current algebra and single particle saturation assumptions. We will show what restrictions these constraints impose on the πK interactions. Next we shall discuss phenomenology of πK scattering using the current algebra analyses. Thus our discussion shall proceed as follows:

a. $(3,3^*) + (3^*,3)$ Breaking and Current Algebra.

b. πK Scattering.

a. $(3,3^*) + (3^*,3)$ Breaking and Current Algebra. Glashow and Weinberg [29] and Gell-Mann, Oakes and Renner [30] have proposed that in the construction of phenomenological Lagrangians that govern the hadronic interactions, the symmetry breaking part be proportional to local field operators that transform like the $(3,3^*) + (3^*,3)$ representation of $SU(3) \otimes SU(3)$. It was shown subsequently by Arnowitt, Friedman, Nath and Suitor [32] that the $(3,3^*) + (3^*,3)$ form of the chiral breakdown arises naturally from a combined action of current algebra conditions and pole dominance of the currents as well as the σ commutators [126]. In addition to the demand of the $(3,3^*) + (3^*,3)$ breaking, the combination of current algebra condi-tions and pole dominance also require the existence of nonets rather than octets of $J^P = 0^\pm$ meson states. Both of the above results are in the nature of predictions of specific representations of the sym-metry breaking interactions, and it is interesting that these deduc-tions arise from dynamical considerations [32].

One of the consequences of the $(3,3^*) + (3^*,3)$ breaking hypo-thesis is that in the $SU(2) \times SU(2)$ sector it implies the existence of a purely $I = 0$ σ-commutator. As discussed in sec. V this assump-tion produces $a_0^{\,0}/a_0^{\,2} = -3.5$, a result which is consistent with the present experiment. For the $K_{\ell 3}$ decay, the breaking scheme leads to a rather small value for the ξ parameter. The experimental situation for ξ is not yet fully settled, though the most recent experimental determination favors a small ξ [31].

The $(3,3^*) + (3^*,3)$ breaking scheme when applied to the array of $\pi K \varkappa$ interaction structure puts stringent constraints and determines the coupling constants in terms of the interpolating constants F_π, F_K and $F_\varkappa$ and the meson mass spectrum. In addition, the breaking scheme forbids $F_\varkappa m_\varkappa$ to lie in certain domains. One has [29]

$$\left| F_\varkappa m_\varkappa \right| > \left| F_K m_K \right| + \left| F_\pi m_\pi \right|$$

or

$$\left| F_\varkappa m_\varkappa \right| < \left| F_K m_K \right| - \left| F_\pi m_\pi \right| . \tag{6.2}$$

There are also constraints from the $K_{\ell 3}$ experiment on the interpolat-ing constants. From the $K_{\ell 3}$ decay one determines that the f_+ form factor obeys

$$F_K / (F_\pi f_+(0)) = 1.28 . \tag{6.3}$$

The hard meson current algebra calculations show $f_+(0)$ to be [127]

$$f_+(0) = 2(F_\pi F_K)^{-1}(F_\pi^2 + F_K^2 - F_\mu^2) . \qquad (6.4)$$

Using Eqs. (6.2)-(6.4) and an m_μ = 1325 MeV as present experiment indicates [128], one finds that F_K/F_π is constrained to lie in the domain

$$1.15 \leq F_K/F_\pi \leq 1.32 . \qquad (6.5)$$

We note that F_μ is now no longer independent but determined in terms of F_K/F_π from Eqs. (6.3) and (6.4). The πK scattering is then essentially determined in terms of just one parameter F_K/F_π which also must lie in the rather limited domains of Eq. (6.5).

We next discuss πK scattering on the basis of the current algebra principles discussed above.

b. πK Scattering. Before proceeding to the hard meson analysis of πK scattering, we shall discuss the soft pion threshold results. Analogous to the $\pi\pi$ case, the πK s-wave scattering lengths are also determined by expansions near threshold in momenta and imposition of soft pion current algebra constraints. For the I = 1/2 and I = 3/2 s-wave scattering lengths $a_o^{1/2}$ and $a_o^{3/2}$ one obtains [127]

$$a_o^{1/2} = 2L(1 + m_\pi/m_K)^{-1} = 0.13 \ m_\pi^{-1}$$

$$a_o^{3/2} = -L(1 + m_\pi/m_K)^{-1} = -0.065 \ m_\pi^{-1} \qquad (6.6)$$

where $L = m_\pi(8\pi F_\pi^2)^{-1} = 0.0825 \ m_\pi^{-1}$. The experimental value of $a_o^{1/2}$ given by Barbero-Galtieri et al. [128] is

$$a_o^{1/2} = (0.22 \pm 0.04) \ m_\pi^{-1} \qquad (6.7)$$

which is somewhat larger than the current algebra prediction of Eq. (6.6). Pond [130] has given a hard meson current algebra analysis of πK scattering. As mentioned already all the amplitudes are essentially determined in terms of one parameter F_K/F_π. The I = 1/2 s-wave scattering length is found to be insensitive to the variations in F_K/F_π and differs from the soft pion values by 10-15%. In contrast, the I = 3/2 scattering length varies quite rapidly as F_K/F_π varies and may differ from the soft pion value by as much as 40%.

In order to extract the phase-shifts from current algebra one proceeds by imposing two body unitarity in its domain of validity in each partial wave. Theoretical results from I = 1/2 s-wave current algebra amplitudes are given in Fig. 23. The experimental phase-shift results are those of Bingham et al. [131], Mercer et al. [132], Yuta et al. [133], and Firestone et al. [134]. The agreement with the data over the entire range of energy up to 1.5 GeV is excellent with some deviations appearing above this energy. I = 3/2 current algebra results are plotted in Fig. 24. The experiment is taken from refs. 135 and 136 (see ref.131). The agreement here is not so good as in the I = 1/2 case. We note that an additional attraction from a stronger iso-scalar contribution in the cross channel would produce a better agreement and also improve further the agreement with experiment in the $I = 1/2$ s-wave at low energy.

The theoretical results for $I = 1/2$, P-wave are plotted in Fig. 25. The experimental data is from Mercer et al. [132]. The agreement with experiment of the current algebra results is almost perfect over all the available data (which presently is up to about 1.2 GeV). We note that in this channel the background in the current algebra amplitude is negligible and the unitarized amplitude is well approximated by the K^* pole in the s channel. For $I = 3/2$ P-wave, current algebra predicts a very small phase-shift consistent with experimental analyses. Thus overall the hard meson current algebra analysis gives a good description of πK interaction over all the available data. Finally, we remark that limited treatments of πK scattering have been given by other authors [137].

VII. CONCLUSION

We have attempted to review the present status of low and inter-mediate energy $\pi\pi$ and πK scattering. As a result of the recent major experimental efforts, especially from CERN [49], pion-pion phase-shift analyses up to 2 GeV in some partial waves are now available. All of the old up-down ambiguities are now resolved below 1 GeV and phase-shift analyses using different techniques are generally compatible though perhaps new ambiguities may have arisen in the above 1 GeV domain. A similar situation exists for πK scattering. At the same time, however, a trend of the data at threshold in the pion-pion P-wave appears unexpected, in that the P-wave scattering lengths appear to be larger than theoretical predictions.

The theoretical schemes discussed in the preceding sections (which were confronted with experiment) are those based on rigorous constraints, field theory models using Pade approximants and super-propagator technique and current algebra.

Among the phenomenologies based on rigorous constraints, those using forward dispersion relations and Roy equations have proven to be the more useful ones, since they do not require extrapolations into the unphysical domain in contrast to calculations using Roskies relations and Martin inequalities. One result that emerges from the above analyses is that the low energy amplitudes are not uniquely determined by the principles of crossing, analyticity and unitarity. For the case of $\pi\pi$ scattering a five-fold ambiguity remains after all the constraints are satisfied and one must specify in addition five independent parameters such as $a_0^{\,0}$, $a_0^{\,2}$, m_ρ, Γ_ρ and m_ε (defined by $\delta_0^{\,0}(m_\varepsilon) = 90^\circ$) to uniquely determine the amplitudes. After doing this, the Roy equations appear to give a good description of the present data, up to their range of applicability, $s = 110 \, m^2$.

Pade approximants have been useful in understanding the generation of resonant states within field theory models though the fits to the data are generally poor. In addition, the technique is not on a secure foundation and in any case the method is applicable only to the class of renormalizable Lagrangians. The super-propagator method shows how unambiguous results may be obtained for a scattering amplitude starting from a

Lagrangian conventionally regarded as non-renormalizable. Applied to the nonlinear pion Lagrangian, the super-propagator technique correctly generates a ρ meson but no $I = 0$ s-wave resonance is generated. The resulting s-wave phases are in poor agreement with experiment.

An alternate procedure then is the current algebra method which with appropriate unitarization can fit all the known data for $\pi\pi$ and πK scattering. The current algebra analyses, of course, may be extended to higher energies even beyond 2 GeV if appropriate account is taken of all the hadron states that influence the energy region under consideration.

The survey we have presented makes no pretense at completeness. One significant omission has been a discussion of finite energy sum rules and duality. No discussion of hadron interactions could be complete without appropriate regard to the duality principle (see, e.g., Quigg's discussion at this conference). It is rather remarkable, however, that in the low and intermediate energy domain where the hard meson current algebra accurately governs hadron interactions, duality and current algebra apparently co-exist! Current algebra achieves this co-existence through the phenomenon of cancellations that arise between pole and sea-gull diagrams over large domains of energy. This cancellation serves then to regulate in a very precise way the amount of backgrounds allowed in different isospin and angular momentum channels. Thus in the $I = 1$, P-wave $\pi\pi$ channel the cancellation between the crossed pole contributions and sea-gull terms is almost exact all the way up to about 2 GeV, allowing this channel to be dominated essentially by the ρ pole. However, the backgrounds are not so inessential in other channels. For the $I = 0$, s-wave $\pi\pi$ channel, for example, there is a significant amount of background at all energies. The prediction of the correct backgrounds to be included along with the resonant contributions in different channels for phenomenological analyses is undoubtedly one of the major triumphs of the current algebra method.

Finally, we wish to remark that there are many non-peripheral pion production experiments, and perhaps the more efficient (though more difficult) way to confront theory and experiment would be to apply present methods directly to the processes $\pi N \rightarrow \pi\pi N$ and $KN \rightarrow \pi K N$.

ACKNOWLEDGMENTS

I would like to express my thanks to Professor R. Arnowitt for many fruitful discussions. I also wish to thank Professor W. Männer for discussions of $\pi\pi$ experimental results. Compliments are also due to the Conference Organizing Committee for making the Conference both pleasant and instructive.

REFERENCES

1. Experimental Meson Spectroscopy - 1974: Proceeding of this Conference.

2. The International Conference on ππ Scattering and Associated Topics, Tallahasse, 1973.

3. Experimental Meson Spectroscopy - 1972. Edited by A. H. Rosenfeld and K. W. Lai.

4. R. E. Diebold, XVI International Conference on High Energy Physics, Chicago-Batavia, 1972.

5. M. Gourdin, Weak and Electromagnetic Form Factors of Hadrons in International Symposium on Electrons and Photon Interactions at High Energies, BONN, August 27-31, 1973.

6. General formalism for the analysis of $K_{\ell 4}$ decay is given by N. Cabibbo and A. Maksymowicz, Phys. Rev. 137, B348 (1965); F. A. Berends, A. Donnachie and G. C. Oades, Phys. Rev. 171, 1457 (1968); A. Pais and S. B. Trieman, Phys. Rev. 168, 1868 (1968). Also see G. E. Kalmus in Proceedings of the Conference on ππ and Kπ Interactions, Argonne National Laboratory 1969. Discussion of recent results from $K_{\ell 4}$ decay is given in sec. V.

7. The formalism for the K → 2π decay is given by T. T. Wu and C. N. Yang, Phys. Rev. Letters 13, 380 (1964); T. D. Lee and C. S. Wu, Ann. Rev. Nucl. Sci. 16, 511 (1966). Also see G. E. Kalmus in ref. 6.

8. M. Jacob, C. H. Llewellyn Smith and Pokorski, Nuovo Cimento 63A, 574 (1969).

9. L. M. Chounet, J. M. Gaillard and M. K. Gaillard, Physics Reports C4, No. 5, 199-324 (1972).

10. A. Martin, Nuovo Cimento 47A, 265 (1967); 58A, 303 (1968); 63A, 167 (1969). G. Auberson, G. Mahoux, O. Brander and A. Martin, Nuovo Cimento 65A, 743 (1970).

11. A. P. Balachandran and J. Nuyts, Phys. Rev. 172, 1821 (1968).

12. R. Roskies, Nuovo Cimento 65A, 467 (1970).

13. S. M. Roy, Physics Letters 36B, 353 (1971).

14. G. Mahoux, S. M. Roy and G. Wanders, Nuclear Physics B70, 297-316 (1974).

15. D. Morgan and G. Shaw, Phys. Rev. D2, 520 (1970); Nuclear Physics B10, 261 (1969).

16. J. L. Basdevant, C. D. Froggart and J. L. Petersen, NORDITA Preprint, September 1973; Physics Letters 41B, 173, 178 (1972).

17. M. Gell-Mann and M. Levy, Nuovo Cimento 16, 705 (1960).

18. J. L. Basdevant and B. W. Lee, Phys. Rev. D2, 1680 (1970).

19. H. Lehman, Physics Letters 41B, 529 (1972). H. Lehman and H. Trute, Nuclear Physics B52, 280 (1973).

20. J. Honnerkamp, Nuclear Physics B36, 130 (1972). G. Ecker and J. Honnerkamp, Physics Letters 42B, 252 (1972).

21. G. Ecker and J. Honnerkamp, Nuclear Physics B52, 211 (1973).

22. M. K. Volkov, TMΦ, 6, 21 (1971).

23. V. N. Pervushin and M. K. Volkov, Preprint, Joint Institute for Nuclear Research, Dubna, 1974. E2-7661.

24. M. Gell-Mann, Physics 1, 63 (1964).

25. S. Weinberg, Phys. Rev. Letters $\underline{18}$, 507 (1967). The s-wave scattering lengths obtained by Weinberg are given by $32\pi a_o^{\,o} = 7m_\pi F_\pi^{-2}$ and $32\pi a_o^{\,o} = -2m_\pi F_\pi^{-2}$. Using the Goldberger-Trieman relation for F_π, Weinberg obtains $a_o^{\,o} = 0.2\ m_\pi^{-1}$ and $a_o^{\,2} = -0.06\ m_\pi^{-1}$. However, the experimental value of $F_\pi = 94$ MeV gives $a_o^{\,o} = 0.16\ m_\pi^{-1}$ and $a_o^{\,2} = -0.046\ m_\pi^{-1}$. We think these latter values are the more correct soft pion predictions.

26. R. Arnowitt and P. Nath, Lectures in Theoretical Physics, Volume XIA, edited by Kalyana T. Mahanthappa, Wesley E. Brittin and O. Barut, published by Gordon and Breach.

27. R. Arnowitt, Proceedings of the Conference on $\pi\pi$ and πK Interactions, Argonne National Laboratory 1969.

28. M. G. Olsson, Phys. Rev. $\underline{162}$, 1338 (1967).

29. S. Glashow and S. Weinberg, Phys. Rev. Letters $\underline{20}$, 224 (1968).

30. M. Gell-Mann, R. O akes and B. Renner, Phys. Rev. $\underline{175}$, 2195 (1968).

31. G. Donaldson, D. Fryberger, D. Hitlin, J. Liu, B. Meyer, R. Piccioni, A. Rothenberg, D. Uggala, S. Wojcicki and D. Dorfan, Phys. Rev. Letters $\underline{31}$, 337 (1973).

32. R. Arnowitt, M. H. Friedman, P. Nath and R. Suitor, Phys. Rev. Letters $\underline{26}$, 104 (1971); Phys. Rev. $\underline{D3}$, 594 (1971).

33. See, e.g., J. L. Petersen, Physics Reports 2C, 155 (1971).

34. R. Arnowitt and P. Nath, to be submitted for publication.

35. We normalize the amplitudes so that for the process of Fig. (1), we have

$$\langle q_3 c q_4 d | q_1 a q_2 b \rangle = -i(2\pi)^4 \delta^4 (q_1 + q_2 - q_3 - q_4)$$
$$\times [16\omega_{q_1} \omega_{q_2} \omega_{q_3} \omega_{q_4} (2\pi)^{12}]^{-\frac{1}{2}}\ M_{cd,ab}(s,t,u).$$

Iso-spin invariance gives

$$M_{cd,ab} = \delta_{ab}\delta_{cd} A(s,t,u) + \delta_{ac}\delta_{bd} B(s,t,u) + \delta_{ad}\delta_{bc} C(s,t,u).$$

The s-channel iso-spin amplitudes are given by

$$A^o(s,t,u) = 3A(s,t,u) + B(s,t,u) + C(s,t,u),$$
$$A'(s,t,u) = B(s,t,u) - C(s,t,u),$$
$$A^2(s,t,u) = B(s,t,u) + C(s,t,u).$$

36. Crossing symmetric properties of $A^I(s,t,u)$ may be deduced from the crossing properties of the amplitudes A, B and C, which are

$$A(s,t,u) = B(t,s,u), \qquad A(s,t,u) = C(u,t,s).$$

37. A. Martin, Nuovo Cimento $\underline{42}$, 930 (1966). H. Lehman, Nuovo Cimento $\underline{10}$, 578 (1958).

38. Thus for the s-wave alone one has only two sum rules. These are

$$\int_0^{4m_\pi^2} ds (4m_\pi^2 - s)(4m_\pi^2 - 3s)(A_o^o(s) + 2A_o^2(s)) = 0$$

and

$$\int_0^{4m_\pi^2} ds(4m_\pi^2-s)(2A_o^o(s)-5A_o^2(s)) = 0 \ .$$

There are three additional relations if one includes the p-waves and so on. For a more complete discussion see G. Wanders, Springer Tracts in Modern Physics, Vol. 57, 1970.

39. The constraints on s-wave $\pi^o\pi^o$ amplitudes are given by the following inequalities

$$A_o^{oo}(4m_\pi^2) > A_o^{oo}(s), \quad \text{for } 0 \leq s < 4m_\pi^2,$$

$$A_o^{oo}(0) > A_o^{oo}(3.155),$$

$$A_o^{oo}(3.205) > A_o^{oo}(0.2134) > A_o^{oo}(2.9863),$$

where the amplitude $A_o^{oo} = (\frac{1}{3}A_o^o + \frac{2}{3}A_o^2)$. There are also derivative constraints on the amplitudes,

$$\frac{d}{ds} A_o^{oo}(s) < 0 , \qquad 0 \leq s \leq 1.127 \, m_\pi^2,$$

$$\frac{d}{ds} A_o^{oo}(s) > 0 , \qquad 1.7 \, m_\pi^2 \leq s \leq 4 \, m_\pi^2,$$

$$\frac{d^2}{ds^2} A_o^{oo}(s) > 0 , \qquad 0 \leq s \leq 1.7 \, m_\pi^2 .$$

40. O. Piguet and G. Wanders, Phys. Letters 30B, 418 (1969). A. P. Balachandran and M. L. Blackman, Phys. Letters 31B, 655 (1970).

41. The amplitudes A_t^I with t-channel i-spin I are related to A^I which have s-channel i-spin I by $A_t^o = 6^{-1}(2A^o+6A^1+10A^2)$, $A_t^1 = 6^{-1}(2A_o^o+3A_o^1-5A_o^2)$ and $A_t^2 = 6^{-1}(2A_o^o -3A_o^1+A_o^2)$.

42. Using the I = 2 component, Morgan and Shaw (Ref. 15) have derived a sum rule for $(2a_o^o+a_o^2)$. However, the sum rule converges badly since it is rather sensitive to small variations in the partial waves saturating the sum rule. For example, a 10 MeV variation in $\Gamma(g \to 2\pi)$ is found to produce an almost hundred percent variation in $(2a_o^o+a_o^2)$ (Ref. 43).

43. E. P. Tryon, Phys. Rev. D4, 1216 (1971).

44. A cubic polynomial in ω is assumed which allows two subtraction constants for each i-spin amplitude. The six subtraction constants are chosen to satisfy six consistency conditions corresponding to the input values of Re $A^{o,1,2}$ at $\sqrt{s} = m_\rho$, the first derivatives of Re $A^{o,1}$ evaluated at $\sqrt{s} = m_\rho$ and the threshold value for A^1.

45. L. J. Gutay, F. T. Meire and J. H. Scharenguivel, Phys. Rev. Letters 23, 431 (1969).

46. D. Cline, K. J. Braun and F. R. Scherer, Nuclear Physics B18, 77 (1970).

47. Gutay et al. give a value of $a_0^0/a_0^2 = -3.2 \pm 0.1$. Correction due to possible quadratic terms gives $a_0^0/a_0^2 = -3.2 \pm 1.0$ (Ref. 14). Cline et al. give a value of $a_0^0 = -3.2 \pm 1.1$.

48. S. D. Protopopescu, M. Alston-Garnjost, A. Barbaro-Galtieri, S. M. Flatté, J. H. Friedman, T. A. Lasinski, G. R. Lynch, M. S. Rabin and F. T. Solmitz, Phys. Rev. $\underline{D7}$, 1279 (1973). This is a 7.1 GeV/c experiment for the reaction $\pi^+ p \to \Delta^{++}\pi^+\pi^-$.

49. For a recent review of the 17.2 GeV/c CERN experiment $\pi^- p \to \pi^-\pi^+ n$ see B. Grayer, B. Hyams, C. Jones, P. Schlein, P. Weilhammer, W. Blum, H. Dietl, W. Koch, E. Lorenz, G. Lütjens, W. Männer, J. Meissburger, W. Ochs and U. Stierliu, CERN Preprint (submitted to Nuclear Physics).

50. A. $M_{\pi\pi}$ range 500-1500 MeV/c^2. G. Grayer, B. Hyams, C. Jones, P. Schlein, W. Blum, H. Dietl, W. Koch, E. Lorenz, G. Lütjens, W. Männer, J. Meissburger, W. Ochs, U. Stierliu, and P. Weilhammer, Proc. 3rd Philadelphia Conf. on Experimental Meson Spectroscopy, Philadelphia, 1972 (American Institute of Physics, New York 1972), p. 5.

51. B. $M_{\pi\pi}$ range 600-1900 MeV/c^2. B. Hyams, C. Jones, P. Weilhammer, W. Blum, H. Dietl, G. Grayer, W. Koch, E. Lorenz, G. Lütjens, W. Männer, J. Meissburger, W. Ochs, U. Stierliu and F. Wagner, to be published in Nuclear Physics. Also AIP Conf. Proc. $\underline{13}$, 206 (1973).

52. C. $M_{\pi\pi}$ range 440-1400 MeV/c^2. P. Estabrooks, A. D. Martin, G. Grayer, B. Hyams, C. Jones, P. Weilhammer, W. Blum, H. Dietl, W. Koch, E. Lorenz, G. Lütjens, W. Männer, J. Meissburger and U. Stierliu, AIP Conf. Proc. $\underline{13}$, 37 (1973).

53. D. $M_{\pi\pi}$ range 900-1120 MeV/c^2. G. Grayer, B. Hyams, C. Jones, P. Schleu, P. Weilhammer, W. Blum, H. Dietl, W. Koch, E. Lorez, G. Lütjens, W. Männer, J. Meissburger, W. Ochs and U. Stierlin, AIP Conf. Proc. $\underline{13}$, 37 (1973).

54. E. $M_{\pi\pi}$ in range 600-920 MeV/c^2. G. Grayer, B. Hyams, C. Jones, P. Schlein, P. Weilhammer, W. Blum, H. Dietl, W. Koch, E. Lorenz, G. Lütjens, W. Männer, J. Meissburger, W. Ochs and U. Stierlin, Paper No. 768 contributed to the 16th International Conference on High-Energy Physics, Batavia, 1972.

55. W. D. Apel, J. S. Ausländer, H. Müller, G. Sigurdsson, H. M. Standenmaier, U. Stier, E. Bertolucci, I. Mannelli, G. Perazzini, P. Rehak, A. Scribano, F. Sergiampietri and M. L. Vincelli, Phys. Letters $\underline{41B}$, 542 (1972).

56. J. P. Baton, G. Laurens and J. Reignier, Physics Letters $\underline{33B}$, 525 and 528 (1970).

57. C. Colton, E. Malamud, P. Schlein, A. D. Johnson, V. S. Stenger and P. G. Wohlmut, Phys. Rev. $\underline{D3}$, 2028 (1971).

58. Contribution to the 16th International Conference on High-Energy Physics, Batavia, 1972. Data from the following authors: O. R. Sander, J. P. Prokop, J. A. Poirier, C. A. Rey, A. J. Lennox, N. N. Biswas, N. M. Cason, W. D. Shephard, U. P. Kenney,

R. D. Klem and I. Spion.

M. Baubillier, B. Durusoy, R. George, M. Goldberg, A. M. Touchard, N. Armenise, M. T. FogliMuciaccia and A. Silvestri: Paper No. 489.

R. D. Baker et al.

59. D. Cohen, T. Ferbel, P. Slattery and B. Werner, Phys. Rev. $\underline{D7}$, 661 (1973).

60. W. D. Walker, J. Carrol, A. Garfinkel and B. Y. Oh, Phys. Rev. Letters $\underline{18}$, 630 (1970).

61. W. Hoogland, G. Grayer, B. Hyams, C. Jones, P. Weilhammer, W. Blum, H. Dietl, W. Koch, E. Lorenz, G. Lütjens, W. Männer, J. Meissburger and U. Stierlin, Nuclear Physics $\underline{B69}$, 266-278 (1974). This is an I = 2 analysis from $\pi^+ p \rightarrow \pi^+ \pi^+ n$ experiment at 12.5 GeV/c.

62. J. L. Basdevant, J. C. LeGuillou and H. Navelet, Nuovo Cimento $\underline{7A}$, 363 (1972).

63. M. R. Pennington and S. D. Protopopescu, Phys. Rev. $\underline{D7}$, 1429, 2591 (1973).

64. B. Bonnier and P. Gauron, Nuclear Physics $\underline{B52}$, 506 (1973).

65. Equations (2.3) or (3.5) arise from considerations of s-u crossing and must be supplemented by additional relations expressing t-u crossing (refs. 62, 66) to exhaust the full content of crossing symmetry. However, the new relations do not constrain the s and p waves.

66. R. Roskies, Phys. Rev. $\underline{D2}$, 247, 1649 (1970). G. Wanders, Nuovo Cimento $\underline{63A}$, 108 (1969).

67. A detailed discussion of this topic is given in Ref. 16. Since the cutoff Λ^2 extends up to 110 m_π^2 the inelastic unitarity condition involving the inelasticity parameter η_ℓ^I must be used. Experimentally the $\pi\pi$ system is very elastic up to the $K\bar{K}$ threshold and $\eta_\ell^I(s) \simeq 1$ for s below $4m_K^2$. Above $4m_K^2$, $\eta_\ell^I(s)$ is fitted from the data.

68. This number is one less than for Morgan and Shaw (ref. 15) and arises due to a more complete treatment of crossing in ref. 16.

69. W. Männer, Proceedings of this conference and private communication.

70. B. Bonnier and N. Johannesson, CERN-Preprint. Ref. TH.1856-CERN.

71. The $C_\pi(s)$ given by Eq. (3.7) is derivable from a single s-channel closed loop with two pion intermediate states.

72. The calculation of ref. 70 is subject to this criticism.

73. J. S. Kang and B. W. Lee, Phys. Rev. $\underline{3D}$, 2814 (1971).

74. Partial wave dispersion relations of the N/D type are used with the assumption of no CDD poles (ref. 75) present.

75. L. Castillejo, R. H. Dalitz and F. J. Dyson, Phys. Rev. $\underline{101}$, 453 (1956).

76. Even though the Roskies sum rules and Martin inequalities are well satisfied, the model has been shown to violate analyticity by Tryon, Phys. Rev. $\underline{D5}$, 1039 (1972).

77. Another calculation using the unphysical region constraints but in a different formalism was attempted by LeGuillou, Morel and Navelet (ref. 78). They constructed parametrizations of the low energy amplitudes which were then constrained to satisfy unitarity in the physical region and all the exact constraints in the unphysical region $0 \leq s \leq 4m_\pi^2$. Using the experimental P-wave phase-shifts between 500 and 1100 MeV as input (and extrapolations of these below 500 MeV), they claimed a unique prediction for the s-wave phase-shift. However, subsequent calculations (refs. 16, 79) have shown that this conclusion is incorrect and that the constraints of analyticity, crossing symmetry and unitarity even when supplemented by the assumption of a correct P-wave amplitude, leave the s-wave amplitude essentially arbitrary.

78. J. C. LeGuillou, A. Morel and H. Navelet, Nuovo Cimento 5A, 659 (1971). Earlier attempts in similar spirit have been made by G. Wanders and O. Piguet, Nuovo Cimento 56A, 417 (1968). B. Bonnier and P. Gaugron, Nuclear Physics 21B, 465 (1970).

79. O. Piguet and G. Wanders, Nuclear Physics B46, 295 (1972).

80. As an illustration, the Brown-Goble model is known to violate both the Roskies relations and the Martin inequalities.

81. L. Brown and R. Goble, Phys. Rev. Letters 20, 346 (1968).

82. R. Dolen, D. Horn and H. Schmid, Phys. Rev. 166, 1768 (1968); C. Schmid, Phys. Rev. Letters 20, 689 (1968).

83. G. Veneziano, Nuovo Cimento 57A, 190 (1968).

84. C. Lovelace, Proceedings of the Conference on $\pi\pi$ and $K\pi$ Interactions, Argonne National Laboratory 1969.

85. See, however, the comment following these equations in sec. IIIb.

86. A number of other schemes for the unitarization of the Veneziano amplitude have also been considered. See, e.g., E. P. Tryon, Phys. Rev. D4, 1202 (1974) and other references quoted therein.

87. The resonance parameters of S^* are:

Expt.	Sheet II pole position (MeV)
Hyams et al. (ref. 51)	$1007 \pm 20 - i (15 \pm 5)$
Binnie et al. (ref. 88)	$987 \pm 6 - i (24 \pm 6)$
Estabrook et al. (ref. 52)	$997 - i\ 5$
Protopopescu et al. (ref. 48)	$997 - i\ 27$

Binnie et al. also give a strong coupling of the S^* to $K\bar{K}$ with $g_{S^* \to K\bar{K}} / g_{S^* \to \pi\pi} = 3.8 \pm 1.0$.

88. D. M. Binnie, J. Carr, N. C. Debenham, A. Duane, D. A. Garbutt, W. G. Jones, J. Keyne and I. Siotis

89. E. P. Tryon, Phys. Rev. Letters 20, 769 (1968).

90. A possible exception is the $I = 1$ $\pi\pi$ partial wave amplitude which for the Veneziano model has been shown to satisfy unsubtracted dispersion relation. See R. T. Park and B. R. Desai, Phys. Rev. D2, 786 (1970).

91. E. P. Tryon, Phys. Rev. D4, 1221 (1971). It is shown here that the s-wave Veneziano amplitude may be approximated to within ten percent below 700 MeV by functions which satisfy twice subtracted dispersion relations.

92. The diagonal Pade approximants have been shown to converge for the case of potential scattering (R. Chisholm, J. Math. Phys. $\underline{4}$, 12 (1963)) and for the anharmonic oscillator (J. J. Loeffel, A. Martin, B. Simon and A. S. Wightman, Phys. Letters $\underline{30B}$, 656 (1969)). No proof of convergence for field theory exists.

93. M. Gell-Mann, Phys. Rev. $\underline{125}$, 1067 (1962).

94. B. W. Lee, Nucl. Phys. $\underline{B9}$, 649 (1969); J. Gerrais and B. W. Lee, ibid, $\underline{B12}$, 627 (1969). Also see K. Symanzik, Nuovo Cimento Letters $\underline{11}$, 1 (1969). Commun. Math. Phys. $\underline{16}$, 48 (1970).

95. If S_ℓ^I satisfy the unitarity condition $S_\ell^I(\lambda)S_\ell^{I*}(\lambda) \equiv 1$ perturbatively, the Pade approximants satisfy the condition $S_\ell^I[N,M]S_\ell^I[N,M]* \equiv 1$. For a review of the Pade approximant method and recent developments in this area see J. Zinn-Justin, Phys. Reports $\underline{1C}$, 55 (1971).
Pade Approximants and Their Applications, Edited by P. R. Graves-Morris, Academic Press 1973.

96. The authors treat F_π as a parameter. The values of the coupling constants for which the phase-shift are exhibited are $F_\pi = 125$ MeV (experimentally $F_\pi = 94$ MeV) and $\lambda_V = 5.63$. The resonance parameters which correspond to these coupling constants are $m_\rho = 780$ MeV, $\Gamma_\rho = 35$ MeV, $m_{fo} = 1115$ MeV, $\Gamma_{fo} = 180$ MeV. The $I = 2$ exotic resonance appears in the D-wave at a mass of 1335 MeV, approximately 200 MeV higher than m_{fo} whereas for the $\lambda\varphi^4$ theory the two masses are essentially degenerate.

97. J. L. Basdevant and J. Zinn-Justin, Phys. Rev. $\underline{D3}$, 1865 (1971). K. S. Jhung and R. S. Willey, International Conference on $\pi\pi$ Scattering and Associated Topics, Tallahasse, 1973.

98. In the Weinberg calculation the pion and the σ field belong to (1/2, 1/2) representation of SU(2) x SU(2). For (N/2, N/2) representation, the Weinberg results are:

$$a_o^o = m_\pi(32\pi F_\pi^2)^{-1}(N(N+2) + 4)$$

$$a_o^2 = \frac{2}{5} m_\pi(32\pi F_\pi^2)^{-1}(N(N+2) - 8).$$

We note that the p-wave scattering length a_1^1 is independent of N, and therefore of the assumption on the sigma commutator.

99. H. J. Schnitzer and S. Weinberg, Phys. Rev. $\underline{164}$, 1828 (1967).

100. I. S. Gerstein and H. J. Schnitzer, Phys. Rev. $\underline{170}$, 1638 (1968).

101. R. Arnowitt, M. H. Friedman and P. Nath, Phys. Rev. Letters $\underline{19}$, 1085 (1967).

102. R. Arnowitt, M. H. Friedman and P. Nath, Phys. Rev. $\underline{174}$, 1999 (1968).

103. R. Arnowitt, M. H. Friedman, P. Nath and R. Suitor, Phys. Rev. $\underline{175}$, 1802 (1968).

104. S. G. Brown and G. B. West, Phys. Rev. Letters $\underline{19}$, 812 (1967); Phys. Rev. $\underline{174}$, 1786 (1968); D. Geffen, Phys. Rev. Letters $\underline{19}$, 770 (1967); T. Das, V. Mathur and S. Okubo, Phys. Rev. Letters $\underline{19}$, 900 (1967).

105. J. Schwinger, Phys. Letters $\underline{24B}$, 473 (1967); Phys. Rev. $\underline{167}$, 1432 (1968). J. Wess and B. Zumino, Phys. Rev. $\underline{163}$, 1727 (1967); B. W. Lee and H. T. Lee, Phys. Rev. $\underline{166}$, 1507 (1968).

106. R. Arnowitt, M. H. Friedman, P. Nath and R. Suitor, Phys. Rev. Letters $\underline{20}$, 475 (1968); Phys. Rev. $\underline{175}$, 1820 (1968).

107. The total $\sigma \rightarrow 2\pi$ decay width is given by
$\Gamma(\sigma \rightarrow 2\pi) = 3m_\sigma^3 (128\pi F_\pi^2)^{-1}(1 - 4\epsilon_\sigma)^{\frac{1}{2}}\lambda^2$ where $\epsilon_\sigma = m_\pi^2/m_\sigma^2$.
The $\rho \rightarrow 2\pi$ decay width is determined by λ_A;
$\Gamma(\rho \rightarrow 2\pi) = m_\rho^3 (96\pi F_\pi^2)^{-1}.(1 - \frac{1}{4}\lambda_A)^2$. The present experimental determination of $\Gamma_\rho \simeq 150\pm 10$ MeV are consistent with $\lambda_A \simeq 0$.

108. P. Nath and S. S. Kere, Proceedings of the LAMPF Summer School, July 1973, LA-3443-C.

109. R. Arnowitt and P. Nath, to be submitted for publication.

110. E. W. Beier, D. A. Buchholz, A. K. Mann, S. H. Parker and J. B. Roberts, Phys. Rev. Letters $\underline{30}$, 399 (1973).

111. A. Zylbersztejn, P. Basile, M. Bourguin, J. P. Boymond, A. Diamant-Berger, P. Extermann, P. Kunz, R. Mermond, H. Suter and R. Turlay, Phys. Letters $\underline{38B}$, 457 (1972).

112. W. Schweinberger, D. Bertrand, M. Czejthey-Barth, P. Van Binst, W. L. Knight, J. Lemonne, C. D. Esveld, F. Bobisut, M. Mattili and G. Miari, Phys. Letters $\underline{36B}$, 246 (1971).

113. R. P. Ely, Jr., G. Gidal, V. Hagopian, G. E. Kalmus, K. Billing, F. W. Bullock, M. J. Esten, M. Govan, C. Henderson, W. L. Knight, F. Russel Stannard, O. Treutler, U. Camerini, D. Cline, W. F. Fry, H. Haggerty, R. H. March and W. J. Singleton, Phys. Rev. $\underline{180}$, 1319 (1969).

114. E. P. Tryon, Preprint.

115. J. A. Jones, W. W. M. Allison and D. H. Saxon, contributions to this conference. S. A. Bunyatov, contribution to this conference. The analysis of Bunyatov, gives $a_0^2 = -(0.07 \pm 0.01)m_\pi^{-1}$.

116. The current algebra effective ranges are given by (see ref.106)
$$2\pi m_\pi (a_0^0)^2 r_0^0 = -F_\pi^{-2}[1+\epsilon_\sigma(3\lambda^2+\lambda_2\lambda)-2\epsilon_\rho(1-\tfrac{1}{4}\lambda_A)^2]$$
$$4\pi m_\pi (a_0^2)^2 r_0^2 = F_\pi^{-2}[1+\epsilon_\sigma\lambda\lambda_2 - 2\epsilon_\rho(1-\tfrac{1}{4}\lambda_A)^2].$$

117. B. Gobbi et. al., Phys. Rev. Lett. $\underline{22}$, 682, (1969).

118. The dip rather than a sharp rise of the phase-shift through 270° is typical of an inelastic resonance.

119. P. K. Williams, Phys. Rev. $\underline{D6}$, 3178 (1972).

120. P. Nath, R. Arnowitt and M. H. Friedman, Phys. Rev. $\underline{D6}$, 1572 (1972).
$f^o \rightarrow \pi\pi$ coupling from scale invariance have also been discussed by B. Renner, Phys. Letters $\underline{33B}$, 559 (1970) and K. Raman, Phys. Rev. $\underline{D3}$, 2900 (1971).

121. H. J. Schnitzer, Phys. Rev. Letters $\underline{24}$, 1384 (1970); Phys. Rev. $\underline{D2}$, 1621 (1970).

122. J. Borges, Nucl. Phys. $\underline{B51}$, 189 (1973).

123. J. Iliopoulos, Il Nuovo Cimento, Vol. LIII A, 552 (1968). The author attempted to incorporate unitarity in Weinberg-like soft pion calculation by expanding in powers of the momenta $(4m_\pi^2-s)^{\frac{1}{2}}$, $(4m_\pi^2-t)^{\frac{1}{2}}$ and $(4m_\pi^2-u)^{\frac{1}{2}}$. He obtained four independent solutions (including the Weinberg solution) all satisfying soft pion constraints and Martin inequalities. The

procedure is limited to a small domain in the vicinity of the threshold.

124. A. Amatya, A. Pagnamenta and B. Renner, Phys. Rev. 172, 1755 (1968); R. Rockmore, Phys. Rev. Letters 24, 541 (1970); J. J. Brehm, E. Golowich and S. C. Prasad, Phys. Rev. Letters 23, 666 (1969); S. C. Prasad and J. J. Brehm, Phys. Rev. D6, 3216 (1972); J. J. Brehm and E. Golowich, ibid., D9, 2064 (1974).

125. This condition is the natural extension of the partial conservation condition for the axial vector current A^μ_a and the strangeness changing axial vector current C^μ where $\partial_\mu A^\mu_a = F_\pi m_\pi^2 \pi_a$ and $\partial_\mu C^\mu = F_K m_K^2 K$, and F_K analogous to F_π is the kaon decay constant.

126. Similar conclusions were also reached by L. K. Pande, Phys. Rev. Letters 20, 224 (1968); Phys. Rev. 184, 1683 (1969).

127. See Ref. 26 and R. Arnowitt, M. H. Friedman and P. Nath, Nucl. Phys. B10, 578 (1969). Similar treatments have also been given by I. S. Gerstein and H. J. Schnitzer, Phys. Rev. 175, 1876 (1968); L. K. Pande, Phys. Rev. Letters 23, 353 (1969); P. N. Chang and Y. C. Leung, Phys. Rev. Letters 21, 122 (1968). P. Auvil and N. Deshpande, Phys. Rev. 183, 1463 (1969).

128. A. Barbaro-Galtieri, M. J. Matison, M. Alston-Garjoust, S. M. Flatte, J. H. Friedman, G. R. Lynch, M. S. Rabin and F. J. Solmit, International Conference on $\pi\pi$ and Associated Topics, Tallahasse, 1973. Also see Phys. Rev. D9, 1872 (1974).

129. The $\pi_a + K \to \pi_b + K$ scattering amplitude M_{ba} is defined analogous to the $\pi\pi$ case (see ref. 35). Crossing symmetry and iso-spin invariance gives

$$M_{ba}(s,t,u) = \delta_{ba} A^+(s,t,u) + i\epsilon_{bac}\tau_c A^-(s,t,u)$$

where $A^{+(-)}$ are symmetric (antisymmetric) under the s $\leftrightarrow$ u crossing. The I-spin amplitudes are $A^{\frac{1}{2}} = A^+ + 2A^-$ and $A^{3/2} = A^+ - A^-$. The partial phase-shifts are given by

$$A^I(s,t) = (8\pi\sqrt{s}/k) \cdot \sum_\ell (2\ell+1)(e^{i\delta_\ell}\sin\delta_\ell)P_\ell(\cos\theta)$$

which gives $8\pi(m_\pi + m_K)a_o^I = A^I((m_\pi + m_K)^2, 0)$.

130. P. Pond, Phys. Rev. D3, 2210 (1971).

131. H. H. Bingham, W. M. Dunwoodie, D. Drijard, D. Linglin, Y. Goldsschmidt-Clermont, F. Muller, T. Trippe, F. Grard, P. Herquet, J. Naisse, R. Windmolders, E. Colton, P. E. Schlein, W. E. Slater and Data from International K^+ Collaboration, Nuclear Phys. B41, 1 (1972).

132. R. Mercer, P. Antich, A. Callahan, C.-Y. Chien, B. Cox, R. Carson, D. Denegri, L. Ettlinger, D. Feiock, G. Goodman, J. Haynes, A. Pevsner, R. Sekulin, V. Sreedhar and R. Zdanis, Nucl. Phys. B32, 381 (1971).

133. M. Yuta, M. Derrick, R. Engelmann, B. Musgrave, F. Schweingruber, B. Forman, N. Gelfand and H. Schulz, Phys. Rev. Letters 26, 1502 (1971).

134. A. Firestone, G. Goldhaber and D. Lissaner, Phys. Rev. Letters 26, 1460 (1971).

135. A. M. Bakker, W. Hoogland, J. C. Kluyver, G. Giacomelli, P. Lugaresi-Serra, A. M. Rossi, D. Merril, J. C. Sheuer, G. Lamidey, A. Rouge and U. Karshon, Nucl. Phys. B24, 211 (1970).

136. B. Jongejans and K. Voorthuis, Amsterdam- Nijmegen Collaboration, contribution to the International Conference on Meson Resonances and Related Electromagnetic Phenomenon, Bologna (1971).

137. G. J. Komen, CERN Preprint. TH.1643-CERN. R. W. Griffiths, Phys. Rev. 176, 1705 (1968). There is very little of the dispersion phenomenology of the πk system. For a summary see e.g. J. L. Petersen, Lectures given at XII Internationale Universitat-swochen für Kernphysik, Schladming, Febr. 1974.

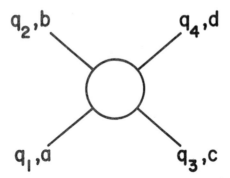

Fig. 1. Pion-pion scattering diagram. a,b,c and d label iso-spin indices. s,t,u are the usual variables $s = -(q_1 + q_2)^2$, $t = -(q_1 - q_3)^2$ and $u = -(q_4 - q_1^2)$. Mass shells are $q_i^2 = -m_\pi^2$.

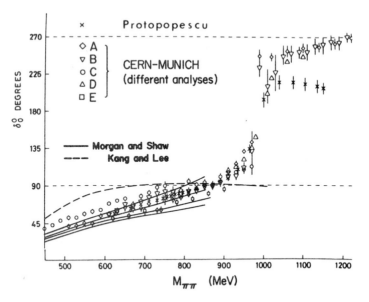

Fig. 2. I = 0, s-wave $\pi\pi$ phase-shift given by the analyses of refs. 15 and 73. See the text sec. III for discussion.

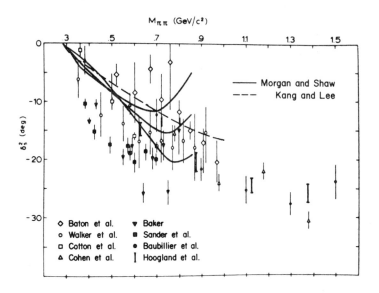

Fig. 3. Same analyses as in Fig. 2 but for I = 2, s-wave ππ phase-shift.

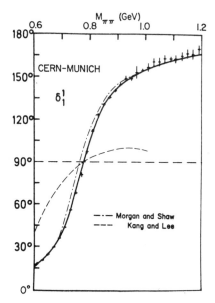

Fig. 4. Same analyses as in Fig. 2 but for I = 1, P-wave ππ phase-shift.

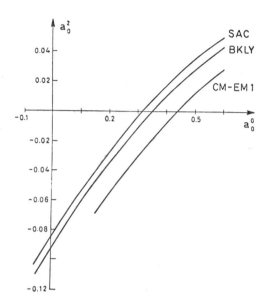

Fig. 5. Universal curves taken from the analysis of ref. 16
showing the s-wave scattering lengths a_o^0 and a_o^2 for the Saclay
(SAC), Berkeley (BKLY) and CM-EMI phase-shift. See the text
(sec. III) for discussion.

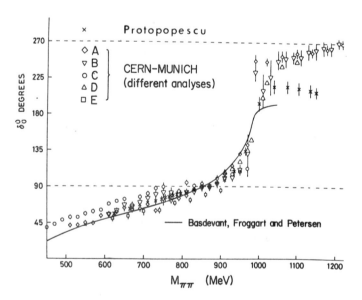

Fig. 6. I = 0, s-wave phase-shift from ref. 16.

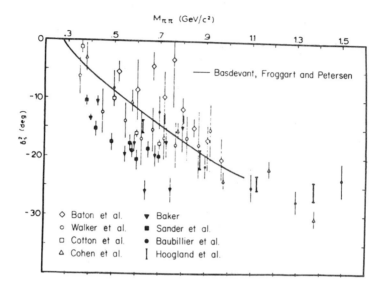

Fig. 7. I = 2, s-wave phase-shift from ref. 16.

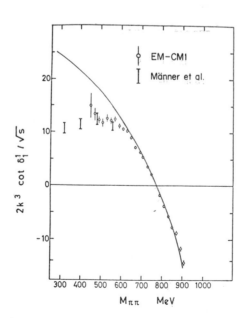

Fig. 8. The theoretical curve (solid line) is taken from ref. 16 and corresponds to $a_o^o = 0.17 \, m_\pi^{-1}$. The $2k^3 \cot\delta_1^1/\sqrt{s}$ is in units of m_π^2. The data is from refs. 52 and 69.

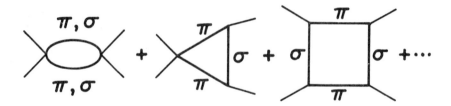

Fig. 9.   Non-pole one-loop diagrams of the σ model.

Fig. 10.   Super-propagator diagrams for ππ scattering.

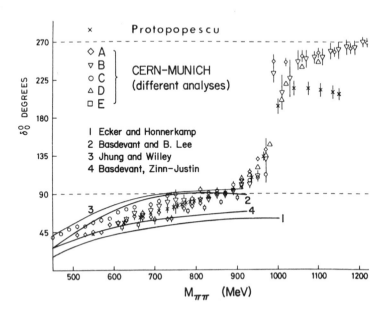

Fig. 11.   I = 0, s-wave ππ phase-shift from Pade approximant and super-propagator methods (see refs. 18, 21 and 97).

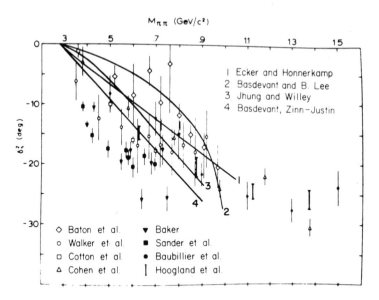

Fig. 12. Same analyses as in Fig. 11 but for I = 2, s-wave ππ phase-shift.

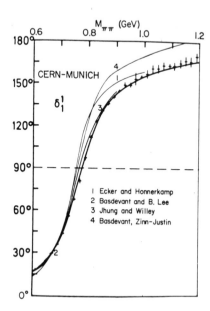

Fig. 13. Same analyses as in Fig. 11 but for I = 1, P-wave ππ phase-shift.

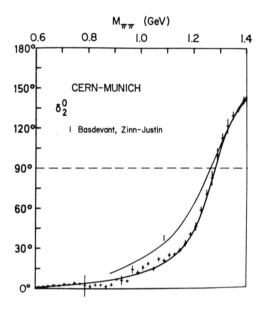

Fig. 14. I = 0, D-wave ππ phase-shift from Basdevant and Zinn-Justin in ref. 97.

$$\text{(diagram)}$$

Fig. 15. Contribution to ππ scattering from pole and sea-gull terms as demanded by current algebra.

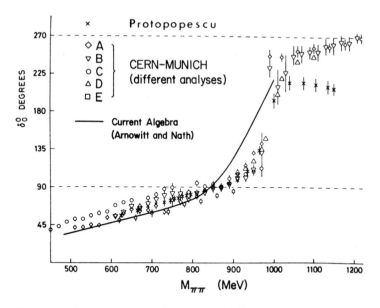

Fig. 16. I = 0, s-wave ππ phase-shift from current algebra, see ref. 34.

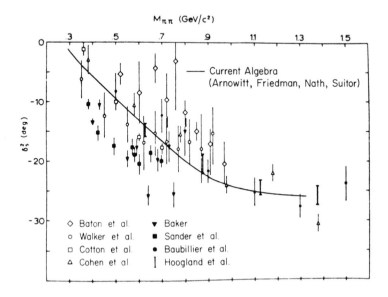

Fig. 17. I = 2, s-wave ππ phase-shift from current algebra, see refs. 34 and 106.

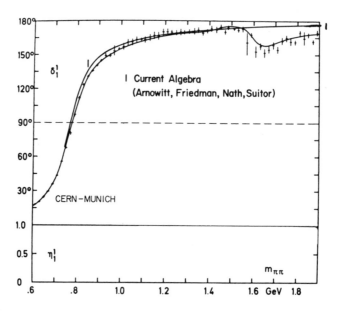

Fig. 18. I = 1, P-wave ππ phase-shift from current algebra, see refs. 34 and 106.

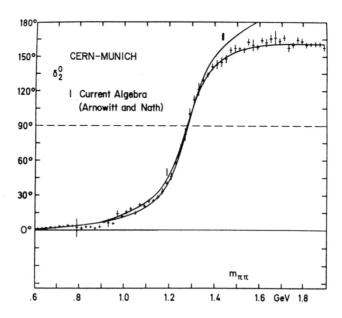

Fig. 19. I = 0, D-wave ππ phase-shift from current algebra, see ref. 34.

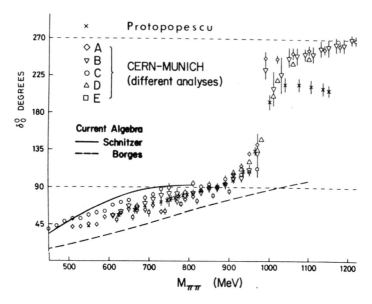

Fig. 20. I = 0, s-wave ππ phase-shift from current algebra with unitarizations of refs. 121 and 122.

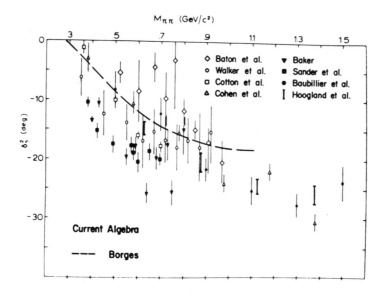

Fig. 21. I = 2, s-wave ππ phase-shift from current algebra with unitarizations of refs. 121 and 122.

234

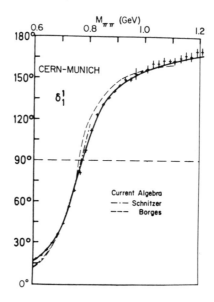

Fig. 22. I = 1, P-wave ππ phase-shift from current algebra with unitarizations of refs. 121 and 122.

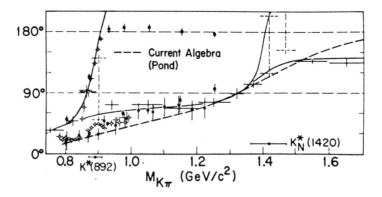

Fig. 23. I = 1/2, s-wave πK phase-shift from current algebra calculation of Pond (see ref. 130) but with the present experimental value of $m_\varkappa$ (see text).

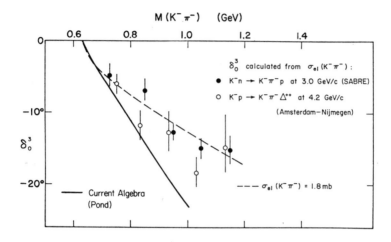

Fig. 24. Results from the same analysis as in Fig. 23 but for I = 3/2, s-wave phase-shift.

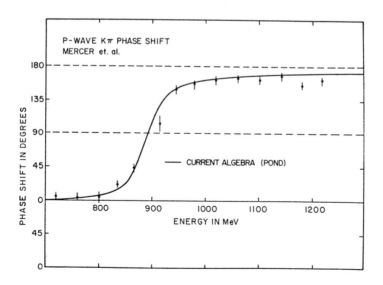

Fig. 25. Results from the same analysis as in Fig. 23 but for I = 1/2, P-wave phase-shift.

CHAIRMAN'S COMMENTS

D.R.O. Morrison
CERN, Geneva Switzerland

In his usual stimulating style, Dr. Quigg reviewed the complete field of meson production. He showed a graph demonstrating that the number of well-established resonances has ceased to increase in recent years (it may even have decreased). On the other hand theory is predicting large number of resonances. The indication is that a few more resonances at lower masses may be found with present experimental techniques. But to answer the questions that theorists are asking, it is probable that considerable improvements in experimental techniques are required - this is due mainly to the great density expected of high mass states.

It is interesting to read again what was said in the last EMS Conference in 1972. Then Chris Quigg in his chairman's comments, said that Geoff Fox, in his epic poem, had shown that "The Deck model, which has come and gone as frequently as the phantom A_2 splitting, seems now to be laid to rest for the last time". At this conference many results have been presented on Partial Wave Analysis of three-meson systems. At this session Dr. Lorella Jones has presented the results of a Deck model calculation - but a Reggeized one - which gives predictions for the reaction $\pi^-p \rightarrow (\pi^-\pi^+\pi^-)p$ which are in reasonable (though not perfect) agreement with the experimental data. Thus the Reggeized Deck model must be considered to have arisen like Phoenix from the ashes and should be considered seriously again.

Then there is the question of A_2 production in the reaction $\pi p \rightarrow (3\pi)p$, which appears anomalous with respect to the diffractive processes. Dr. Jones described how it could be included in their Reggeized Deck model to give reasonable agreement with experiment. A_2 production has not previously been observed in coherent production on heavy nuclei. However, at this session Dr. Russ described how a Partial Wave Analysis of the $(3\pi)^-$ system produced in the reaction $\pi^-Z \rightarrow (3\pi)^-Z$ where Z is a nucleus, gives clear evidence for A_2 production.

Dr. Winkelman presented results on inclusive rho-meson production in pp and πp collisions. In the former the ρ's are mainly produced centrally while in π^-p reactions they are mainly produced forwards in the c.m. At 205 GeV/c, about a quarter of pions are associated with rho production and this is a feature not normally incorporated in models.

COHERENT NUCLEAR PRODUCTION OF MESONS

J. S. Russ

Carnegie-Mellon University, Pittsburgh, Pennsylvania 15213

ABSTRACT

The coherent production of $(3\pi)^-$ systems from nuclear targets is considered, with emphasis on results from a high statistics (130,000 reconstructed events) experiment at 23 GeV/c by the Carnegie-Mellon-Northwestern-Rochester collaboration. Features of diffractive dissociation in the coherent process are compared with diffractive data from hydrogen. The momentum dependence of the coherent yield and the multiparticle absorption cross sections in nuclear matter are discussed. Coherent nuclear A_2 production is seen.

This talk is to review the present state of coherent production of mesons from nuclei and will focus upon the specific case of the coherent dissociation process $\pi \to 3\pi$ on nuclear targets. The reasons for considering coherent production on nuclear targets have been appreciated for many years, from the early experiments in deuterium and in heavy liquid or Ne fills in bubble chambers.[1],[2] This type of process allows one to do the following:

(A) Isolate small t behavior and obtain a high-statistics event sample for $t' \lesssim 0.1 (GeV/c)^2$. This is a major aid for studies of production mechanisms.

(B) Select t-channel exchange quantum numbers characteristic of the vacuum: $\Delta I = 0$; $J^P = 0^+, 1^-, 2^+, \ldots$ One expects in such cases to have a dominant contribution from Pomeron exchange, but without momentum-dependence analyses, other allowed exchanges cannot be ignored. Indeed, they are almost certainly there, perhaps suppressed by the coherence condition. This point is not at all well understood and is currently receiving much attention.

(C) Examine mass spectra of diffractive processes with large statistics and restricted degree of complication from competing production mechanisms. Especially interesting in the 3π case, of course, is the question of the nature of the A_1 and A_3 enhancements. Here the Illinois partial wave fitting is especially useful.

(D) Within the framework of a nuclear model, examine the total cross section for multihadron states at short distances in nuclear matter. The idea, simply stated,

is that the outgoing wave function for the dissociated
state is attenuated by the remaining nuclear matter
within the nucleus in which the interaction occurs. By
varying nuclear dimensions, i.e., by changing A, one
can vary the average path length in nuclear matter for
the dissociated state and hence become sensitive to the
cross section σ (multihadron, nucleon).

The mechanism for detecting coherent scattering in the multi-
GeV region relies on the characteristic forward peaking of $d\sigma/dt$,
due to nuclear form factor effects. This peaking arises simply
from the fact that for all nucleons to contribute coherently to
the final state, energy must not be transferred to the internal
degrees of freedom of the scatterer, here, nuclear excitation,
breakup, or particle production at the nuclear vertex. The nuclear form factor expresses the probability that at momentum
transfer t the nucleus will remain unaltered and will recoil as a
whole. Given an apparatus with good angular resolution, this momentum transfer dependence is much more readily detectable in
counter experiments than the unaltered nature of the target nucleus, which is unobservable in nearly all cases involving heavy
nuclei.[3] Examples of angular distributions for $\pi^- \to (3\pi)^-$ at
22.5 GeV/c are seen in fig. 1. The level of the incoherent
background may be estimated from the data for $|t| > 0.2(\text{GeV/c})^2$.
Signal to background ratios at t'=0 are $\gtrsim$ 20/1.

Previous work, chiefly bubble chamber studies, has been
summarized in recent reviews.[1],[2] An ETH(Zurich)/Milan/CERN/
Imperial College collaboration has made a high statistics study
chiefly of $\pi^- \to (3\pi)^-$, $(5\pi)^-$ at 8.9-15.1 GeV/c.[4],[5] These
data, together with new data from the CMU/Northwestern/Rochester
collaboration at 15 and 22.5 GeV/c provide the high energy, high
statistics samples which allow one to extract significant physics
information from coherent dissociation of pions.

These physics ideas are usually set within the framework of
eikonal-type optical models, exploiting the strong limits on
transverse momentum ($p_\perp$) in coherent scattering. These models
have been discussed extensively, first by Glauber[6] and in the
context of coherent dissociation processes by Kölbig and
Margolis and others.[7] The success of these ideas in describing coherent processes in heavy nuclei has been demonstrated
by their validity in describing ρ^0 photoproduction. Data from
some very lovely experiments by groups at Cornell, DESY and SLAC
have been analyzed successfully by a number of authors, using
similar formalisms.[8] The same type of model works admirably
in describing all significant features of the coherent dissociation processes in nuclei from Be to U, with attendant variation in nuclear diffraction slopes from 50 $(\text{GeV/c})^{-2}$ to
500 $(\text{GeV/c})^{-2}$. This range of predictive power gives one some
confidence that model-dependent information, such as absorption
cross sections for the outgoing state, is in fact correct.

General Features of Coherent Dissociation
_____

The general features of coherent dissociation processes in general, without particular dependence on incident particle or the detailed nature of the outgoing final state, are summarized as follows:

(A) The angular distribution for coherent events sharpens as A is increased with a nearly exponential t slope. The slope parameter agrees roughly with an absorbing disc prediction: b ~ 2R(A). The exact dependence of R on A has been studied in some detail in ρ^0 photo-production[8] but not in exclusively hadronic inter-actions with equivalent sensitivity. Our group plans to do this both for π^- and p dissociation.

(B) High mass final states are suppressed due to t_{min} effects. For high momenta, producing a final state of mass M_x from a beam particle of mass m_B at momentum p_B requires

$$\sqrt{-t_{min}} \doteq (M_x^2 - m_B^2)/2p_B$$

in a coherent process, assuming the recoiling nucleus carries off momentum, but negligible kinetic energy.

(C) Based on analysis of $\pi \to 3\pi$, 5π by the ETH/Milan/CERN/IC group and $\pi \to 3\pi$ and $p \to p\pi\pi$ by the CMU/Northwestern/Rochester group, the multiparticle final states have nuclear absorption cross sections very similar to the particle that produces them. The extra pions in the dissociated state appear not to contribute to the at-tenuation of the outgoing wave for any 3π mass, sug-gesting a general coherence of hadronic matter when interacting in nuclei regardless of the ultimate final state - resonant or not.

(D) As expected in a diffractive process, the dissociated mass spectrum is independent of the target nucleus, aside from the t_{min} effects which cause additional suppression of high mass states in heavier elements. Figure 2 shows the 3π mass distribution for carbon and lead, not corrected for efficiency, integrated over $|t'| < 0.05$, for 22.5 GeV/c $\pi^- \to (3\pi)^-$. The details of the distributions are identical, and the form-factor suppression at high masses accounts for the small additional fall-off in $M_{3\pi}$ from Pb near 1.8 GeV/c^2.

Aside from these phenomena which are more or less consistent with one's expectations, the EMCI group's analysis of the momen-tum dependence of the dissociation cross section shows a growth from 9-15 GeV/c, consistent with an exponent n in the relation

$$\frac{d^2\sigma}{dt\,dM} = A\,p^n$$ of +0.2.[5] They note that such a growth had been suggested previously in hydrogen data, integrated over a large t

bite, in which diffractive events were flagged by Longitudinal Phase Space selection of three forward-going pions with $-t' < 0.1$ $(GeV/c)^2$.[9] The effect has rather limited statistical support but is seen at all masses.

Another potential surprise from that data is a rather pre-liminary report at this conference based on a separation of the 0^- substate of the 3π system from the dominant 1^+ wave by an analysis of the moments of the normal to the decay plane.[10] The $0^-/1^+$ ratio is said to be A-dependent, indicating a possible difference in the absorption cross sections for the two partial waves. This is a potentially interesting topic, since the 0^- is predominantly $\epsilon\pi$ compared to the $\rho\pi$ 1^+ state, but in my opinion, due to the small admixture of 0^- in the 3π state, a careful par-tial wave analysis is required. Such efforts employing the Illinois Partial Wave Analysis Program[11] are underway on our data, and on the CERN data.

Coherent Nuclear Dissociation at 22.5 GeV/c

In view of what was already known, the new data from the CMU/Northwestern/Rochester group, consisting of R. M. Edelstein, E. Makuchowski, C. Meltzer, E. Miller, J. Russ, B. Gobbi, J. Rosen, H. Scott, S. Shapiro, and L. Strawczynski, was aimed at pursuing the momentum dependence by going to 22.5 GeV/c, ex-ploiting the excellent systematic stability and angular reso-lution available in the Lindenbaum/Osaki Mark I spectrometer to do high resolution studies of $d\sigma/dt$, and at improving one's knowledge of the low mass region in the $M_{3\pi}$ spectrum. To con-trast this experiment's capabilities to those of previous work, the data sample at 22.5 GeV/c is $4\times$ the CERN sample at 15 GeV/c, with angular resolution of 0.4 mrad on the 3π vector momentum compared with 1,2 mrad. Our $M_{3\pi}$ acceptance is highest at $M_{3\pi} \sim 0.9$ GeV/c^2 (~80%) while the CERN acceptance curves favor large 3π masses. Hence, the two experiments complement one another in developing a picture of the mass spectrum. Because of the good angular resolution of this apparatus, resolution smearing of $d\sigma/dt$ near t=0 is a 5% effect even in Pb, for which the slope is ~500. The resolution was dominated by multiple Coulomb scattering in the nuclear target, giving constant reso-lution in $q_\perp$, with $\sigma_{q_\perp} = .009$ GeV/c.

The apparatus is shown in fig. 3. To enhance the coherent signal, incoherent events were suppressed by using a box of veto counters, sandwiches of Pb and scintillator, which surrounded the target region except for a beam entrance hole and the 3π exit hole. This veto box distorts the incoherent spectrum but, by the nature of the coherent process, has no effect on it

(except for δ rays.)[12] The incoherent background under the coherent peak is strongly suppressed, and using a simple exponential
extrapolation of the incoherent background, one gets coherent/
incoherent ratios at t'=0 of 20/1 or better. There is considerable controversy in the literature about the shape of the incoherent background at small t'. The Pauli principle requires
that at t'=0, the density of states vanish. However, for t' non-
zero but very small, there are many nuclear states that can be
reached by collective movement, or single-particle excitation
without isospin change. Moreover, such incoherent nuclear states
are unlikely to produce a deexcitation particle (primarily soft
γ rays that are very hard to detect) that would veto the event.
Hence, the incoherent background may even continue to rise at
small t'.

The only available information on the subject comes from a
SLAC experiment on π^+ photoproduction on complex nuclei.[13] In
such an experiment the isospin state of the nucleus is changed,
reducing the density of states at small t' much more drastically
than in the equivalent t' range in dissociation processes. Their
results indicate a suppression by a factor of 2-3 at t' < .02
compared with t' ~ .25 $(GeV/c)^2$ but certainly not zero. Therefore, in cases that don't involve isospin changes of the nuclear
state, one may expect much less suppression.

In view of the complexity of the situation, one can only
say that he expects a non-zero incoherent contribution at small
t'. It is likely that finite resolution will completely obscure any effects due to the Pauli principle exactly at t'=0.
In any case, Monte Carlo reproduction of all the counterbalanc-
ing effects is hopelessly complicated and not to be trusted.
Fortunately, the subtraction is not large, and one can test the
results of any particular background subtraction by the sensitivity of nuclear parameters, especially the nuclear radius, to
his assumptions.

Characteristics of $\pi^- \to (3\pi)^-$ at 22.5 GeV/c

Referring again to the 3π mass distributions in fig. 2, it
is striking that despite excellent detection efficiency for
$M_{3\pi} < 0.7$ GeV/c^2, fewer than 1% of all recorded events in coherent dissociation $\pi^\pm \to (3\pi)^\pm$ have $M_{3\pi} < 0.7$ GeV/c^2. Indeed,
the dissociation cross section rises only when
$M_{3\pi} \sim m_\rho + m_\pi - \Gamma_\rho \sim .75$ GeV/c^2, then climbs very steeply through
the A_1 region.

The importance of the strong two-particle states like the
ρ in the dissociation process can be seen from the submass plots
$M_{\pi^+\pi^-}$ and $M_{\pi^-\pi^-}$ in fig. 4. Again, there is no particular A-
dependence in these characteristics; carbon and Pb show similar
distortions of phase space. As one studies the higher 3π masses
the f^0 plays a strong role for $M_{3\pi} > 1.4$ GeV/c^2, competing with
the ρ to dominate the $M_{\pi^+\pi^-}$ distribution for the high mass

sample. These features are essentially the same as those seen in the hydrogen data.[14]

These data have been analyzed for the absorption cross section, following the method of Kölbig and Margolis.[7] Exploiting the high statistics and good angular resolution, these have been fit by extrapolating $\frac{d^2\sigma}{dt\,dM}$ to t'=0 and comparing to the model predictions for this same quantity, written as $\left(\frac{d^2\sigma}{dt\,dM}\right)_{nucleon}$ x $|F_A(t',M_x,\sigma_2)|^2$. Since $\left(\frac{d^2\sigma}{dt\,dM}\right)_{nucleon}$ depends on t' and M_x but not on A, all the target element dependence and, in particular, σ_2 dependence is contained in the form factor. By virtue of the extrapolation to t'=0, which is insensitive to the diffraction minimum, sensitivity to resolution smearing is reduced, compared to the integration over the central diffraction maximum used by the EMCI group, and the eikonal approximation for $|F_A|^2$, neglecting transverse displacements, is nearly exact. The model prediction for F_A is, using normalization A = $\int d^3r\ \rho(r)$,

$$F_A(t,M_x,\sigma_2) = \int e^{i\ \vec{q}_\perp\cdot\vec{b}}\ b\ db\ d\phi \int dz\ e^{i\ q_{\shortparallel}z}\ \rho(b,z)\ \ \ x$$

$$e^{-\frac{1}{2}(1-i\alpha_1)T_1(b,z)\ -\ \frac{1}{2}(1-i\alpha_2)T_2(b,z)}$$

with $T_1(b,z) = \int_{-\infty}^{z} \sigma_1\,\rho(b,z)dz$ and $T_2(b,z) = \int_{z}^{\infty} \sigma_2\,\rho(b,z)dz$. From π^-p scattering data, σ_1 = 25 mb, α_1 = -0.14 at 23 GeV/c.[15] We have varied α_2 between 0 and -0.2 and find the results rather insensitive to α_2. A good measurement of α_2 will be difficult. In agreement with the EMCI findings, we find the 3π absorption cross sections are small, of order 20 mb ± 5 mb for all masses. The curves and best cross sections are shown in fig. 5(a). These are hand-drawn curves, not computer fits yet and as such are indicative of the size of σ_2, not yet definitive of its value. In particular, sensitivity to nuclear parameters and to the effects of Coulomb phase shifts have not yet been studied fully. They will not modify the general conclusion, however.

For comparison, fig. 5(b) shows the EMCI results at 15 GeV/c.[5] As the data summary in fig. 5 shows, the results are in quite good agreement, indicating little, if any, momentum dependence of the multiparticle-nucleon cross sections, just as there is little variation in the pion-nucleon cross sections in this momentum region.

One can extract the diffractive π-nucleon differential cross section for $\pi^- \to (3\pi)^-$ at t=0, once σ_2 is known, and compare to results in hydrogen, where the diffractive signal again is defined by the requirement that all three pions go forward. Figure 6 shows the 15 GeV/c results, and the new 22.5 GeV/c results have been added. The agreement is reasonably good, but again, our data are still being refined, so that the errors are artificially large. The rather large value of C_o from hydrogen for $\bar{M}_{3\pi} = 1$ GeV/c^2 suggests non-diffractive production may be strong at low masses in hydrogen.

In general, the features of the single-nucleon interaction as deduced from studying coherent dissociation agree quite well with and provide much higher statistics than available hydrogen data. A comparison of mass spectra corrected for t_{min} effects and efficiency compiled in ref. 5 is shown in fig. 7. These comparisons in figs. 6 and 7 illustrate the power of coherent dissociation studies to provide clean, high statistics samples of diffractive data on nucleons. This technique may well be employed at NAL energies to study the characteristics of high mass diffractive states, especially in lighter nuclei.

The CMU/Northwestern/Rochester experiment does permit a direct comparison of π^+ and π^- dissociation yields in the same mass region. The data at 15 GeV/c, shown in fig. 2(c) and 2(d), have a cross section ratio from Carbon such that

$$R = \frac{d^2\sigma(\pi^+ \to (3\pi)^+)/dtdM}{d^2\sigma(\pi^- \to (3\pi)^-)/dtdM} = 0.93\pm0.08$$

for $1.2 < M_{3\pi} < 1.4$ GeV/c^2. For Cu, in the range $1.0 < M_{3\pi} < 1.2$ this same ratio $R = 0.82\pm.12$. This is suggestive of some suppression of π^+ relative to π^-, but more refined analysis is required.

One can also calculate momentum dependence of the diffractive yields from 15 to 23 GeV/c in these data.

Based on data from carbon and copper, we see no decrease in yield between 15 GeV/c and 23 GeV/c. In the lowest mass bin, there is no change within errors, while the regions $1.0 < M_{3\pi} < 1.2$ and $1.2 < M_{3\pi} < 1.4$ show a small increase in yield. In terms of the momentum power law $d\sigma/dt = A\ p^n$, we find the following:

	$M_{3\pi}$	.8-1.0	1.0-1.2	1.2-1.4
carbon	n	-.08±.15	.16±.15	-.11±.17
copper	n	.22±.17	0.0±.16	0.6±0.4

At this time the strongest statement one should make on the basis of these data is that we would be very hard-pressed to accommodate a $p^{-0.4}$ power law as is usually applied to inelastic diffractive processes in hydrogen.

In order to apply the powerful tool of Partial Wave Analysis to gain a better understanding of the 3π system, the CMU/NW/Rochester group has collaborated with U. Kruse and T. Roberts of Illinois to study the details of coherent dissociation. The partial wave intensities as a function of 3π mass are shown in fig. 8 for data from π^- carbon $\to (3\pi)^-$ carbon. Results from other elements - Aℓ, Cu, Ag - show similar features. For comparison the hydrogen results at 40 GeV/c, for $.3 > -t' > .04$ (GeV/c)^2 are shown in fig. 9.[16] One sees in both cases that the 1^+ partial wave dominates, representing ~75% of the yield. The 0^- wave is significant at low mass, while the 2^- rises in the A_3

region.

The analysis shows two features that are unexpected. In the 3^+ wave there is a sharp rise at 1.7 GeV/c^2, a narrow structure that doesn't correspond to any known resonance. Unfortunately, this occurs just at the edge of the acceptance, so its significance is somewhat questionable at this time.

An effect which is unquestionable, however, is the presence of a coherent A_2 signal in the 2^+D wave.[17] This new feature of coherent production is seen as a small structure in the mass spectrum plots, but the characteristic peaking at small t and the slope variation with A are the unmistakable signature of coherent production. The angular distributions of the 1^+S and 2^+D waves in the A_2 mass region 1.2 < $M_{3\pi}$ < 1.4 are displayed in fig. 10. Note that the angular distribution of the 2^+ wave rises sharply at t' ~ 0 to peak at small t' and then falls smoothly into the incoherent background at -t' > .1 (GeV/c)^2. A detailed investigation of the properties of the 2^+D wave show it is produced with $|\Delta J_z|$ = 1 and 2^+ exchange. A detailed analysis of the A_2 position and width is quite consistent with hydrogen results.

Because of the kinematic suppression due to $|\Delta J_z|$ = 1, the 2^+D angular distribution has been fit to a form

$$\frac{d\sigma}{dt'} = At' e^{\alpha t'} + Bt' e^{\beta t'}$$

The quantities A, α, B, β for the 2^+D wave are compared with those from the 1^+S wave fits in the same mass interval in Table II. As can be seen from fig. 8, this mass region includes 30% of the total 1^+S signal. The coherent slope parameters α_{2+} and α_{1+} are the same within errors. The strength coefficients A_{2+} for each element are large, comparable to those for the 1^+ state. The small size of the A_2 peak in the coherent mass plots is due principally to the kinematic suppression factor t', not due to a small coupling strength. As one can see from fig. 10, it is not possible to fit the 2^+D angular distribution with any simple incoherent background term without any coherent production but including a kinematic constraint factor of t' to satisfy angular momentum conservation.

Given this coherent A_2 signal at high energy one must believe that A_2 production on nucleons has a substantial Pomeron contribution. With a single energy analyzed one cannot, of course, separate coherent f^0 exchange from exchange. It would be, however, quite surprising to find an f^0-exchange amplitude at 23 GeV/c comparable in size to the $I\!P$ amplitude responsible for the A_1 peak in the 1^+S wave. Both our collaboration and the EMCI group will soon have completed PWA of 15 GeV/c data. If the coherent A_2 production were due chiefly to f^0 exchange, the 2^+D wave should be a very prominent feature in the partial wave decomposition at 15 GeV/c.

Other Types of Experiments

In another experiment aimed at studying the mechanisms of coherent production, an Illinois group has reported an experiment to study the coherent dissociation $\pi^- \to (3\pi)^-$ accompanied by excitation of carbon to its 4.33 MeV excited state ($J^P = 2^+$).[18] By detecting the 4.33 MeV deexcitation γ-ray in coincidence with the $\pi^- \to (3\pi)^-$ signal at 6 GeV/c, the group hoped to test the ideas behind the nuclear models used in analyzing dissociation processes. Coherence in this case is maintained by demanding a specific transition, leading to a specific final state, no matter where in the carbon nucleus the dissociation occurred. The summary of the experiment is shown in fig. 11. It is a difficult experiment, and statistical accuracy is limited by the small acceptance. Nonetheless, the features of the mass distribution and dominant partial waves are essentially the same as those for the ground state dissociation process. The t distribution shows a coherent peak damped by the kinematic zero required by angular momentum conservation at t'=0.

In a different vein a SACLAY group has reported at this conference another attempt to study I=0 meson systems by doing a "pickup" reaction, in this case $d+d \to He^4+x^0$ from 2-4 GeV/c.[19] Recall that work had been done with $p+d \to He^3+x$. The cross section for the dd reaction is very small, a few μb, and it was not easy to see even the ω above background. This is further evidence that nucleon-transfer reactions are not good ways to search for new I=0 mesons.

Summary

To summarize, then, the work on coherent dissociation of pions on nuclei has proven to be a fruitful tool in meson spectroscopy. The important results that have emerged are the following:

(1) Valid information about diffractive particle-nucleon interactions can be obtained from studying coherent dissociation on complex nuclei.

(2) Coherent A_2 production on complex nuclei has been observed at 22.5 GeV/c with a strength comparable to that for A_1 production, indicating a strong diffractive component to A_2 production.

(3) Employing Glauber-type nuclear models, one finds a 3π state in nuclear matter is absorbed with a cross section very similar to that of its parent π. This suggests that the 3-particle nature doesn't become significant until a separation exceeding $R_{nucleus}$ is attained.[20]

(4) The momentum dependence of coherent production is very weak, as expected for Pomeron exchange. With weak statistical support, one may conclude the diffractive yield rises from 9 to 15 GeV/c.[5] From 15 to 22.5 GeV/c our data show some tendency for the yields to rise more rapidly than the optical model prediction, but the statistical significance is weak. At present, averaging over all mass intervals in Cu and carbon, one concludes that n =.04±.16, where the error is dominated by systematics. As analysis proceeds, the error on

n should be reduced to .08. From these results it is clear that continuation of these studies to compare different particles, different multiplicities at different mass regions will be a powerful tool for further analyses of meson systems.

REFERENCES

(1) H. H. Bingham, CERN Preprint 70/60 (unpublished).

(2) H. J. Lubatti, Proceedings of the Second Aix-en-Provence Conference on Elementary Particles, Supplément au Journal de Physique, 34, C1 (1973).

(3) Exceptions are Si or Ge, in which the recoil energy can, in principle, be detected down to very small t, or gaseous elements.

(4) C. Bemporad, W. Beusch, A. Melissinos, E. Polgar, D. Websdale, J. Wilson, J. Dufey, K. Freudenreich, R. Frosch, F. Gentit, P. Muhlemann, J. Codling, J. Lee, M. Letheren. G. Bellini, M. Di Corato, and G. Vegni, Nucl. Phys. B 33, 397 (1970); ibid, B 42, 627 (1972). This group is abbreviated as EMCI.

(5) P. Muhlemann, K. Freudenreich, F. Gentit, G. Bellini, M. di Corato, G. Vegni, C. Bemporad, W. Beusch, E. Polgar, D. Websdale, P. Astbury, J. Lee, and M. Letheren, Nucl. Phys. B 59, 106 (1973).

(6) R. J. Glauber, Boulder Lectures in Theoretical Physics (1959).

(7) K. S. Kölbig and B. Margolis, Nucl. Phys. B 6, 85 (1968). See also High Energy Physics and Nuclear Structure, Plenum (1970) - articles by Czyz, Glauber, Gottfried, and Margolis.

(8) S. D. Drell and J. S. Trefil, Phys. Rev. Letters 16, 522 (1966); 832(E). J. Formanek and J. S. Trefil, Nucl. Phys. B 4, 165 (1968). G. von Böchmann, B. Margolis, and L. C. Tang, Phys. Letters 30B, 254 (1969). H. Alvensleben, U. Becker, W. K. Bertram, M. Chen, K. J. Cohen, T. M. Knasel, R. Marshall, D. Quinn, M. Rhode, G. H. Sanders, H. Schubel, and S. C. C. Ting, Nucl. Phys. B 18, 333 (1970). For an analysis of nuclear sizes in p+A coherent elastic scattering, see R. J. Glauber and G. Matthiae, Nucl. Phys. B 21, 135 (1970).

(9) J. Beaupré, M. Deutschmann, H. Grässler, H. Kirk, R. Schulte, U. Gentsch, W. Nowak, G. Bossen, H. Drevermann, Ch. Kanazersky, M. Rost. B. Stöcker, K. Böckmann, J. Campbell, V. Cocconi, G. Kellner, W. Kittel, D. Morrison, H. Schiller, D. Sotiriou, and H. Wahl, Phys. Letters 41B, 393 (1972) and Nucl. Phys. B46, 1 (1972).

(10) Paper 8, submitted to this conference.

(11) D. Brockway, Ph.D. thesis, Univ. of Illinois Report No. COO-1195-197.

(12) I thank R. Diebold for a useful comment on this point.

(13) C. Boyarski, R. Diebold, S. Ecklund, G. Fischer, Y. Murata, B. Rechter, and M. Sands, Phys. Rev. Letters 23, 1343 (1969).

(14) G. Ascoli, et al., Phys. Rev. D 7, 669 (1973).

(15) K. Foley, R. S. Jones, S. J. Lindenbaum, W. A. Love, S. Ozaki, E. D. Platner, C. A. Quarles, and E. H. Willen, Phys. Rev. 181, 1775 (1969).

(16) Yu. Antipov, G. Ascoli, R. Busnello, M. Kienzle-Focacci, W. Kienzle, R. Klanner, A. Lebedev, P. Lecomte, V. Roinishvili, A. Weitsch, and F. Yotch, Nucl. Phys. B 63, 153 (1973). See also ref. 14.

(17) U. Kruse, T. Roberts, R. M. Edelstein, E. Makuchowski, C. Meltzer, E. L. Miller, J. Russ, B. Gobbi, J. Rosen, H. Scott, S. Shapiro, and L. Strawczynski, Phys. Rev. Letters, to be published.

(18) G. Ascoli, T. Chapin, R. Cutler, L. Holloway, L. Koester, U. Kruse, L. Nodulman, T. Roberts, J. Tortora, B. Weinstein, and R. Wojslaw, Phys. Rev. Letters 31, 795 (1973).

(19) J. Banaigs, J. Berger, L. Goldzahl, L. VuHai, M. Cottereau, C. Le Brun, F. Fabbri, and P. Picozza, to be published. Private communication from L. Goldzahl.

(20) K. Gottfried, Phys. Rev. Letters 32, 957 (1974). The ideas raised in this article call into question the simple interpretation of the absorption cross section as employed in the optical model formulation.

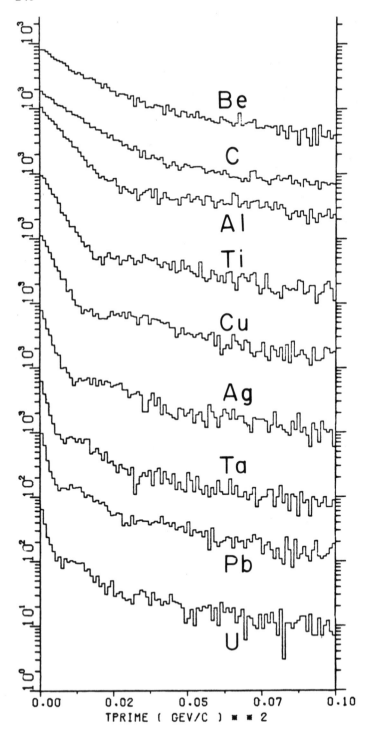

Fig. 1 Angular distributions for $\pi^- A \to \pi^+ \pi^- \pi^- A$ at 22.5 GeV/c for nine nuclear targets

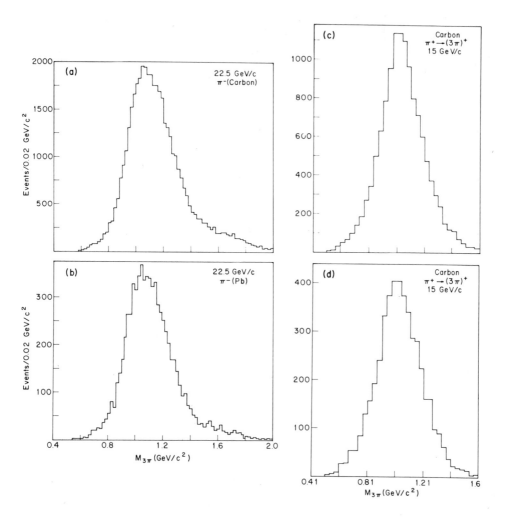

Fig. 2 $M_{3\pi}$ distribution for $-t' < .05(GeV/c)^2$

      (a) Carbon at 22.5 GeV/c
      (b) Pb at 22.5 GeV/c
      (c) Carbon, $\pi^- \ C \to (3\pi)^- C$ at 15 GeV/c
      (d) Carbon, $\pi^+ \ C \to (3\pi)^+ C$ at 15 GeV/c

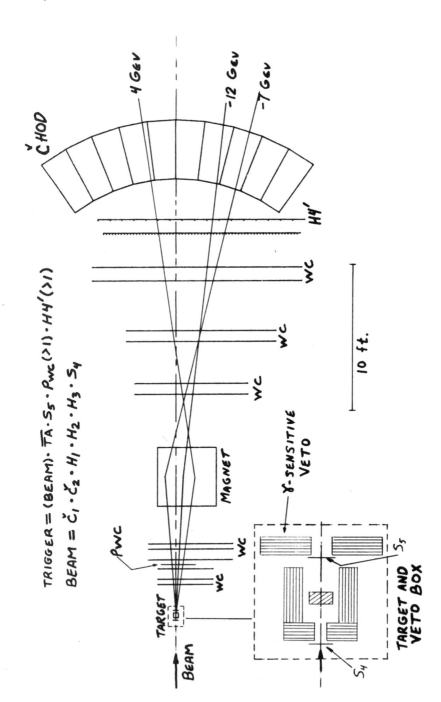

$TRIGGER = (BEAM) \cdot \overline{TA} \cdot S_5 \cdot P_{WC}(>1) \cdot H4'(>1)$

$BEAM = \check{C}_1 \cdot \check{C}_2 \cdot H_1 \cdot H_2 \cdot H_3 \cdot S_4$

Fig. 3 Layout of the Lindenbaum-Ozaki Mark I spectrometer as used in this experiment.

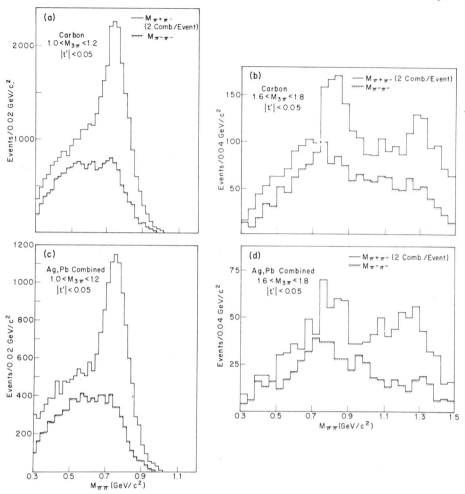

Fig. 4   (a) $M_{\pi^+\pi^-}$ and $M_{\pi^-\pi^-}$ distributions from Carbon for
$1.0 < M_{3\pi} < 1.2$ GeV/c^2

    (b) distributions from Carbon for $1.6 < M_{3\pi} < 1.8$

    (c) distributions summed from Ag and Pb for $1.0 < M_{3\pi} < 1.2$

    (d) sum of Ag and Pb for $1.6 < M_{3\pi} < 1.8$

252

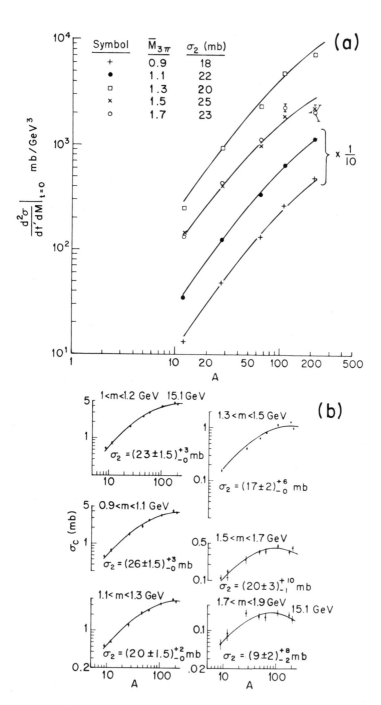

Fig. 5 Diffractive yields from nuclear targets versus atomic weight A for 200 MeV/c^2 mass intervals. The optical model curves have fixed parameters $\sigma_1 = 23$ mb, $\alpha_1 = \alpha_2 = -0.14$ and σ_2 varied for best fit. (a) 22.5 GeV/c intercepts at $t'=0$ (b) 15.1 GeV/c yields integrated to first diffraction minimum. (ref. 5)

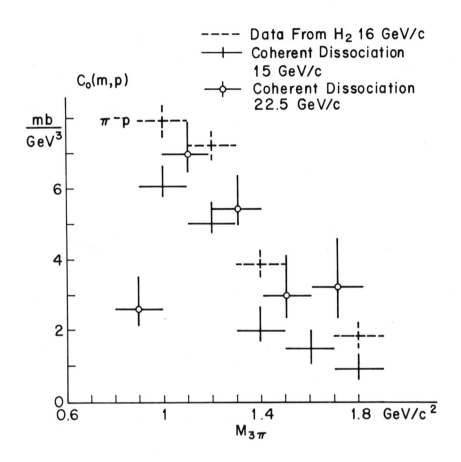

Fig. 6 Effective single nucleon diffractive cross sections at t=0 for 200 MeV/c^2 mass bands from coherent dissociation compared with bubble chamber results from 16 GeV/c $\pi^-p \rightarrow \pi^+\pi^-\pi^-p$, with $-t' < 0.1$ and requirement of all three pions in the forward hemisphere.

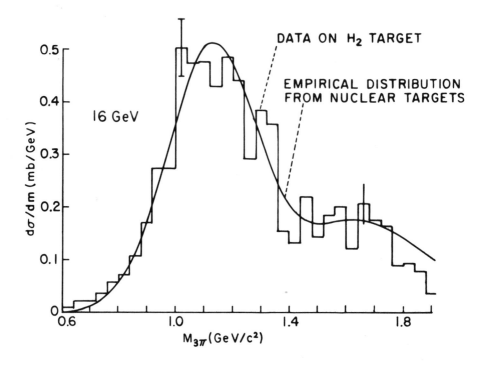

Fig. 7 $M_{3\pi}$ mass spectrum in carbon at 15 GeV/c, corrected for form
factor effects, compared with the $M_{3\pi}$ spectrum from
"diffractive" events selected by longitudinal phase space
cuts in $\pi^-p \rightarrow (3\pi)^-p$ at 16 GeV/c.

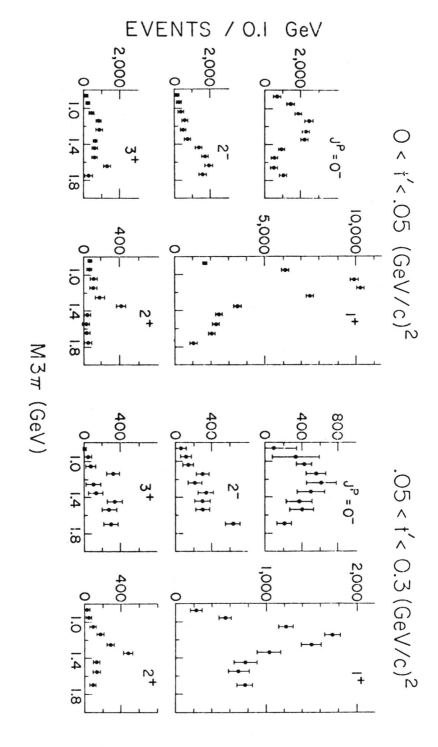

Fig. 8   Event distributions vs. $M_{3\pi}$ for $\pi^- \rightarrow \pi^+\pi^-\pi^-$ on Carbon at 22.5 GeV/c, showing partial wave intensities for the coherent ($-t' < .05$) and incoherent (.05 < $-t'$ < .30) regions.

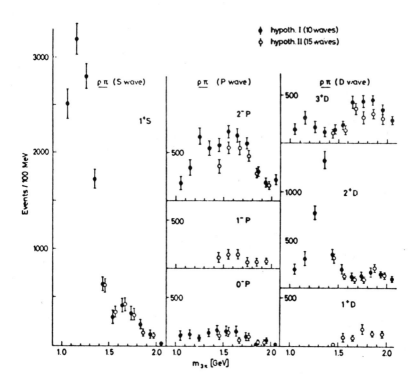

Fig. 9 Same plot as fig. 8 for $\pi^- p \to (3\pi)^- p$ at 40 GeV/c (ref. 16). Events shown have $.04 < -t' < 0.30$ (GeV/c)^2.

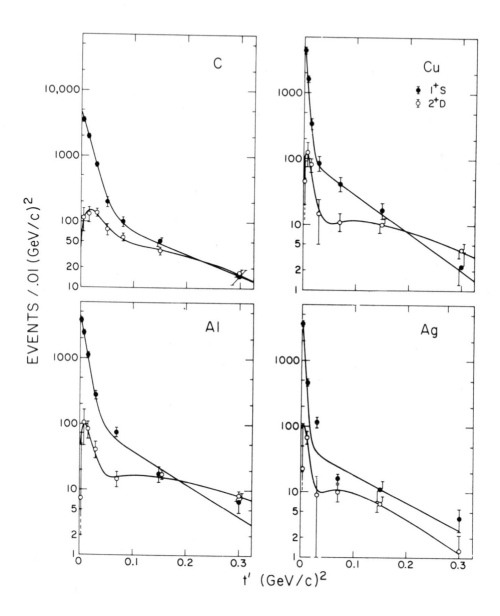

Fig. 10 Angular distributions for 1^+S and 2^+D partial waves
for $1.2 < M_{3\pi} < 1.4$ GeV/c^2 from C, Aℓ, Cu, and Ag.
The curves shown are described in the text.

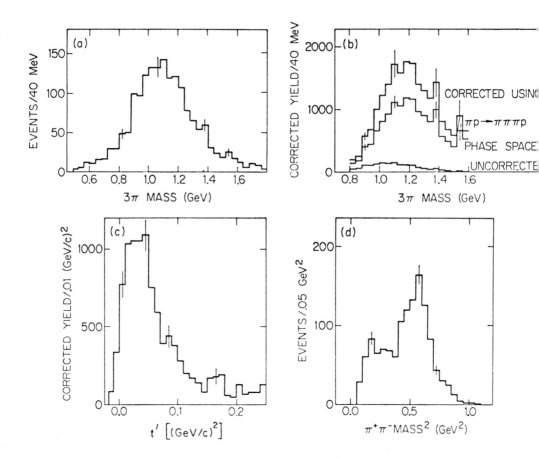

Fig. 11 (a) $M_{3\pi}$ spectrum in $\pi^- C \to (3\pi)^- C^*(4.44)$ at 6(GeV/c).
(b) efficiency-corrected mass spectrum using phase space or (top curve) hydrogen data for the 3π distributions. (c) t' distribution. (d) $M^2_{\pi^+\pi^-}$ distribution.

INCLUSIVE MESON RESONANCE PRODUCTION*

Frederick C. Winkelmann
Lawrence Berkeley Laboratory
University of California
Berkeley, California 94720

ABSTRACT

The status of inclusive meson resonance production is reviewed. New data is presented on inclusive ρ^0 production in 205 GeV/c $\pi^- p$ interactions.

INTRODUCTION

Compared to inclusive single particle production--which has been studied extensively[1] up to ISR energies for a great variety of beams and targets--very little detailed information now exists on inclusive production of meson (or baryon) resonances. Among the reasons for this are (a) the production cross sections for specific resonances are small at high energy (except for the vector mesons); (b) the familiar difficulties of separating resonance from background are accentuated by small signal-to-noise ratios, particularly when high multiplicities are involved; and (c) many states, such as ω^0 and $\rho^\pm$, are difficult to study inclusively because π^0 detection is required.

Nevertheless, resonances--via decay--account for a significant fraction of pion and kaon production, and, for nonzero spin, carry polarization as an additional variable for studying reaction mechanisms. Also, vector mesons may be an important source of lepton pairs.

In this talk I will review what is currently known about inclusive production of meson resonances and will present new data on inclusive ρ^0 production in 205 GeV/c $\pi^- p$ interactions. The data considered come from the experiments (all in bubble chambers) listed in Table I.

Table I. Inclusive Meson Resonance Production Experiments.

	Initial particles	Beam momentum (GeV/c)	Meson resonances studied	Bubble chamber	No. of events	Ref.
(1)	$K^+ p$	8.2	$K^{*+}(890)$, $K^{*+}(1420)$	CERN 80 cm	10.5K	2
(2)	γd	7.5	ρ^0	SLAC 82"	7.6K	4
(3)	pp	12,24	ρ^0, ω^0, $K^{*\pm}(890)$	CERN 2 m	275K	5
(4)	$\pi^- p$	8	ρ^0	BNL 80"	15K	9
(5)	$\pi^- p$	11.2	ρ^0	CERN 2 m U.K. 1.5 m	60K	11
(6)	$\pi^- p$	15	ρ^0, ω^0	SLAC 82"	18.5K	12
(7)	$\pi^- p$	205	ρ^0	NAL 30"	3.2K	13

As can be seen from this Table, except for the $K^*(1420)$, inclusive analyses have so far been done only for vector mesons. In the following, each experiment is described separately. Results are then compared in Table V and general features of inclusive meson resonance production are summarized. The review concludes with a list of basic questions still to be answered.

(1) $\underline{K^+p \rightarrow K^{*+}(890)X \text{ and } K^{*+}(1420)X \text{ at } 8.2 \text{ GeV/c (Ref. 2)}}$

The reaction

$$K^+p \rightarrow K^o_s \pi^+ X \qquad (1)$$

was studied at 8.2 GeV/c to determine the properties of inclusive $K^{*+}(890)$ and $K^{*+}(1420)$ production. Figure 1 shows the $K^o_s \pi^+$ mass distribution for several ranges of the Feynman variable x of the $K^o_s \pi^+$ system in the center of mass. Strong $K^{*+}(890)$ and relatively weaker $K^{*+}(1420)$ production is observed. The resonance fractions for each x-interval were obtained by fitting Breit-Wigners plus polynomial or phase space background to the mass spectra.

The inclusive cross sections for $K^{*+}(890)$ and $K^{*+}(1420)$ production (corrected for unseen K^o decays and including the $K^+\pi^o$ decay mode) are approximately 1.5 mb and 0.46 mb, respectively, compared to an inelastic K^+p cross ection of 13.8±0.3 mb. The inclusive K^o cross section is 5.76±0.33 mb, so that 23% of the K^o's come from $K^{*+}(890)$ or $K^{*+}(1420)$ decay. A triple-Regge analysis of the reaction $K^+p \rightarrow K^oX$, taking into account K^* decay as a source of K^o's, is given in Ref. 3.

Figure 2 compares the inclusive x-distributions for $K^{*+}(890)$ for all K^o, and for K^o from $K^{*+}(890)$ decay. We note the following features:

(a) Both K^o and $K^{*+}(890)$ come off preferentially in the forward direction (x > 0), suggesting that production of these particles is strongly associated with excitation of the incoming K^+ beam.

(b) The sharp peak near x($K^{*+}(890)$) = 1 arises from the highly peripheral quasi-two-body reaction $K^+p \rightarrow K^{*+}\Delta^+$.

(c) The x-dependences for inclusive K^o and $K^{*+}(890)$ production are quite different; however, similar shapes are observed for the x-distributions of K^o from $K^{*+}(890)$ decay and for all K^o.

Figure 3 shows the transverse momentum squared (p^2_T) distributions for K^o and $K^{*+}(890)$ production. A steeper falloff is observed for K^o; the exponential slopes for $0 < p^2_T < 0.5 \text{ GeV}^2$ are 5.5 GeV^{-2} for K^o and 4.3 GeV^{-2} for $K^{*+}(890)$.

To determine the $K^{*+}(890)$ production mechanism the missing mass (M) and momentum transfer (t') behavior of $K^+p \rightarrow K^{*+}(890)X$ was studied, with the following conclusions:

(i) For M < 1 GeV (corresponding to $K^+p \rightarrow K^{*+}(890)p$) and for $|t'| < 0.15 \text{ GeV}^2$, reaction (1) proceeds predominantly via natural parity exchange (ω^o-f^o).

(ii) For $|t'| < 0.15 \text{ GeV}^2$ and M > 1 GeV, π-exchange becomes dominant.

(iii) For $|t'| > 0.4 \text{ GeV}^2$ and M > 1 GeV, $\rho-A_2$ exchange dominates. A similar analysis of the $K^{*+}(1420)$ production mechanism indicates that natural parity exchange dominates for M < 1 (corresponding to $K^+p \rightarrow K^{*+}(1420)p$), whereas for M > 1 pseudoscalar exchange becomes important.

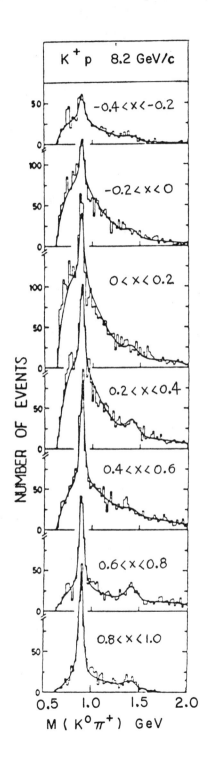

Fig. 1. $K^0\pi^+$ mass distribution
for several ranges of $x(K^0\pi^+)$
in $K^+p \rightarrow K^0\pi^+X$ at 8.2 GeV/c.

262

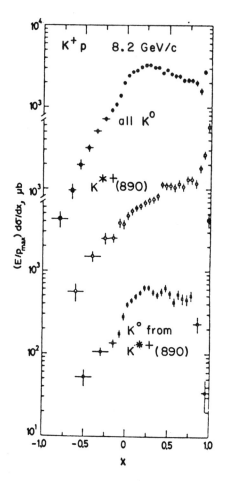

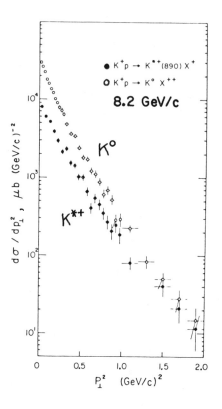

Fig. 2.   x-distribution for
K^{*+}, all K^O, and K^O from
K^{*+} decay in 8.2 GeV/c
K^+p interactions.

Fig. 3. p_T^2 distributions for
inclusive K^{*+} and K^O production in 8.2 GeV/c K^+p
interactions.

(2) $\underline{\gamma d \to \rho^o X \quad \text{at } 7.5 \text{ GeV/c}}$

Inclusive ρ^o photoproduction in the reactions

$$\gamma p \to \pi^+\pi^- X \tag{2}$$

and

$$\gamma n \to \pi^+\pi^- X \tag{3}$$

has been studied in an experiment using the SLAC 82-inch deuterium-filled bubble chamber exposed to a nearly monochromatic 7.5-GeV/c photon beam. Below a lab momentum of 1.3 GeV/c, outgoing protons and π^+'s were identified using ionization. Above 1.3 GeV/c all tracks were assumed to be pions; $K^\pm$ contamination, estimated to be $\sim 3\%$ of the charged tracks, was ignored.

Figure 4 shows the $\pi^+\pi^-$ mass distribution for reactions (2) and (3) combined. A distinct shoulder at the ρ^o mass is observed. The shaded histogram, which shows a more pronounced ρ^o, is restricted to the two- and three-prong topologies. The inclusive ρ^o cross section (which I have estimated by counting events above the hand-drawn background curve in Fig. 4) is ~ 50 μb. This is about 20% of the total γd cross section of 240 μb.

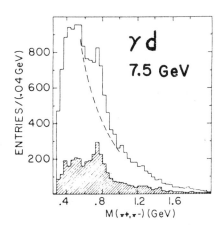

Fig. 4. Inclusive $\pi^+\pi^-$ mass distribution for $\gamma d \to \pi^+\pi^- X$ at 7.5 GeV/c. The shaded histogram is restricted to 2- and 3-prong events.

Figure 5 shows the x-dependence of the $\pi^+\pi^-$ system for all $\pi^+\pi^-$ combinations in the mass band $0.6 < M(\pi^+\pi^-) < 0.85$ GeV, which contains about 25% ρ^o. For comparison, Fig. 5 also shows the x-distribution for inclusive π^+ production in $\gamma p \to \pi^+ X$ (the corresponding distributions for $\gamma p \to \pi^- X$ and $\gamma n \to \pi^\pm X$ are similar). We observe that for $x < 0.7$ the π^+ and ρ^o distributions are similar. Above $x = 0.7$, however, the π^+ distribution continues to fall, whereas the ρ^o distribution rises sharply as x approaches 1. This forward peak comes mainly from the reactions $\gamma p \to \rho^o p$ and $\gamma n \to \rho^o n$, which, according to vector dominance, correspond to ρ^o-nucleon elastic scattering.

The p_T^2 behavior for inclusive ρ^o production is shown in Fig. 6. The exponential slopes below and above $p_T^2 = 0.12$ GeV^2 are 9.1 ± 0.1 and 4.4 ± 0.3 GeV^{-2}, respectively.

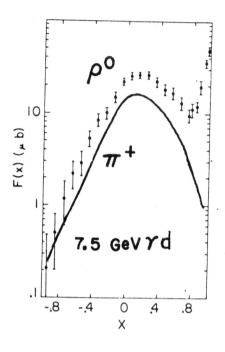

Fig. 5. x-distribution for inclusive ρ^0 production in $\gamma d \rightarrow \pi^+\pi^-X$ at 7.5 GeV/c. Also shown is the x-distribution for π^+ in $\gamma p \rightarrow \pi^+X$ at the same energy.

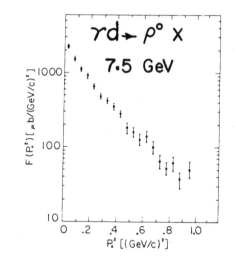

Fig. 6. p_T^2 distribution for inclusive ρ^0 production in $\gamma d \rightarrow \pi^+\pi^-X$ at 7.5 GeV/c. The exponential slopes are 9.1 ± 0.1 GeV^{-2} for $p_T^2 < 0.12$ Gev^2, and 4.4 ± 0.3 GeV^{-2} for $p_T^2 > 0.12$ Gev^2.

(3) <u>pp → ρ⁰X, ω⁰X, and K*±X at 12 and 24 GeV/c (Ref. 5)</u>

Inclusive ρ^0 and $K^*(890)$ production and semi-inclusive ω^0 production were studied in pp collisions at 12 and 24 GeV/c. Figure 7 shows the inclusive $M(\pi^+\pi^-)$ distribution for the reaction

$$pp \rightarrow \pi^+\pi^-X$$

at each beam momentum. (Only π^+'s backward in the pp center-of-mass

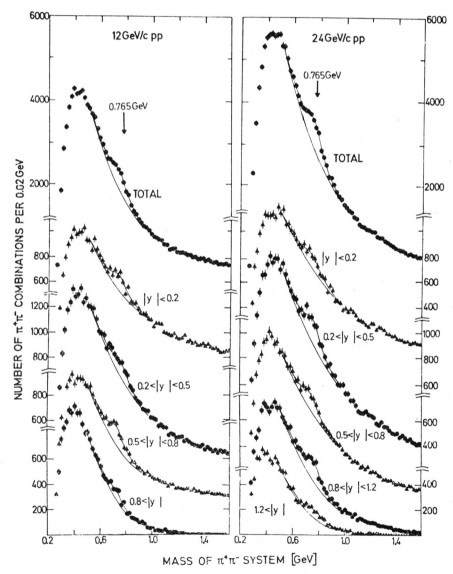

Fig. 7. $\pi^+\pi^-$ mass distributions for $pp \rightarrow \pi^+\pi^-X$ at 12 and 24 GeV/c for all events, and for intervals of c.m. rapidity y of the $\pi^+\pi^-$ system.

were considered, since these π^+ have low enough lab momentum to be
distinguished from protons by ionization.) A distinct shoulder at
the ρ^o mass is observed. Also shown is $M(\pi^+\pi^-)$ for several intervals
of center-of-mass rapidity, y, of the $\pi^+\pi^-$ system. The inclusive ρ^o
production cross section as well as the ρ^o cross section for each y-
interval were obtained by fitting a P-wave Breit-Wigner plus a second-
order polynomial background to the 0.5 to 1.0 GeV $\pi^+\pi^-$ mass region.
In a similar way, cross sections were obtained for inclusive $K^{*+}(890)$
production in the reaction

$$pp \rightarrow K_s^o \pi^{\pm} X \ ,$$

and for semi-inclusive ω^o production in the reaction

$$pp \rightarrow \pi^+\pi^-\pi^o + \text{charged particles,}$$

which was isolated by one-constraint kinematic fitting. Distributions
in y and p_T^2 for ρ^o, ω^o, and K^{*+} are shown in Figs. 8 and 9.

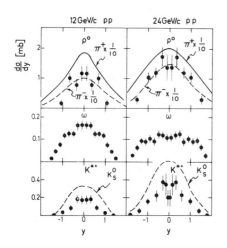

Fig. 8. Distributions in the
center-of-mass rapidity y for
inclusive ρ^o and K^{*+}, and semi-
inclusive ω^o production in 12
and 24 GeV/c pp interactions
(solid circles). The smooth
curves show the inclusive $\pi^{\pm}$
and K_s^o distributions.

Table II summarizes the results at 12 and 24 GeV/c on inclusive
cross sections, polarization, and average value of p_T. From this
Table and from Figs. 8 and 9, we note the following features:

(a) In pp collisions vector mesons are produced centrally. At both
energies the inclusive π^- and ρ^o rapidity distributions have very
similar shapes. The same holds for the K^o vs K^{*+} distributions. At
24 GeV/c, $\rho^o/(\text{all } \pi^-) \approx 10\%$ and $K^{*+}/(\text{all } K_s^o) \approx 50\%$; both of these
ratios are consistent with being independent of rapidity. Similar
ratios are observed at 12 GeV/c.

(b) $d\sigma/dp_T^2$ for ρ, ω, and K^* is consistent with an exponential fall-
off in p_T^2 for $0 < p_T^2 \lesssim 1.2 \text{ GeV}^2$; $\langle p_T \rangle$ increases with the mass of the
produced particle.

(c) The ρ, ω, and K^* polarizations, obtained from decay angular distributions in various frames, are all consistent with zero. This disagrees with the dual resonance model of Fenster and Uretsky,[6] which
predicts strong polarization of the ρ.

(d) The ρ^o and K^{*+} cross sections have risen by a factor of two
between 12 and 24 GeV/c.

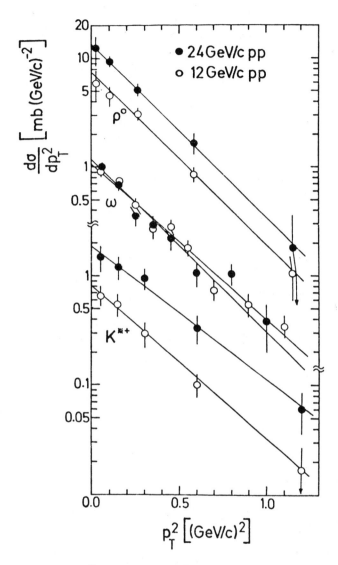

Fig. 9. p_T^2 distributions for inclusive ρ^0 and K^{*+}, and semi-inclusive ω^0 production in 12 and 24 GeV/c pp interactions. The fitted exponential slopes are given in Table II.

Table II. Properties of inclusive vector meson production in pp interactions at 12 and 24 GeV/c.[5]

| Final state | Particle | Inclusive cross section (mb) | | $\langle p_T \rangle$[d] (MeV/c) | Polariza- tion (12 and 24 GeV/c) |
		12 GeV/c	24 GeV/c		
$\pi^- X$	π^-	21.1±0.4	33.8±0.6	320	--
$K_S^0 X$	$K^0, \bar{K}^0$	1.15±0.03[a]	2.51±0.06[a]	405	--
$\pi^+ \pi^- X$	ρ^0	1.80±0.25	3.49±0.42	470	~ 0
$\pi^+ \pi^-$ + charged particles[b]	ρ^0	0.32±0.06	0.30±0.05	--	--
$\pi^+ \pi^- \pi^0$ + charged particles[b]	ω^0	0.32±0.02	0.32±0.03	460	~ 0
$K_S^0 \pi^+ X$	$K^{*+}(890)$	0.25±0.03[c]	0.64±0.06[c]	530	~ 0
$K_S^0 \pi^- X$	$K^{*-}(890)$	0.02±0.02[c]	0.14±0.02[c]	--	--

[a] These are twice the cross sections for K_S^0 production corrected for undetected K_S^0 decays.
[b] Semi-inclusive.
[c] These are twice the cross sections for $K^{*\pm} \to K_S^0 \pi^{\pm}$ production corrected for undetected K_S^0 decays and corrected for the $K^{*\pm} \to K^{\pm} \pi^0$ decay mode.
[d] At 24 GeV/c.

(e) At both energies, $\sigma(K^{*+}) + \sigma(K^{*-}) \approx 0.2 \; \sigma(\rho^0)$[7] and $\sigma(K^{*+}) \gg \sigma(K^{*-})$.

(f) $\sigma(\omega^0, \text{semi-inclusive}) \approx 0.1 \; \sigma(\rho^0)$.

The fraction of π's which come from ρ-decay can be calculated assuming roughly equal ρ^+, ρ^-, and ρ^0 cross sections. This gives $\sigma(\rho) \approx 3 \times \sigma(\rho^0) = 10.5$ mb, which is equivalent to the production of about 0.3 ρ's per inelastic pp collision at 24 GeV/c. Using $\sigma(\pi) \approx$ 144 mb[5] then implies that approximately 1/7 of the produced π^+, π^-, and π^0 come from ρ decay. This is a large enough fraction to affect the details of inclusive single pion distributions.

Vector mesons have small decay branching ratios into lepton pairs [e.g., $(\rho^0 \to e^+ e^-)/(\rho^0 \to \text{all}) = 0.43 \pm 0.05\%$ and $(\rho^0 \to \mu^+ \mu^-)/(\rho^0 \to \text{all}) = 0.67 \pm 0.12\%$]. Thus the vector dominance contribution to the reaction

$$pp \to \ell^+ \ell^- X$$

can be calculated from measured inclusive vector meson cross sections assuming $\sigma(\rho^0) = \sigma(\omega^0)$ and $\sigma(\varphi^0) \ll \sigma(\rho^0)$. The result is that the $\mu^+ \mu^-$ pairs observed[8] in the reaction

$$p + (\text{bound nucleon}) \to \mu^+ \mu^- X$$

cannot be explained by leptonic decay of high-mass Breit-Wigner tails of the ρ, ω, and φ.

(4) $\pi^- p \to \rho^o X$ at 8 GeV/c (Ref. 9)

Inclusive ρ^o production has been studied in $\pi^- p$ interactions at 8 GeV/c. Figure 10 shows the inclusive $\pi^+\pi^-$ mass distribution for the 2-, 4-, and 6-prong topologies of the reaction $\pi^- p \to \pi^+\pi^- X$. A distinct ρ^o signal is observed only in the 2- and 4-prongs (the 2-prongs also show some f^o). The ρ^o cross sections, obtained from Fig. 10 by counting events above a smooth background, are given in Table III. Cross sections for some exclusive ρ^o channels are also given. The total ρ^o cross section--most of which comes from the 4-prong events--is 3.2 mb. This is about 14% of the overall inelastic cross section of 23 mb.

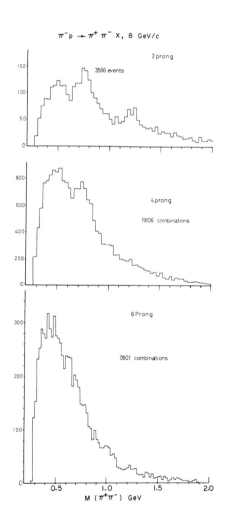

Fig. 10. $\pi^+\pi^-$ mass distributions for the 2-, 4-, and 6-prong topologies of $\pi^- p \to \pi^+\pi^- X$ at 8 GeV/c.

Table III. Inclusive and exclusive ρ^0 cross sections in 8 GeV/c π^-p interactions.[9]

Prongs	Final state	Cross section (mb)	
		Exclusive	Inclusive
2	$\rho^0 n$	0.23±0.02 ⎫	
	$\rho^0 n + m\pi^0$, m ≧ 1	0.44±0.06 ⎭	0.67±0.06
4	$\rho^0 \pi^- p$	0.64±0.05 ⎫	
	$\rho^0 \pi^- p\pi^0$	0.49±0.04 ⎬	1.97±0.16
	$\rho^0 \pi^- p + m\pi^0$, m ≧ 2 ⎫	0.84±0.15 ⎭	
	$\rho^0 \pi^- n\pi^+ + m\pi^0$, m ≧ 0 ⎭		
6	$\rho^0 X$	---	≲ 0.46
≧ 8	$\rho^0 X$	---	~ 0.07
		Total	3.2 mb

Figure 11 shows the x-distribution for inclusive ρ^0 production. For comparison, x-distributions are also given for $\pi^\pm$ in $\pi^- p \to \pi^\pm X$, for ρ^0 in the exclusive reaction $\pi^- p \to \rho^0 n$, and for $K^{*+}(890)$ in the reaction $K^+ p \to K^{*+}(890)X$ at 8.2 GeV/c (discussed above). We note the following features:

(a) ρ^0's are produced mainly in the forward direction (x > 0) with a forward-backward ratio of about 6. This suggests that in $\pi^- p$ collisions at this energy ρ^0's come predominantly from excitation of the incoming beam, in contrast to the situation just considered for pp collisions at 12 and 24 GeV/c, where the ρ^0's are produced centrally.

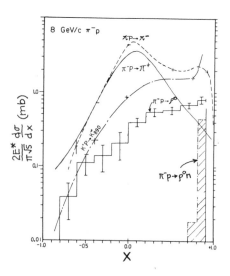

Fig. 11. x-distribution for inclusive ρ^0, π^+, and π^- production in 8 GeV/c $\pi^- p$ interactions, and for inclusive K^{*+} production in 8.2 GeV/c $K^+ p$ interactions.

At x = 0 the inclusive ρ^0/(all π^-) production ratio is ~ 7% compared to ~ 10% for pp collisions at 24 GeV/c.

(b) The highly peripheral reaction $\pi^-p \to \rho^0 n$ contributes strongly near x(ρ^0) = 0.9.

(c) The shapes of the x-distributions for inclusive ρ^0 and $\pi^\pm$ production are quite different.

(d) There is a strong similarity over the entire range of x between $\pi^-p \to \rho^0 X$ at 8 GeV/c and $K^+p \to K^{*+}(890)X$ at 8.2 GeV/c, indicating similar production mechanisms.

Figure 12 compares the x-distributions for all π^+ and π^-, and for those π^+ and π^- coming from ρ^0 decay. The latter $\pi^\pm$ fall mainly in the forward region but somewhat less so than the parent ρ^0's due to the smearing effect of the ρ^0 decay. The decay pion spectra do not show the sharp peaking near x = 0 predicted on the basis of scaling in ρ-production.[10] For x < 0 the decay pion distributions are similar to those for overall π^+ and π^- production.

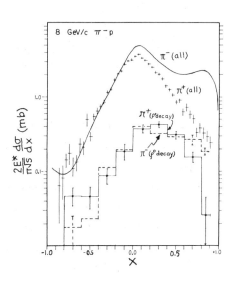

Fig. 12. x-distributions for all $\pi^\pm$ and for $\pi^\pm$ coming from ρ^0 decay in 8 GeV/c π^-p interactions.

The distribution in the momentum transfer t' (= t − t_{min}) between beam and outgoing ρ^0 is shown in Fig. 13. A pronounced change of slope occurs at |t'| = 0.2 GeV^2, below and above which the slopes are approximately 6.9 and 2.0 GeV^{-2}, respectively. It is concluded from this and from the t' dependence of the ρ^0 spin-density matrix elements that ρ^0 production is dominated by π exchange for |t'| $\lesssim$ 0.1 GeV^2 and by A_2 exchange for larger |t'|.

(5) $\pi^-p \to \rho^0 X$ at 11.2 GeV/c (Ref. 11)

The $\pi^+\pi^-$ mass distribution for the reaction $\pi^-p \to \pi^+\pi^-X$ at 11.2 GeV/c is shown in Fig. 14. In order to reduce the background in the ρ^0 region, both π^+ and π^- were required to be forward in the center-of-mass. A clear peak at the ρ^0 mass is observed. The ρ^0 cross section (which I have estimated by counting events above the hand-drawn

272

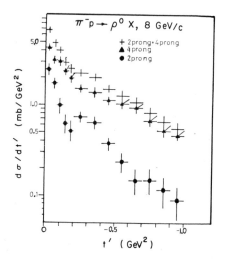

Fig. 13. t'-distributions for $\pi^- p \to \rho^0 X$ at 8 GeV/c. For the combined 2- plus 4-prong data the exponential slopes are ~ 6.9 GeV^{-2} for $|t'| < 0.2$ GeV^2 and ~ 2.0 GeV^{-2} for $|t'| > 0.2$ GeV^2.

background curve in Fig. 14) is $\sigma(\rho^0) \approx 2.0\pm0.5$ mb for $x_{\pi+} > 0$, $x_{\pi-} > 0$. Figure 14 also shows $M(\pi^+\pi^-)$ for $-t(\text{beam},\pi^+\pi^-) < 0.4$ GeV^2. This cut greatly enhances the ρ^0 signal and reveals some indication of the f^0.

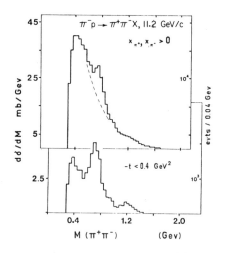

Fig. 14. $\pi^+\pi^-$ mass distribution in $\pi^- p \to \pi^+\pi^- X$ at 11.2 GeV/c with both π^+ and π^- forward in the center of mass. The dashed line is an estimate of the background in the ρ^0 region. The lower histogram has the restriction $-t(\text{beam},\pi^+\pi^-) < 0.4$ GeV^2.

Figure 15 shows the p_T^2 distribution of the $\pi^+\pi^-$ system in the ρ-region $(0.64 < M(\pi^+\pi^-) < 0.88$ GeV), again for $x_{\pi+} > 0$ and $x_{\pi-} > 0$. The observed sharp peak at low p_T^2 can be associated mainly with ρ^0 by examining the $\pi^+\pi^-$ mass spectrum for intervals of p_T^2. The forward slope, indicated by the dashed line in Fig. 15, is 6.4 ± 0.4 GeV^{-2} for $p_T^2 \lesssim 0.2$ GeV^2. This slope (a) is less than that for inclusive π^+ and π^- production in the same experiment $(8.9\pm0.8$ and 9.4 ± 0.4 GeV^{-2}, respectively, for $p_T^2 \lesssim 0.2$ GeV^2 and all x); (b) is close to the t' slope of 6.9 GeV^{-2} for $|t'| < 0.2$ GeV^2 for inclusive ρ^0 production

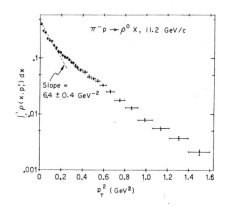

Fig. 15. p_T^2 distribution of the forward $\pi^+\pi^-$ system in the ρ-region for $\pi^-p \to \pi^+\pi^-X$ at 11.2 GeV/c. The dashed line corresponds to a slope of 6.4 ± 0.4 GeV^{-2}.

in π^-p collisions at 8 GeV/c (see Fig. 13); and (c) is about twice that for inclusive ρ^0 production in pp reactions at 12 GeV/c (3.6 ± 0.4 GeV^{-2} for $p_T^2 \leq 1.2$ GeV^{-2}; see Fig. 9).

(6) $\pi^-p \to \rho^0pX$ at 15 GeV/c (Ref. 12)

In a paper submitted to this Conference preliminary results were presented on semi-inclusive ρ^0 production in 15 GeV/c π^-p interactions. In order to reduce the possible number of $\pi^+\pi^-$ combinations and thereby improve the ρ^0 signal-to-background ratio, only events containing an outgoing proton with lab momentum ≤ 1 GeV/c were considered. Figure 16 shows the $\pi^+\pi^-$ mass distribution for these slow proton events. The shaded histogram contains 4-prong events only. Clear ρ^0 production is observed.

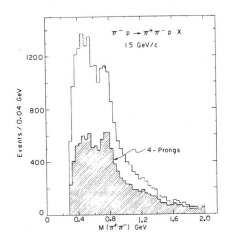

Fig. 16. $\pi^+\pi^-$ mass distribution in $\pi^-p \to \pi^+\pi^-pX$ at 15 GeV/c. The shaded histogram shows 4-prong events only.

Preliminary estimates for the ρ^0 cross sections are 0.24 ± 0.06 mb, 4-prongs; 0.13 ± 0.04 mb, 6-prongs; and 0.05 ± 0.05 mb, 8-prongs (the 2-prong events do not contribute since a proton is required in the final state). The overall cross section for the reaction $\pi^-p \to \rho^0pX$ is

0.43±0.15 mb. About 40% of this cross section comes from the exclusive reaction $\pi^-p \to \rho^0\pi^-p$, which is found to be dominated by fragmentation of the π^- beam into a low-mass $\rho^0\pi^-$ system (A_1, A_2 production).

(7) $\pi^-p \to \rho^0 X$ at 205 GeV/c (Ref. 13)

I would now like to present some new and still preliminary data on inclusive ρ^0 production obtained by the Berkeley-NAL collaboration from an exposure of the NAL 30-inch hydrogen bubble chamber. As shown below, because of poor momentum resolution on the fast tracks coming from beam fragmentation, ρ^0 production can only be studied in the central and backward regions in this experiment.

All charged tracks were taken to be pions, except below 1.4 GeV/c where protons and π^+ were separated in the usual way by means of ionization. From a study of K_S^0 production it is estimated that $\lesssim 10\%$ of the charged tracks are kaons, assuming $\sigma(K_S^0) \approx \sigma(K^+) \approx \sigma(K^-)$. The $K^{*\pm} \to K_S^0\pi^\pm$ signal in our data is so small that K^{*0} contamination in the ρ^0 region (from $K^\pm\pi^\mp$ misidentified as $\pi^\pm\pi^\pm$) is negligible.

Figure 17 shows the inclusive $\pi^+\pi^-$ mass distribution for the following three intervals of center-of-mass rapidity, y, of the $\pi^+\pi^-$ system: $-3 < y < -1$ (backward or target fragmentation region), $-1 < y < 1$ (central region), and $1 < y < 3$ (forward or beam fragmentation region). The backward and central regions show a shoulder at the ρ^0

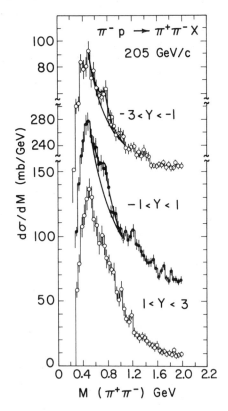

Fig. 17. $\pi^+\pi^-$ mass distribution for three intervals of center-of-mass rapidity y of the $\pi^+\pi^-$ system in $\pi^-p \to \pi^+\pi^-X$ at 205 GeV/c.

mass. There is also some indication of f⁰ in the central region. The forward region, however, shows no distinct ρ^0; this is expected since the $\pi^+\pi^-$ mass resolution here is ±140 MeV--comparable to the ρ width and sufficient to wash out the ρ signal (the mass resolutions in the backward and central regions are ±20 and ±50 MeV, respectively).

The inclusive ρ^0 cross section for backward and central production was obtained by fitting the $0.5 < M(\pi^+\pi^-) < 1.0$ GeV mass region to a P-wave Breit-Wigner (with a fixed mass of 0.765 GeV and variable width) plus a second-order polynomial background. This gives $\sigma(\rho^0) =$ 2.4±0.8 mb for $-3 < y < -1$, and 8.9±2.2 mb for $-1 < y < 1$. The cross section for <u>observable</u> ρ^0's is thus 11.3±2.3 mb. This, of course, should be considered a <u>lower-limit</u> since the beam-fragmentation contribution is not included.

The fraction of π's coming from ρ-decay can be estimated assuming equal cross sections for ρ^+, ρ^- and ρ^0 production. Using $\sigma(\rho) \approx 3 \times \sigma(\rho^0) \gtrsim 34$ mb and $\sigma(\pi) \approx 240$ mb (Ref. 13) then gives that $\gtrsim 25\%$ of all pions are products of ρ decay.

Figure 18 shows the inclusive mass distribution for all $\pi^+\pi^-$ combinations $(-3 < y(\pi^+\pi^-) < 3)$. A fit to the ρ^0 region yields $\sigma(\rho^0) =$ 11.1±2.6 mb, consistent with $\sigma(\rho^0)$ for $y < 1$, indicating that the fit is insensitive to the resolution-smeared ρ^0 signal from beam fragmenta-

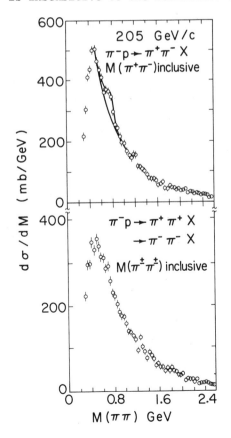

Fig. 18. Inclusive $\pi^+\pi^-$ and $\pi^\pm\pi^\pm$ mass distributions for $\pi^-p \to \pi\pi X$ at 205 GeV/c.

tion. Also shown in Fig. 18 is the structureless inclusive mass spectrum for pions of like charge ($\pi^+\pi^+$ and $\pi^-\pi^-$).

The distributions in y for ρ^0, for all π^+, and for all π^- are shown in Fig. 19. The $\rho^0/(\text{all } \pi^{\pm})$ cross-section ratios for the backward and central region are given in Table IV. In the central region,

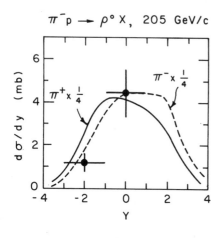

$\pi^- p \rightarrow \rho^0 X$, 205 GeV/c

Fig. 19. Distribution in center-of-mass rapidity y for inclusive production of ρ^0 (solid circles), π^+ (solid curve), and π^- (dashed curve) in 205 GeV/c $\pi^- p$ interactions. For π^+ and π^-, $d\sigma/dy$ has been multiplied by 1/4. Due to poor $\pi\pi$ mass resolution in the forward direction, the ρ^0 cross section has not been determined for $y > 1$.

$-1 < y < 1$, the $\rho^0/(\text{all } \pi^-)$ ratio is 26±6%, which is 2 to 3 times larger than the corresponding ratio in 24 GeV/c pp interactions ($\sim 10\%$ at $y = 0$) or 8 GeV/c $\pi^- p$ interactions ($\sim 7\%$ at $y = 0$). The $\rho^0/(\text{all } \pi^+)$ ratio in the central region is 28±7%, in agreement with the value of 21% obtained by Barnett and Silverman[14] on the basis of a general peripheral calculation applied to the diagrams shown in Fig. 20.

Table IV. Inclusive $\rho^0/(\text{all } \pi^{\pm})$ ratios in 205 GeV/c $\pi^- p$ interactions.[13]

c.m. rapidity interval	$\rho^0/(\text{all } \pi^+)$	$\rho^0/(\text{all } \pi^-)$
$-3 < y < -1$	0.12±0.04	0.16±0.06
$-1 < y < 1$	0.28±0.07	0.26±0.06
$y > 1$	not measurable	

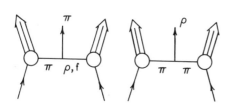

Fig. 20. Diagrams used by Barnett and Silverman[14] to calculate the inclusive ρ/π production ratio at high energy.

COMPARISON

Table V compares some of the features of inclusive meson resonance production in the experiments just described.

Table V. Comparison of inclusive meson resonance (R) production data.

R	Initial particles	p_{beam} (GeV/c)	σ_{inel} (mb)	$\sigma(R)$ (mb)	$\langle n_R \rangle$	$\langle n_{ch} \rangle$	$\dfrac{\langle n_R \rangle}{\langle n_{ch} \rangle}$	R/all π^- ($y_{c.m.} = 0$)	R exponential slope (GeV^{-2})
ρ^0	γd	7.5	0.24[a]	~0.050	0.20	3.4	0.059	--	--
	pp	12	29.8	1.80±0.25	0.06	3.43	0.018	~0.10	9.1±0.1 (0 < p_T^2 < 0.12), 4.4±0.3 (0.12 < p_T^2 ≲ 1)
	pp	24	30.6	3.49±0.42	0.11	4.25	0.026	~0.10	3.6±0.4 (0 < p_T^2 < 1.2)
	pp	8	22.7	~3.2	0.14	3.2	0.044	~0.07	6.9 (0 < |t'| < 0.2), 2.0 (0.2 < |t'| < 1)
	pp	11.2	22.1	2.0[b]	0.09	3.4	0.026	--	6.4±0.4 (0 < p_T^2 < 0.2)
	π^- p	15	21.2	0.42±0.15[c]	0.02	3.9	0.005	--	--
	π^- p	205	20.9	11.3±2.3[d]	0.54	8.0	0.068	0.26±0.06	--
K*+1420	K+p	8.2	13.8	~0.46	0.033	~3.4	0.010	--	2.3 (0 < |t'| < 0.7)
K*+890	K+p	8.2	13.8	~1.5	0.11	~3.4	0.032	--	4.3 (0 < p_T^2 < 0.5)
K*+890	pp	12	29.8	0.25±0.03	0.0083	3.43	0.0024	~0.018	3.4±0.4 (0 < p_T^2 < 1.2)
K*+890	pp	24	30.6	0.64±0.06	0.0209	4.25	0.0049	~0.023	2.8±0.3 (0 < p_T^2 < 1.2)
K*890	pp	12	29.8	0.02±0.02	0.0007	3.43	0.0002	--	--
K*890	pp	24	30.6	0.14±0.02	0.0046	4.25	0.0011	--	--

[a] Total cross section.

[b] Forward hemisphere only ($x_{\pi^+}, x_{\pi^-} > 0$).

[c] ≤1 GeV/c proton required in final state.

[d] Beam fragmentation region ($y_{c.m.}$ (ρ^0) > 1) not included.

SUMMARY

The general properties of inclusive meson resonance production may be summarized as follows:

(a) A limited amount is now known about inclusive vector meson production, particularly below ~ 15 GeV/c, but almost nothing is known about the inclusive properties of other meson resonances (for baryon resonances, only the Δ^{++} has been studied inclusively[15]).

(b) K^*'s in K^+p interactions at 8.2 GeV/c, and ρ^o's in π^-p interactions between 8 and 15 GeV/c come predominantly from beam excitation and are therefore produced mainly in the forward direction in the center of mass. However, for pp collisions, ρ^o, ω^o, and K^* are produced centrally.

(c) For π^-p collisions, the ρ^o cross section is typically 2-3 mb at low energy, and increases to $\gtrsim 11$ mb at 205 GeV/c.

(d) A substantial fraction (15-25%) of outgoing pions come from ρ-decay, producing a significant effect on one- and two-pion inclusive distributions.

(e) In lower-energy π^-p interactions (where ρ^o's come mainly from beam fragmentation) the ρ^o's are strongly polarized and have p_T^2 slopes of typically 6-7 GeV^{-2}, whereas in lower-energy pp interactions (where ρ^o's are produced centrally) the ρ^o's are unpolarized and have smaller p_T^2 slopes of 3-4 GeV^{-2}.

CONCLUSION

Clearly, there are still many basic questions about inclusive meson resonance production still to be answered. I would like to conclude this review with a basic set of questions aimed specifically at ρ production but relevant to the other meson resonances as well.

(a) How does $AB \rightarrow \rho X$ depend, for example, on projectile (A) and target (B); on total energy; on mass and multiplicity of X?

(b) What are the properties of inclusive production of charged ρ's?

(c) How does the average number of ρ's per event, $\langle n_\rho \rangle$, depend on charged multiplicity, n_{ch}? Does $\langle n_\rho \rangle$ increase with n_{ch}?

(d) Does $AB \rightarrow \rho X$ scale?

(e) What are the ρ-π correlations?

(f) How are one- and two-particle inclusive π distributions affected by ρ production?

(g) What is the ρ/π ratio as a function of x, y, and p_T? What is ρ/π at high p_T?

(h) What fraction of the single leptons or lepton pairs observed at high p_T in proton-nucleus collisions[16] come from vector meson decay as opposed to direct production?

(i) What are the ρ production mechanisms? What is the ρ component of beam or target fragmentation; in particular, are ρ's produced in high-mass diffraction dissociation? What is the multiperipheral contribution? How much ρ production is there from decay of higher-mass resonances or particle clusters?

ACKNOWLEDGMENTS

I would like to thank T. Ferbel for his encouragement and R. M. Barnett for discussions on the predicted ρ/π ratio.

REFERENCES

*Work supported by the U. S. Atomic Energy Commission.

1. Single particle inclusive production has recently been reviewed by H. Bøggild and T. Ferbel, Report No. COO-3065-76 (April 1974), to be published in Annual Reviews of Nuclear Science.

2. P. Chliapnikov, O. Czyzewski, Y. Goldschmidt-Clermont, M. Jacob, and P. Herquet, Nucl. Phys. B37, 336 (1972); P. Chliapnikov, L. N. Gerdyukov, B. A. Manyukov, F. Grard, V. P. Henri, P. Herquet, R. Windmolders, F. Verbeure, G. Ciapetti, D. Drijard, W. Dunwoodie, Y. Goldschmidt, A. Grant, S. Nielsen, L. Pape, Z. Sekera, J. Tuominiemi, N. Yamdagni, and V. A. Yarba, Inclusive K^O, K^{*+}(892), π^+ Production in K^+p Interactions at 8.2 GeV/c and Quark Models, CERN/D.Ph.II/PHYS 73-20 (1973); J. V. Beaupre et al., Nucl. Phys. B30, 381 (1971) [8.2 GeV/c K^+p].

3. M. Hontebeyrie, Y. Noirot, and M. Rimpault, Resonance Decay Contributions and Regge Trajectory in $K^+p \rightarrow K^O X^{++}$, Université de Bordeaux Report No. PTB-54 (December 1973).

4. J. Gandsman, G. Alexander, S. Dagan, L. D. Jacobs, A. Levy, D. Lissauer, and L. M. Rosenstein, Nucl. Phys. B61, 32 (1973); G. Alexander et al., Phys. Rev. D8, 712 (1973) and Nucl. Phys. B68, 1 (1974) [7.5 GeV/c γd].

5. V. Blobel, H. Fesefeldt, H. Franz, B. Hellwig, U. Idschok, J. W. Lamsa, D. Mönkemeyer, H. F. Neumann, D. Roedel, W. Schrankel, B. Schwarz, F. Selonke, and P. Söding, Phys. Letters 48B, 73 (1974); V. Blobel et al., Nucl. Phys. B69, 237 (1974); V. Blobel et al., Multiplicities, Topological Cross Sections, and Single Particle Inclusive Distributions from pp Interactions at 12 and 24 GeV/c, DESY 73/36 (1973) [12 and 24 GeV/c pp].

6. S. Fenster and J. Uretsky, Phys. Rev. D7, 2143 (1973).

7. This suggests that $\sigma(K^{*O}) \ll \sigma(\rho^O)$. If, on the other hand, $\sigma(K^{*O})$ were comparable to $\sigma(\rho^O)$, then $K^-\pi^+$ or $K^+\pi^-$ combinations from the K^{*O} could seriously distort the true ρ^O signal if $K^\pm$ were misidentified as $\pi^\pm$. S. Tovey points out that this kind of reflection is a significant problem in studying inclusive ρ^O production in 14.3 GeV/c K^-p interactions.

8. J. H. Christenson, G. S. Hicks, L. M. Lederman, P. J. Limon, B. G. Pope, and E. Zavattini, Phys. Rev. Letters 25, 1523 (1970).

9. T. Kitagaki, K. Abe, K. Hasegawa, A. Yamaguchi, T. Nozaki, K. Tamai, and R. Sugahara, An Inclusive Study of ρ^O Production in π^-p Interactions at 8 GeV/c, submitted to the XVIth International Conference on High-Energy Physics, Batavia (1972); and T. Kitagaki, private communication. [8 GeV/c π^-p]

10. L. Brink, W. N. Cottingham and S. Nussinov, Phys. Letters 37B, 192 (1971).

11. P. Borzatta, L. Liotta, S. Ratti, P. Daronian, and A. Daudin, Nuovo Cimento 15A, 45 (1973); C. Caso et al., Nuovo Cimento 66A, 11 (1970) [11.2 GeV/c π^-p].

12. A. Levy, D. Brick, M. Hodous, F. Hulsizer, V. Kistiakowsky, P. Miller, A. Nakkasyan, I. Pless, V. Simak, P. Trepagnier, J. Wolfson, and R. Yamamoto, Vector Meson Production in Semi-Inclusive π^-p Interactions at 15 GeV/c, submitted to this Conference; A. Levy, private communication [15 GeV/c π^-p].

13. Berkeley-NAL Collaboration: H. H. Bingham, D. M. Chew, B. Y. Daugéras, W. B. Fretter, G. Goldhaber, W. R. Graves, A. D. Johnson, J. A. Kadyk, L. Stutte, G. H. Trilling, F. C. Winkelmann, G. P. Yost, D. Bogert, R. Hanft, F. R. Huson, D. Ljung, C. Pascaud, S. Pruss, and W. M. Smart; results on total and elastic cross sections and charged particle multiplicities are given in D. Bogert et al., Phys. Rev. Letters $\underline{31}$, 1271 (1973) [205 GeV/c $\pi^- p$].

14. R. M. Barnett and D. Silverman, The Relative Production of ρ and π in the Central Plateau, University of California Irvine Report No. 74-27 (June 1974); and R. M. Barnett, private communication.

15. See, for example, F. T. Dao et al., Phys. Rev. Letters $\underline{30}$, 34 (1973); D. Brick et al., Phys. Rev. Letters $\underline{31}$, 488 (1973); and J. Gandsman et al., Ref. 4.

16. See, for example, J. P. Boymond, R. Mermod, P. A. Piroué, R. L. Sumner, J. W. Cronin, H. J. Frisch, and M. J. Shochet, Observation of Large Transverse Momentum Muons Directly Produced by 300 GeV Protons, University of Chicago Report No. EFI-74-24 (1974), and references therein.

VECTOR MESON PRODUCTION
IN SEMI-INCLUSIVE
π - p INTERACTIONS*

A. Levy†, D. Brick††, M. Hodous,
R. Hulsizer, V. Kistiakowsky,
P. Miller, A. Nakkasyan‡,
I. Pless, V. Simak§, P. Trepagnier,
J. Wolfson and R. Yamamoto
Laboratory for Nuclear Science
Massachusetts Institute of Technology
Cambridge, Massachusetts 02139

ABSTRACT

We have studied a sample of events taken in the SLAC 82"
chamber and measured by PEPR. These events contain an
identified proton with momentum less than 1 GeV/c. We
have isolated the produced ρ^0 and ω in these events and
have been able to classify them as to production mechanism.
We find that these vector mesons are mainly decay products
of diffractively produced mesons.

INTRODUCTION

Single particle inclusive spectra of π, K, p, Λ have been extensively
studied these past five years [1]. However, much less has been learned
about inclusive production of resonances [2-6]. The main difficulty in
studying resonances stems from the fact that they usually lie on a high
background. This difficulty is solved in exclusive studies by techniques
which are relatively easy to apply for few body final states. Such techniques are e.g. the Longitudinal Phase Space (LPS) technique [7] and the
prism plot technique [8]. However, such techniques are for the time
being not applicable to inclusive or even semi-inclusive studies.

The inclusive study of the Δ^{++}(1236) resonance [2-4] is in some way
easier, especially at high energies, than the study of mesonic resonances,
e.g. the ρ-meson [4, 6]. The Δ(1236) is close to the phase space threshold
and therefore at high enough energies where the pπ mass distribution
extends to high values, there is not much background left underneath the
Δ(1236) resonance. Another advantage of this baryonic resonance over the
ρ resonance is that it involves less combinations than for the vector meson.
In, say, an 8 prong event in π - p interaction one has 3 combinations for
the Δ^{++} resonance compared to 12 combinations for the ρ^0.

*   Work supported in part by the U. S. A. E. C.
†   On leave of absence from the Tel-Avic Univ., Israel.
†† Present address: L. P. C., Strasbourg.
‡   Present address: CERN, Geneva, Switzerland.
§   Present address: Inst. of Physics, SCAV, Prague, Cz.

Inclusive measurements carried out so far of the ρ^0 were performed in γp collisisions [4], where one has a strong leading particle effect and in pp interactions [6], where the ρ^0 comes off centrally. In both cases the ρ is defined via a mass cut only, carrying along a high background.

In this paper we study ρ and ω production in π - p interactions at 15 GeV/c. In order to overcome the problem of the large background we make a semi-inclusive study of these resonances by measuring the reaction

$$\pi - p \rightarrow p\pi - X^0 \qquad (1)$$

Such a semi-inclusive study after a certain x cut of the π - , provides a sample with a large signal to noise ratio in the region $M_{X^0} \sim m_\rho$. Having isolated what we believe to be a relatively clean sample of ρ and ω events, we try to answer the question to what extent these observed vector mesons are secondary products from primarily produced clusters of well-defined spin and iso-spin. In an inclusive study of Δ^{++} (1236) production, we showed [3] that most of the Δ^{++} are decay products of higher mass N*'s, which are diffractively produced. We come to a similar conclusion in the present study, namely, that most of the observed vector meson production are secondary products of A mesons decaying into $\rho\,\pi$ and B meson decaying into $\omega\,\pi$.

DATA

The data used in this analysis consists of 18518 events which are observed in the SLAC 82-inch HBC which was exposed to a π - beam of 15 GeV/c momentum. These events were selected from a sample of 56616 events measured with precision encoding and pattern recognition (PEPR). All 18518 events have one proton of momentum less than 1 GeV/c, guaranteeing the positive identification of protons and thus also of the π^+'s produced in the slow-proton events. All negative particles were assumed to be π -. Additional details on the selection and event processing are described elsewhere [9].

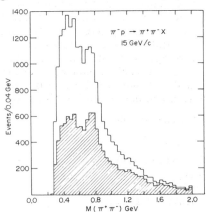

Fig. 1. The inclusive $\pi^+\pi$ - mass distribution in the reaction π - p $\rightarrow \pi^+ \pi$ - + X, where X includes a proton with momentum less than 1 GeV/c. The shaded area represents 4 prong events only.

Figure 1 shows the $\pi^+ \pi$ - mass distribution in the reaction

$$\pi - p \to \pi^+ \pi - X'^0 \qquad (2)$$

where X'^0 includes one identified proton. The distribution has a maximum about 450 MeV and shows no structure except the ρ^0. As evident from this figure the signal to noise ratio is very small in the ρ region and stays small even when selecting only the 4-prong events, where one has only two combinations per event (shaded area of Figure 1). In order to get a more purified ρ sample, we studied the semi-inclusive reaction (1) for topologies of $\geqslant 4$ prongs. Figure 2 shows the missing mass distribution recoiling against the p π -. A strong enhancement in the ρ region is evident, where now the signal to noise ratio is about 2/3. This enhancement includes of course also ω events, which as will be shown below, can be easily separated from the ρ events. As can be seen from the shaded area of Figure 2, all the events in the ρ region come from 4 prong events. We shall restrict therefore our semi-inclusive study of the ρ^0 and ω^0 to the 4 prong part of reaction (1). These two resonances are easily separable by studying the invariant mass of the $\pi^+ \pi$ - included in X^0. The ω events have to fulfill the relation $M (\pi^+ \pi -) < M (X^0)$, while for the ρ^0 events one has an equality of these two masses.

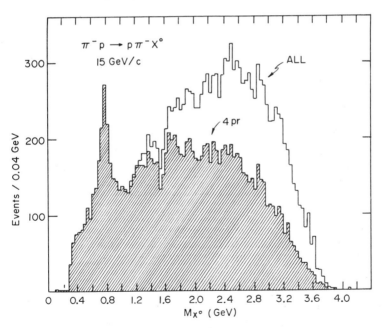

Fig. 2. Invariant mass distribution of X^0 in the reaction
$\pi - p \to \pi - p + X^0$. The shaded area represents 4 prong events only.

284

In order to further reduce the background underneath the vector mesons, we studied the x distribution of the π - in the 4 prong sample of reaction (1), x being the Feynmann variable defined as the ratio of the cms longitudinal momentum of the π - and half the overall cms energy of the initial π - p system. This x distribution displayed in Figure 3 has a maximum near x = 0, i.e. the pionization region. In addition there is an accumulation of events near x~0.9 which is the beam fragmentation region. This peak is due to the leading particle effect and corresponds to events in which the remaining pX⁰ mass distribution exhibit a shape which is reminiscent of the diffraction dissociation

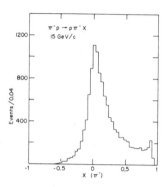

Fig. 3. x distribution of π - produced in the reaction $\pi - p \rightarrow \pi - p + X^0$, for the 4 prong sample.

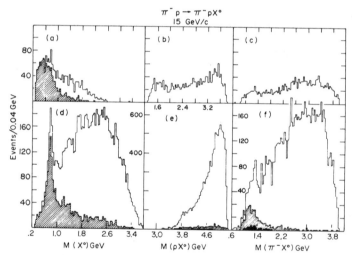

Fig. 4. Mass distributions in the reaction π - p → π - p + X⁰ for the 4 prong sample, using events which satisfy x (π -) > 0.5: (a) M (X⁰), (shaded area represents events which satisfy the condition M $(\pi^+ \pi -) = M (X^0)$), (b) M (pX⁰) and (c) M (π - X⁰). Figures (d), (e) and (f) have events which satisfy the condition x (π -) < 0.5: (d) M (X⁰), (e) M (pX⁰) and (f) M (π - X⁰). The shaded area in (d) represents events which satisfy the condition M $(\pi^+ \pi -) = M (X^0)$. The lightly shaded parts in (e) and (f) are events for which X⁰ is in the ρ⁰ region and the dark parts are those events where X⁰ is in the ω region (see text for definition of ρ and ω events).

of the target proton. One may avoid the leading particle effect by imposing a cut on $x_{\pi -}$. Figure 4a and 4b show the leading particle effect
($x_{\pi -} > 0.5$) on the invariant mass of X^0 and pX^0, respectively. The ρ signal seen earlier in Figure 2 seems to have disappeared almost completely,
(events which satisfy $M (\pi^+ \pi -) = M (X^0)$ appear in the shaded area of
Figure 4a), while the (pX^0) shows a diffraction-like low mass enhancement. A study of these $M (pX^0) < 2.4$ GeV sample shows that most of the
π^+'s included in X^0 form a strong Δ^{++} (1236) signal (not shown). Figure
4c shows that the $\pi - X^0$ mass distribution prefers high masses, which
is indeed expected if constituents of X^0 travel in the same hemisphere
as the proton. We conclude, therefore, from Figures 4a, 4b and 4c that
there seems to be no vector meson production in the case where the produced $\pi -$ is in the beam fragmentation region.

Figures 4d, 4e and 4f are mass distributions of events for which
$x_{\pi -} < 0.5$. This region is mainly the pionization region (there are no
events where the $\pi -$ is produced in the target fragmentation region). The
vector meson (ρ or ω) signal to noise ratio has increased from about $\frac{2}{3}$
(Fig. 2) to ~ 2 (Fig. 4d). The shaded area represents events which
satisfy the condition $M (\pi^+ \pi -) = M (X^0)$. A clear ρ signal is apparent.
We will, therefore, define our ρ events as those which fulfill the following
criteria: (a) $x_{\pi -} < 0.5$ (b) $M (X^0) = 0.66 - 0.86$ GeV and
(c) $M (\pi^+ \pi -) = M (X^0)$. The definition of ω events is (a) $x_{\pi -} < 0.5$,
(b) $M (X^0) = 0.74 - 0.83$ GeV and (c) $M (\pi^+ \pi -) < M (X^0)$. There is no
Δ^{++} (1236) signal in the $p \pi^+$ mass distribution for those events where the
π^+ is part of the ρ or the ω as defined above. Figure 4e shows the pX^0
mass distribution. This distribution is very different from the one in
Figure 4b; there is an absence of the low mass enhancements and a
peaking at high masses. The lightly shaded areas are events satisfying
our criteria as being ρ events and the dark area are ω events. Figure 4f
shows the invariant mass of $(X^0 \pi -)$ for events where $x_{\pi -} < 0.5$. A mass
concentration around ~ 1.2 GeV is an indication of A and B production.
This peak is further enhanced by taking $\pi - \rho^0$ (lightly shaded) and
$\pi - \omega$ (dark area) mass combinations. The $\pi - \rho^0$ mass distribution shows
both A_1 and A_2 production, while the B is seen decaying into $\pi \omega$.

DISCUSSION

Our way of analyzing the semi-inclusive reaction (1) together with a
cut on $x_{\pi -}$, results in a vector meson signal with a relatively low background underneath. Having defined our ρ and ω events in the previous
section, we want to find their origin. Is there any direct vector meson
production or are they decay products? In order to be able to answer this
question we studied the mass distribution of $p \pi -$ recoiling against the
vector meson. Figure 5a shows this distribution for ρ events and 5b

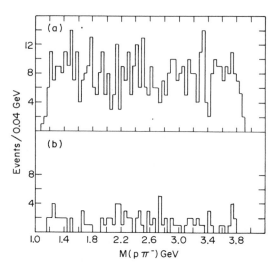

Fig. 5. Invariant mass distribution of (p π -) produced in the reactions: (a) π - p → π - p ρ⁰ (b) π - p → π - p ω.

for ω events. As can be seen very little if any structure is seen near the Δ⁰ (1236) mass. Thus if the quasi-two body reaction π - p → V⁰ Δ⁰ exists, it is produced with a very low cross-section, as expected from an I = 1 exchange reaction at 15 GeV/c. However, studying the π - ρ and π - ω mass distribution (Figure 4f) we conclude that the vector meson is almost always produced in association with a π - as part of a low-mass system, which has the characteristics of diffraction dissociation of the incoming projectile. There is clear A_1 - A_2 signal in the ρ π - and a B signal in the ω π - mass distribution.

We studied the t' = | t - t_{min} | behavior of the A region. For this study we used the shaded part of Figure 4f, namely the ρ⁰ π - events. An exponential fit for events having M (ρ⁰ π -) ◁. 4 GeV yields a slope of ~9 $(GeV/c)^{-2}$. Events above this mass region have a shallower slope (~6 $(GeV/c)^{-2}$). A breakup of the low mass ρ⁰ π - region into A_1 (1.0 - 1.2 GeV) and A_2 (1.25 - 1.35 GeV) regions showed the A_1 to have a much steeper t' distribution than the A_2 region: ~12 $(GeV/c)^{-2}$ for the A_1, compared to ~6 $(GeV/c)^{-2}$ for the A_2. A similar behavior was reported in a study of pion diffraction dissociation reactions $π^± p → (π^± π^+ π -) p$ at 16 GeV/c [10].

We conclude, therefore, that the vector mesons seen here in this semi-inclusive study are in fact decay products of higher mass clusters which are diffractively produced.

SUMMARY AND CONCLUSION

In a semi-inclusive study of π^- p interactions at 15 GeV/c, we have isolated ρ and ω events. We achieved a relatively clean sample of these vector mesons with a 2:1 signal to noise ratio by studying reaction (1) and making a cut of $x_{\pi^-} < 0.5$. We have shown that there exists very little vector meson production in the target diffraction sample.

Most of the vector meson observed in this experiment are shown to be decay products of a $(V^0 \pi)$ system diffractively produced off the projectile.

REFERENCES

(1)   M. E. Law et al, "A Compilation of Data on Inclusive Reactions", LBL-80
(2)   F. T. Dao et al, Phys. Rev. Letters 30, 34 (1973),
(3)   D. Brick et al, Phys. Rev. Letters 31, 488 (1973),
(4)   J. Gandsman et al, Nucl. Phys. B61, 32 (1973),
(5)   P. Chliapnilcov et al, Nucl. Phys. B37, 336 (1972),
(6)   V. Blobel et al, DESY 73/51 October 1973,
(7)   W. Kittel et al, Nucl. Phys. B30, 333 (1971),
(8)   J. E. Brau et al, Phys. Rev. Letters 27, 1481 (1971),
(9)   D. Brick, Ph. D. thesis, M. I. T. (unpublished).
(10) J. V. Beaupre et al, Phys. Letters 41B, 393 (1972),

THE DECK MODEL REVISITED

Lorella M. Jones
University of Illinois, Urbana, Ill. 61801

ABSTRACT

Phases between various 3π partial waves in
$\pi^- p \rightarrow \pi^- \pi^- \pi^+ p$ can be understood if the Reggeized Deck
effect is used for non-resonant 3π production, and a nor-
mal Regge model is used for A_2 production.

Partial wave analyses of the reaction $\pi^- p \rightarrow \pi^- \pi^- \pi^+ p$ have shown
that the A_1 effect does not have resonant phase variation. This
makes a Deck effect explanation of this bump likely; and we have recently shown[1] that all the mass and angle distributions in the A_1
region can be reasonably well fit with an improved version of this
model.

Briefly, our model has on-mass-shell $\pi\pi$ and πN scattering amplitudes connected by a Reggeized pion propagator, and is symmetrized
in the identical π^- particles (Fig. 1). The $\pi\pi$ scattering is calculated from phase shifts and elasticities given by $\pi\pi$ scattering
people; the πN scattering is calculated using phase shifts at low
energies and a Regge model at high energies. The Reggeized pion
propagator looks like

$$\frac{\left(\frac{S_{3\pi} - U_{3\pi}}{2}\right)^{\alpha_\pi(t_R)} e^{-\frac{i\pi\alpha_\pi(t_R)}{2}}}{t_R - m_\pi^2} \qquad (1)$$

As my time is limited, I will refrain from showing slides of all
possible projections of the final state momenta. Instead, I want to
focus on one feature of the model--the phases of the 3π partial
waves.

Once one is committed to Reggeizing the pion exchange, which
Berger showed vastly improved the shape of the 3π mass distribution,
the signature phase of Eq. 1 will tend to make different partial
wave amplitudes have different phases. We should compare the magnitudes of these with the corresponding relative phases obtained from
the data to see whether the model accurately reproduces the data.

I wish to make 3 basic points about these phases:

I. The fitting program used on the data gives phases for the
model which agree with explicit partial wave analysis of the model.

If we examine only the part of the amplitude with $\pi^- \pi^+$ scattering, it is straightforward to explicitly compute partial wave amplitudes by multiplying by assorted D functions and doing angular inte-
grations.[2] Hence, the phases can be definitely found.

One can then generate events according to this model (the two
diagram model) and fit them with the fitting routine.

The two methods are compared in Fig. 2. Notice that agreement
is quite good, especially in the A_1 region. This shows that, even

though the fitting program makes some assumptions which are more valid for resonance production than for Deck models, it is capable of getting the right answers.

II. Relative phases between various non-resonant partial waves are in reasonable agreement with the data.

In Fig. 3 we give relative phases computed from the full model; these are to be compared with the experimental phases in Fig. 4.

Note especially the 1^+s-1^+p and 1^+s-2^-p phase differences, whose constancy was important for the decision that the A_1 was a Deck effect. The mysterious 90° phase difference between the 1^+s and 1^+p waves can be shown to arise almost wholly from the pion signature phase.

III. The A_1-A_2 interference phase can be understood in a straightforward manner if this Deck model is used for A_1 production and a Regge model is used for A_2 production.

We can compute the phase of the 1^+s ($\rho\pi$) wave at $m_{3\pi} = m_{A_2}$ from explicit partial wave analysis of the Deck amplitude. Sample values of this phase are given in Table I. Note that in the forward direction this phase is advanced by about 60° over the 90° phase expected in elastic scattering; this extra phase comes from the partial waving of the signature factor.

Table I

Deck Model Phase for A_1 Production at $M_{3\pi} = M_{A_2}$

P_{lab} \ t-t_{min}	0	-.1	-.2	-.3 GeV^2
6 GeV	146°	158°	170°	179°
16	146°	156°	166°	175°
26	146°	156°	165°	173°
36	146°	155°	164°	172°

A_2 production is not included in the Deck model and must be computed by a different method. The data for natural parity exchange A_2 production fall off too slowly to be fit by ρ and f exchange; one must include a trajectory with a higher intercept.

We have fit A_2 natural parity exchange differential cross section data over the range 5-40 GeV/c with an $f + P$ Regge model.[3/] In Fig. 5 we show the energy dependence of the integrated cross section for two fits of equally good χ^2--the solid line shows results with f and Pomeron residue functions of the same sign, the dashed line is for residues with opposite signs. From the fitted parameters, we can predict an A_2 production amplitude phase.

We compute $\varphi_{A_2} - \varphi_{A_1}$ and compare with data (Fig. 6). The phases from the fit with Pomeron and f contributions of the same sign agree with the data to within 30° over the region measured. This is quite adequate, considering all approximations made. We do not think it is possible to get agreement if a phase near 90° for A_1 production is assumed.

SUMMARY

Phases between various 3π partial waves in $\pi^- p \to \pi^- \pi^- \pi^+ p$ can be understood if the Reggeized Deck effect is used for non-resonant 3π production, and normal Regge models are used for A_2 production.

REFERENCES

1. G. Ascoli, R. Cutler, L. M. Jones, U. Kruse, T. Roberts, B. Weinstein, and H. W. Wyld, "A Deck Model Calculation of $\pi^- p \to \pi^- \pi^+ \pi^- p$", to appear in the Ap.1 edition of Phys. Rev. D (1974).
2. G. Ascoli, L. M. Jones, B. Weinstein, and H. W. Wyld, Phys. Rev. D8, 3894 (1973).
3. G. Ascoli, L. M. Jones, R. Klanner, U. E. Kruse, and H. W. Wyld, "The A_2-A_1 Interference Phase", University of Illinois preprint ILL-(TH)-74-5. Submitted for publication.

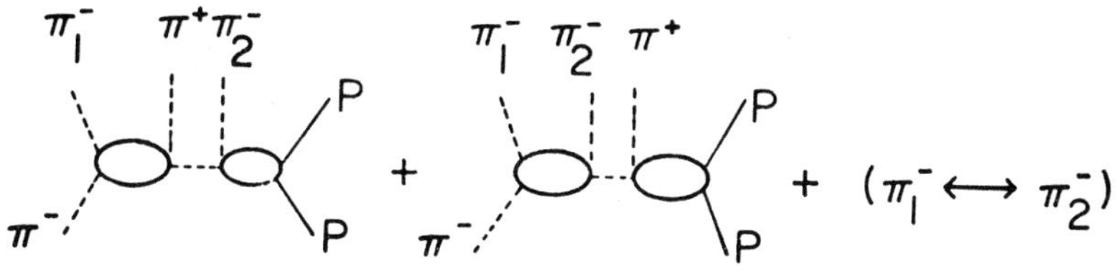

Figure 1   Our Deck Model for $\pi^- p \to \pi^- \pi^+ \pi^- p$.

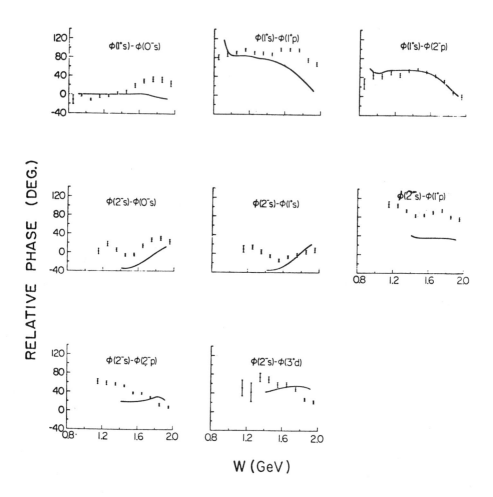

Figure 2   Relative phases of the partial waves important in A_1 and A_3 regions. Solid lines indicate calculated values; points with error bars were obtained by applying the program FIT to a sample of Monte Carlo events.

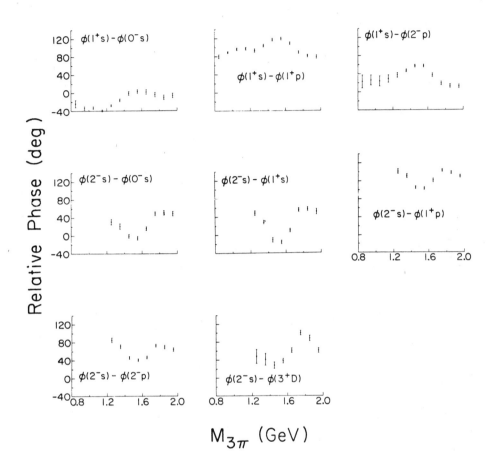

Figure 3   Relative phases between partial wave amplitudes obtained
by applying FIT to the Deck model.

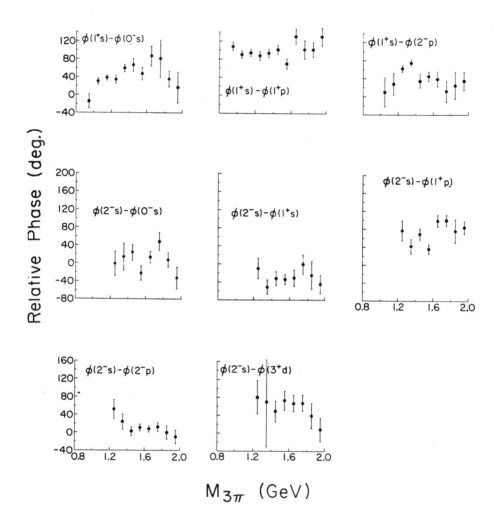

Figure 4 Relative phases between partial wave amplitudes obtained by applying FIT to the data.

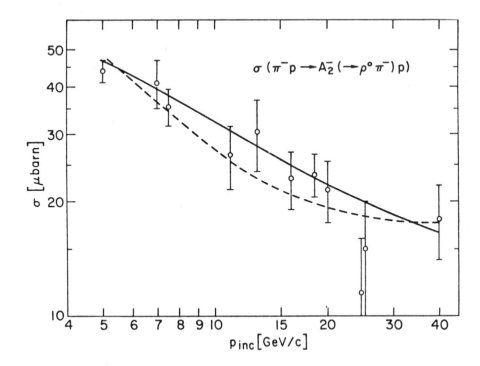

Figure 5   Measured and fitted integrated cross sections for $\pi^- p \to A_2^-$
($\to \rho^o \pi^-$)p for $1.2 \leq M_{3\pi} \leq 1.4$ GeV and $0 < |t-t_{min}| < 0.7$
$(\text{GeV/c})^2$, as a function of beam momentum.   Solid curve for
fit with $K > 0$ and dashed curve for fit with $K < 0$, where
K is the ratio of f and Pomeron residues.

296

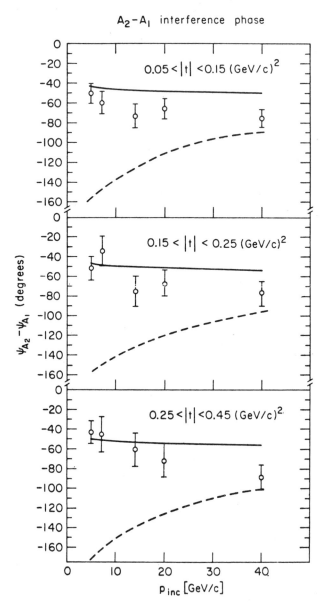

$A_2 - A_1$ interference phase

Figure 6 Comparison of measured A_2-A_1 interference phase to the
phase predicted from Regge and Deck model calculation.
Dependence on incident momentum is shown for different
momentum transfer intervals (note: the data at 40 GeV/c
extend only to momentum transfer of -0.3 $(GeV/c)^2$). Solid
(dashed) curve for fit with K > 0 (K < 0), where K is
ratio of f and Pomeron residues.

MESON SPECTROSCOPY AND THE PHENOMENOLOGY OF
HIGH ENERGY COLLISIONS

C. Quigg*
Institute for Theoretical Physics
State University of New York
Stony Brook, Long Island, New York 11794

ABSTRACT

Properties of high-energy scattering amplitudes are
deduced from recent $\pi\pi$ phase shift analyses. A brief review is given of the duality issues surrounding diffractively produced states. In this context, finite missing mass
sum rules are introduced. The triple-Regge expansion is
discussed, and its relationship to the Chew-Low formula is
emphasized.

INTRODUCTION

With the invitation to speak at this Conference came an admoni-
tion to emphasize aspects of my subject most relevant to meson spectroscopy. To make certain that I would comply, I reviewed the recent
progress in the field by reading through the proceedings of EMS-'68,
'70, and '72, together with other relevant documents. The results of
my survey[1] are summarized in Fig. 1. To a good first approximation,
no new mesons have been established since Meson Conferences began! I
have three comments on these data: (1) At the very least, the meetings
in this series have served the valuable purpose of deflating wild
claims. (2) More and more, the methods of experimental meson spectroscopy and of experimental baryon spectroscopy have merged, and there
is less justification for separate gatherings on mesons and baryons.
(3) I trust that no one seriously believes that because no new mesons
have been found, none exist. Many arguments exist in favor of additional states. These are based (separately) on duality[2], on quark[3],
Regge[4], and Veneziano[5] requirements for daughters, on the possibility[6]
that exotic mesons exist[7], and on theorists' dreams[8].

Given the absence of newly-established states, I need make no
further apology for taking the global point of view that we want to
understand everything -- spectrum and interactions -- at once, and
that instead of merely cataloging states, we must try to discover how
they fit into a dynamical picture. In this spirit, there are two
very useful approaches that I have elected not to follow. The first
is that of the "canonical EMS talk" on the relation of production and
decay characteristics to resonance properties, which has been largely
exhausted (for the moment) by Fox[9] and by Kane[10]. The second involves recent work[11] by Estabrooks, Hoyer, Martin, and others on the
computation of the external-mass dependences of scattering ampli-

*Research supported in part by the National Science Foundation under
Grant No. GP-32998X.

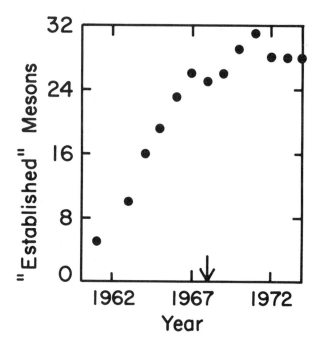

Fig. 1: The number of established mesons tabulated by the Particle Data Group and its forerunners, as a function of time. The arrow indicates the date of the first Meson Conference.

tudes by triple-Regge techniques. If this new tool lives up to its initial promise, it will surely revivify the study of (quasi-) two body reactions.

I will instead discuss what inclusive phenomenology has to ask of, and say to, meson spectroscopy. Specifically, I want to treat the long-standing problems of duality and diffractively-produced states, the applications of finite missing-mass sum rules, and the measurement of $\pi\pi$ inclusive cross sections at high energies. To place these in proper context, let us begin by reviewing the analogous concepts in the case of two-body reactions. In the course of this review, we shall come across a few new results.

DUALITY AND THE $\pi\pi$ SYSTEM

One of the most valuable insights into the systematics of hadronic interactions is provided by the connection between low-energy behavior and high-energy amplitudes first demonstrated by Dolen, Horn, and Schmid[12]. At a qualitative level, this is exhibited by the conjunction of the Legendre zeroes in resonance amplitudes near the location of the zeroes in high-energy amplitudes. The classic illustration of this cooperation between resonances in the πN system is shown in Figs. 2 and 3. Figure 2 shows the first zeroes in the contributions of prominent resonances to the s-channel nonflip and flip ampli-

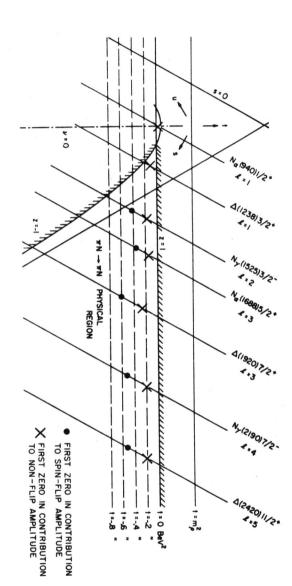

Fig. 2: First zeroes of the prominent resonances on the Mandelstam plot for the πN problem (from Ref. 12).

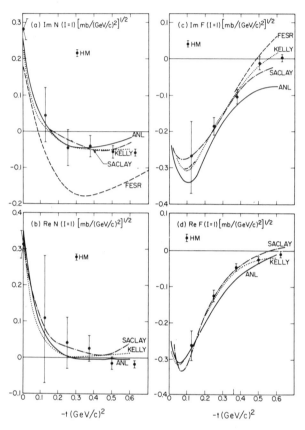

Fig. 3: Results of πN amplitude analyses at 6 GeV/c for I=1 t-channel exchange. N and F refer to s-channel nonflip and flip amlpitudes respectively. The four analyses come from References 13 (ANL), 14 (HM), 15 (Kelly), and 16 (Saclay). For clarity, we omit the error estimates on all except the HM analysis.

tudes, which cluster about $-t=0.2$ and 0.5 $(GeV/c)^2$, respectively. High energy amplitudes for t-channel I=1 exchange, determined by amplitude analysis at 6 GeV/c, are shown in Fig. 3. The imaginary parts have zeroes which coincide roughly with the locations of resonance zeroes. Double zeroes in the real parts, although difficult to locate with any precision, seem to occur at $-t \approx 0.5$ $(GeV/c)^2$ in the flip amplitude, and somewhere in the interval $0.3 < -t < 0.6$ $(GeV/c)^2$ in the nonflip amplitude. The comparison of these results with Regge model expectations is well-known and has been reviewed elsewhere.[17] Legendre zeroes for the prominent ππ resonances are plotted in Fig. 4. We may expect that the conjunction of first zeroes near $-t=0.3$ $(GeV/c)^2$ will be reflected in the structure of the high-energy amplitudes.[18]

The precise form of the duality relationship between low energy and high energy amplitudes is given by the finite energy sum rules. If we assume Regge asymptotic behavior at fixed t

$$\text{Im } T \sim \sum_i r_i \, \nu^{\alpha_i(t)}, \qquad (1)$$

where $\nu = \frac{1}{2}(s-u)$, then analyticity leads to the finite energy sum rule (FESR)

$$\int_o^{\bar{\nu}} d\nu \, \nu^N \text{ Im } T = \sum_i r_i \, \frac{\bar{\nu}^{\alpha_i(t)+N+1}}{\alpha_i(t)+N+1} . \qquad (2)$$

In eq. (2), N is an even (odd) integer for an invariant amplitude T that is odd (even) under crossing $\nu \to -\nu$. If the upper limit $\bar{\nu}$ is chosen so that the left-hand-side can be evaluated from low-energy phase shifts, the right-hand-side constrains the properties of the

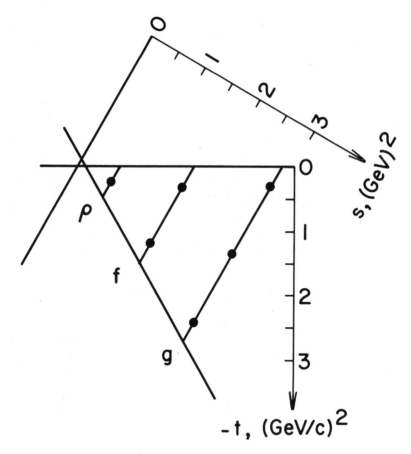

Fig. 4: Zeroes of the prominent resonances on the Mandelstam plot for the $\pi\pi$ problem.

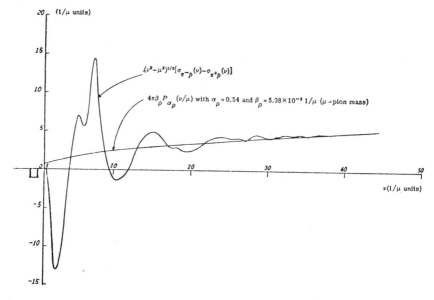

Fig. 5: The integrands for the two sides of the finite energy sum rule (2) for the crossing-odd forward scattering amplitude in πN scattering, with N=0 (from Ref. 19).

Reggeon exchanges assumed to dominate the high-energy amplitudes. Figure 5 shows the low-energy and high-energy integrands of (2) for the t=0 crossing-odd amplitude in πN scattering. The low-energy contribution, which is given in terms of the $\pi^{\pm}p$ total cross section difference, shows prominent resonance wiggles, whereas the high-energy side, which is specified by the exchange of a ρ Regge pole, is a smooth interpolating function. A similar "semilocal" interpolation for the t-channel isospin = 1 amplitude in forward ππ scattering is shown in Fig. 6, in which the prominent resonance bumps are those of

Contribution to $I_t = 1$ Sum Rule

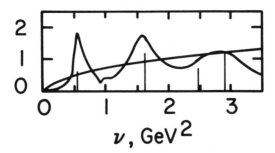

Fig. 6: The integrands for the two sides of (2) for the crossing-odd forward scattering amplitude in ππ scattering, with N=0.

the ρ, f, and g, and the interpolation is provided by a schematic ρ-exchange amplitude, with $\alpha_\rho(0) = \frac{1}{2}$.[20]

Considerable evidence supports the idea[21] that ordinary meson Reggeon exchanges are driven by direct-channel resonances (as opposed to "background"). The contributions of the ρ, f, ρ', and g resonances contained in the Hyams phase shift solution are indicated in Fig. 6 as spikes at the resonance positions. As expected, the resonance contributions dominate the sum rule. This circumstance provides one argument in favor of the existence of higher-mass resonances. If the Regge pole amplitude is a suitable asymptotic form at some high energy ν_H, the resonance-dominated FESR connection requires direct channel resonances, at least up to the energy ν_H.[22] In other words, if the Regge pole amplitude is an interpolator of resonances, it must have resonances to interpolate.

The $I_t=0$ amplitude corresponds to the exchange of the Pomeranchuk singularity and the f° (or P') Regge pole. The integrand of the forward, first-moment $I_t=0$ FESR is shown in Fig. 7 together with a Regge interpolation. The dashed curve is the contribution of the f°-

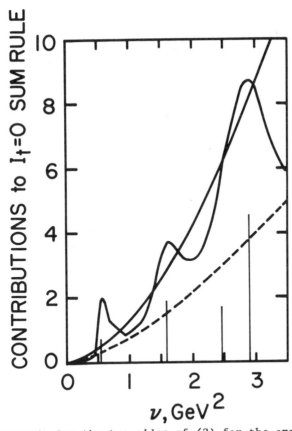

Fig. 7: The integrands for the two sides of (2) for the crossing-even $I_t=0$ forward scattering amplitude in $\pi\pi$ scattering, with N=1.

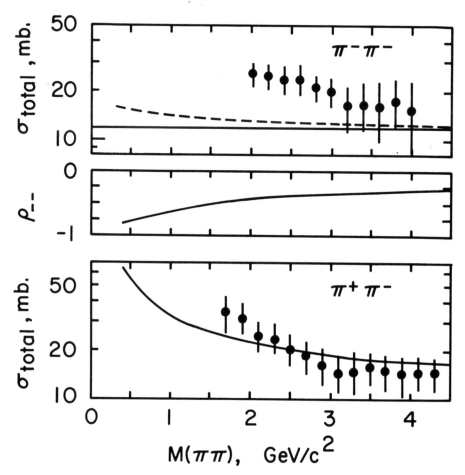

Fig. 8: ππ total cross sections determined using the optical theorem (from Ref. 25). The various curves are explained in the text.

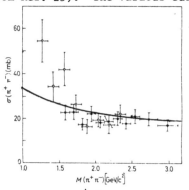

Fig. 9: Total $\pi^+\pi^-$ cross sections from Ref. 26.

exchange amplitude, which I define to be equal to the ρ-exchange amplitude determined above. What remains after subtracting the f^0 contribution from the sum rule is to be identified with Pomeron exchange. Indeed, the residual contribution does have roughly the expected ν^1 energy dependence. The fact that we are dealing with a first-moment sum rule means that the highest-energy phase shifts, which are the least certain, are weighted most heavily, so we are faced with considerable uncertainties in any deductions we might make about high-energy behavior. Nevertheless, a best-eyeball-fit to the I_t=0 and I_t=1 sum rules leads to a high-energy cross section

$$\sigma_{total}(\pi^+\pi^-) \simeq (7 + 30\ s^{-\frac{1}{2}})\ mb., \tag{3}$$

which (also by eye) carries an uncertainty of ∿50%, due largely to the uncertainty of the I_t=0 integrand at high $\pi\pi$ masses.[23]

The 7 mb. asymptotic total cross section found here is somewhat lower than the values of about 12 - 16 mb. deduced from factorization and Regge pole fits to πN and NN scattering. Figures 8 and 9 show the comparison of one such Regge pole prediction[24] with total cross sections extracted from data on $\pi N \to \pi\pi N$ by means of (rather long-distance) pole extrapolation and the optical theorem. There is order-of magnitude agreement between the model and the two data sets. In the experiments, the real part of the forward elastic scattering amplitude has been neglected. The exchange-degenerate Regge pole fits suggest that for $\pi^-\pi^-$ scattering, in which the ρ+f exchange contribution is purely real, the ratio $\rho \equiv Re\ T(0)\ /\ Im\ T(0)$ becomes quite appreciable at low energies. This is shown on the middle graph of

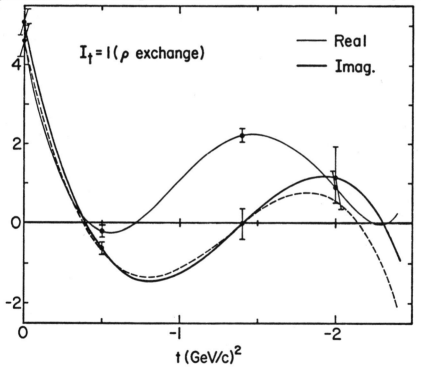

Fig. 10: The I_t=1 amplitude at high energy, determined from a FESR analysis (from Ref.27). The broken line is the imaginary part computed from low-energy resonances alone. In this analysis, the high energy cutoff was $\sqrt{s}$ = 1.4 GeV.

Fig. 8. The dashed curve superimposed upon the $\pi^-\pi^-$ data in Fig. 8 shows the cross section that would be deduced via the optical theorem from the theoretical forward elastic cross section, upon neglect of the real part. In the energy range spanned by the measurements, the error committed is not gross, especially when compared with the uncertainties inherent in the pole extrapolation.

The high-energy amplitudes can be evaluated away from the forward direction and, with the aid of the so-called continuous moment sum rules (CMSR) one can deduce the properties of the real part as well. The results of such a calculation[27] are shown in Fig. 10. As we anticipated in the discussion of Fig. 4, there is a zero in the imaginary part near $-t = 0.4$ $(GeV/c)^2$, as well as a nearby double zero in the real part. It remains to be verified directly that the experimental high-energy amplitudes have this structure. The dashed line in Fig. 10 shows the imaginary part computed from the low-energy resonances alone. The good agreement with the full calculation supports the idea that there is a duality between meson Regge poles and direct-channel resonances.

Let us now turn to the sum rule for the $I_t=2$ amplitude. This amplitude corresponds to an exotic exchange at high energies, so should vanish in first approximation. With the higher energy $\pi\pi$ phase shifts now becoming available, this is the first instance in which we can study how the suppression of a forbidden exchange is accomplished. Because the $I_s=0$ and 1 resonances contribute with opposite signs to the $I_t=2$ amplitude, it was expected in ancient times that the alternation of $I_s=0$ and $I_s=1$ resonances on the leading trajectory would cause the $I_t=2$ amplitude to vanish. However, the fact that the sum rule is a first-moment (N=1) FESR means that the contributions of higher-mass resonances are weighted more strongly. Hence, instead of a cancellation one appears to be faced with a divergent alternating series, in the narrow resonance limit.[28] This is shown in Fig. 11 in which the contributions of ρ, f, g are indicated as spikes for two different sets of resonance parameters. In contrast, in the dual resonance model[5], an exact cancellation takes place at each resonance mass between the highest-spin resonance and its daughters, which alternate in isospin. The enforcement of this cancellation is one of the ingredients which fixes the relative partial widths into $\pi\pi$ of the many resonant states in the model. Returning to Fig. 11 we see that the cancellation of the $I_t=2$ amplitude, which is nearly exact if one integrates the Hyams amplitudes past the g-region, occurs through a combination of these two mechanisms. There is an important cancellation between resonance towers as well as the competition between parent waves and daughter waves within each tower. I look forward with considerable anticipation to the extension and refinement of this analysis as better phase shifts become available at still higher energies.

I have already referred twice to the two-component duality[21] of Freund and Harari, which is represented schematically in Fig. 13. While it is supported by resonance-saturated sum rules, as we saw in Fig. 10, the most graphic illustration of the idea is the one invented by Harari and Zarmi[29]. By forming the contributions of s-channel partial-wave amplitudes to amplitudes of definite t-channel isospin

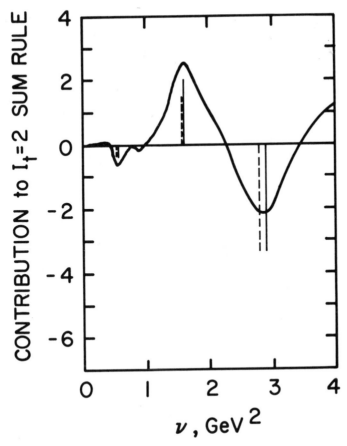

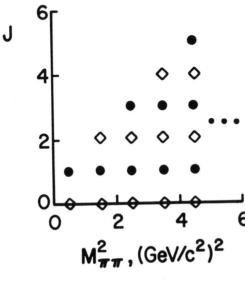

Fig. 11 (above): The integrands for the two sides of (2) for the $I_t=2$ forward amplitude, with N=1. Resonance parameters are from the Hyams solution (——) and from the 1973 wallet cards (- - -).

Fig. 12 (left): Spectrum of the Veneziano model for $\pi\pi$ scattering, showing the $I_s=0$ (◊) and $I_s=1$ (●) states.

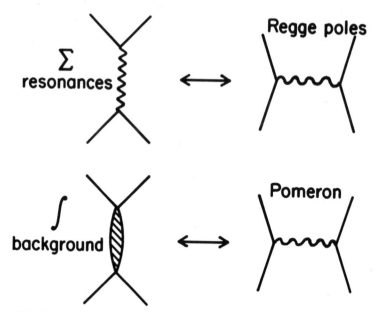

Fig. 13: Two-component duality.

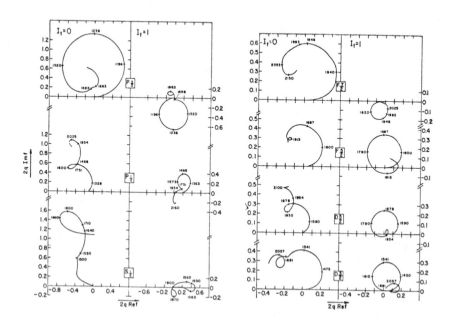

Fig. 14: Argand diagrams for linear combinations of $I_s=\frac{1}{2}$ and $I_s=3/2$ partial-wave amplitudes for πN scattering corresponding to $I_t=0,1$.

they showed (Fig. 14) that the $I_t=1$ (ρ-exchange) partial waves are relatively background-free, i.e. correspond to resonances, whereas the $I_t=0$ (P+f exchange) partial waves clearly contain nonresonant background, in addition to the resonant circles. It is clearly of interest to examine the even partial waves in $\pi\pi$ scattering for similar behavior[30]. This can now be done meaningfully only for the s-wave, with the results shown in Fig. 15. There is a strong background component in the $I_t=0$ partial wave, as expected, and much less background is apparent in the $I_t=1$ partial wave. The $I_t=2$ partial wave is nearly background-free[31]. One wave does not settle anything, of course, but this may be taken as an indication of things to come and, I hope, as motivation to push the phase shift analysis. In particular, the very puzzling problem of duality for high-energy exotic exchanges should become approachable.

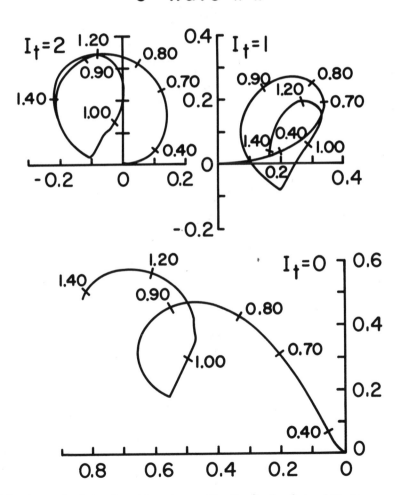

Fig. 15: Argand plots for the $\pi\pi$ s-wave $I_t=0$, 1, 2 amplitudes.

THE TRIPLE-REGGE EXPANSION[32]

Consider the missing-mass reaction depicted in Fig. 16(a),

$$a + b \rightarrow c + X \text{ (of mass } M), \qquad (4)$$

in the regime $M^2/s \ll 1$, $t/s \ll 1$. It is natural to treat (4) as a two-body peripheral exchange process, as indicated in Fig. 16(b). To compute the cross section for such a reaction, we square the matrix element of Fig. 16(b), and sum over all possible final states X with the specified mass M. The sum over final states, which is represented in Fig. 16(c), is [by the optical theorem] simply a flux factor times the total cross section for scattering the virtual particle i (a Reggeon) from the real incoming particle b. Thus

$$\frac{d\sigma}{dtdM} = \frac{\pi^2 |\beta_{ac}^i(t)|^2 |\xi_i(t)|^2}{16\pi[s-(M_a+M_b)^2][s-(M_a-M_b)^2]} \left(\frac{s}{M^2}\right)^{2\alpha_i(t)} \times \text{Flux} \times \sigma_{ib}(M^2,t), \quad (5)$$

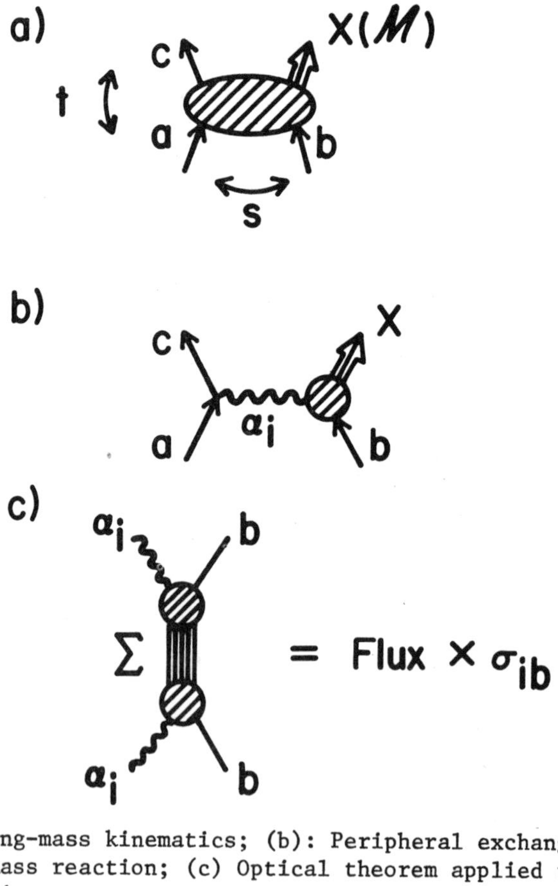

Fig. 16(a): Missing-mass kinematics; (b): Peripheral exchange picture for the missing-mass reaction; (c) Optical theorem applied to Reggeon -particle scattering.

where β occurs at the $\alpha c i$ vertex in Fig. 16(b), $\xi_i(t)[s/M^2]^{\alpha_i(t)}$ is the Regge propagator, and $\xi_i(t) = [\tau + \exp(-i\pi\alpha_i(t))]/\sin\pi\alpha_i(t)$ is the signature factor for a Reggeon with signature τ. If we now parameterize σ_{ib} in terms of Regge poles

$$\sum_{\bar\alpha} \quad \begin{array}{c} \alpha_i(t) \\ \overline{\alpha} \\ \alpha_i(t) \end{array} \quad \begin{array}{c} b \\ b \end{array}$$

i.e. as

$$\sigma_{ib}(M^2,t) = A(t) + B(t)/M + \cdots \tag{6}$$

we obtain a *triple-Regge expansion* for $d\sigma/dtdM^2$.

Now let us specialize to the case of pion exchange, and consider for definiteness the reaction

$$\pi^- + p \to n + X, \tag{7}$$

for which

$$\beta(t) = \sqrt{-t}\ F(t)\ \sqrt{2}\ g/\sqrt{4\pi}, \tag{8}$$

where $F(t)$ is a form factor satisfying $F(\mu^2)=1$, and

$$\xi(t) = \frac{1 + \exp(-i\pi\alpha)}{\sin\pi\alpha} \to \frac{2/\pi}{t-\mu^2} \tag{9}$$

for $\alpha\approx0$. Assembling all the pieces of (5), we obtain

$$\frac{d\sigma}{dtdM^2} = \frac{1}{4\pi[s-(M+\mu)^2][s-(M-\mu)^2]}\ \frac{2g^2|F(t)|^2}{4\pi}\ \frac{(-t)}{(t-\mu^2)^2}$$
$$\times[M^2(M^2-4\mu^2)]^{\frac{1}{2}}\ \sigma_{\pi\pi}(M^2,t), \tag{10}$$

an equation which appears in the classic paper of Chew and Low[33]. [Obviously, anyone who professes faith in the Chew-Low formula now is morally bound to believe with equal fervor in the triple-Regge expansion!]

An obvious check of the ideas involved in the derivation of (5) is the measurement of a known cross section by the Chew-Low/Triple-Regge techniques. For example, in the reaction

$$p + n \to p_{slow} + X, \tag{11}$$

the π^-p total cross section enters. Figure 17 compares a calculation by Field and Fox[34] with *preliminary* data of the Rutgers/ICL/UICC Collaboration[35] taken in the Internal Target Laboratory at NAL at $s=110$ GeV^2 and $t=-0.16$ $(GeV/c)^2$. There is also rough agreement[36] between an OPE calculation and the NAL/UCLA data[37] on the reaction

$$p + p \to \Delta^{++} + X \tag{12}$$

at 303 GeV/c.

These tentative successes suggest exploratory measurements of

312

$\sigma_{\pi\pi}$ at NAL and of $\sigma_{\pi p}$ at the CERN ISR[38]. The energy regime in which the $\pi\pi$ total cross section can be explored in the reaction

$$\pi N \rightarrow N + X \tag{13}$$

at NAL may be deduced from the kinematical maps in Fig. 18. Apparently measurements up to $M^2 \sim 150$ GeV^2 are practical. Since one of the largest sources of uncertainty in the traditional Chew-Low technique arises from the extrapolation to the pion pole, it is worthwhile to make the point strongly that from the perspective of high-energy phenomenology, results would be of interest even before the pole extrapolation is attempted. Specifically, the energy (M^2) dependence of $\sigma(M^2,t)$ at fixed values of t is an important object of study.

DUALITY AND DIFFRACTIVELY-PRODUCED STATES

Duality has acquired something of a bad name in spectroscopic circles because it has been invoked repeatedly to moot the question, "Is the A_1 a resonance?" without contributing to a resolution of the basic issue[39]. That basic issue, which is still unresolved and is once again receiving considerable attention, is represented in Fig. 19:

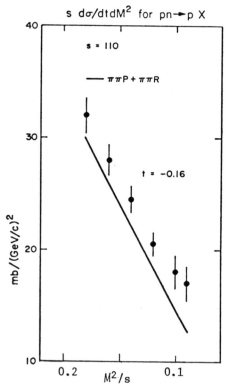

Fig. 17: Data[35] and a triple-Regge prediction for reaction (11) [from Ref. 34]. Theory uses the measured $\sigma_{\pi p}$, and $F(t) \equiv 1$.

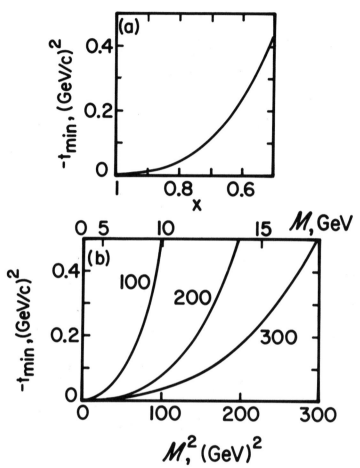

Fig. 18: Kinematics of the reaction πN→NX in terms of the scaling variable x≡1−M^2/s, and in terms of M^2, at NAL energies.

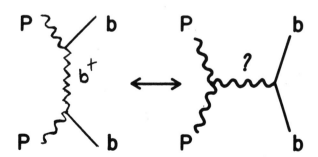

Fig. 19: The "basic issue" concerning duality and diffractively-produced states.

to what are diffractively-produced states dual?

The first guess[40], made during the infancy of triple-Regge analysis, was that two-component duality should hold in its original form for Pomeron-hadron scattering [see again Fig. 13]. If s-channel resonances were dual to t-channel Regge exchanges, we should expect

$$\left.\frac{d\sigma}{dtdM^2}\right|_{resonance} \propto \frac{1}{M^3} , \qquad (14)$$

a constant (in s) cross section at high energies. If background were dual to Pomeron exchange and if, in addition, the triple-Pomeron coupling would be negligible at all (small) values of t [41], we should have

$$\left.\frac{d\sigma}{dtdM^2}\right|_{background} \propto s^{-1}, \qquad (15)$$

which vanishes at high energies. The claim was made that any cross section persisting to high energies would correspond to the production of resonant states. I have never had any sympathy for this proposal because I think it amounts to a definition of a resonance which seems unconnected with any preëxisting notion of what a resonance is. This is, of course, a matter of taste. A more objective objection is that all triple-Regge analyses now seem to indicate[42] that the triple-Pomeron coupling is relatively strong. Therefore, even if normal duality does apply to Pomeron-hadron scattering, the proposed definition of a resonance is lost.

A second try [remember that the first one is not disproved] was made by Einhorn, Green, and Virasoro[43]. An analysis of the dual resonance model suggested to them that two-component duality should be "abnormal" for Pomeron-hadron scattering, i.e. that s-channel resonances should be dual to t-channel Pomeron exchange. In this circumstance, we should have

$$\left.\frac{d\sigma}{dtdM^2}\right|_{resonance} \propto \frac{1}{M^2} , \qquad (16)$$

independent of the incident energy.

One test between these two alternatives has been proposed by Roberts and Roy[44] in a paper submitted to this Conference. Although I do not find their arguments entirely satisfactory, the evidence they cite is worth reviewing. They compared the reactions

$$\pi N \to (\pi, A_1, A_3)N \qquad (17)$$

with the corresponding K-induced reactions

$$KN \to (K, Q, L)N. \qquad (18)$$

They then argued that the triple-Regge diagrams [Fig. 20(a)] corresponding to Pomeron and f°-Reggeon exchange are distinguished by the different couplings of P and f° to external mesons, namely

$$\beta^P_{\pi\pi}/\beta^P_{KK} \simeq 1 \quad versus \quad \beta^f_{\pi\pi}/\beta^f_{KK} \simeq 2. \tag{19}$$

The data[45] indicate that the π-induced and K-induced cross sections are essentially equal. This, presumably, is a fact; but what of the interpretation? Roberts and Roy want to conclude that abnormal duality holds for Pomeron-hadron scattering. Such an interpretation is based upon two important assumptions.

(i) The states A_1^-, A_3, Q, L are resonances.
[That this is assumed is of course stated in
Ref. 44.]
(ii) The possible meson Regge exchanges in
Pomeron-hadron scattering are the familiar
ones with all their known idiosyncrasies,
e.g. the large f-f' splitting which gives
rise to the possibility of high-energy
SU(3) breaking.

Ordinarily, one would regard (ii) as perfectly reasonable, and would

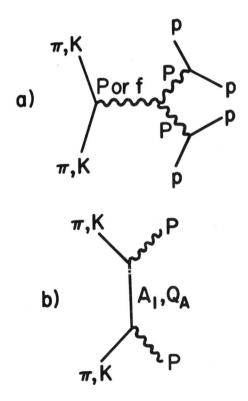

Fig. 20: Diagrams referred to in the discussion of the analysis by Roberts and Roy.

not go looking for trouble. However, since we are concerned with quite low missing masses, for which neglect of f'-exchange is not obviously justified, let us illuminate (ii) by means of an alternative (and equally strong) assumption. Consider the Pomeron-hadron s-channel of Fig. 20(b). Assume that π and K are members of an SU(3) octet, and that A_1 and Q are likewise members of an octet. Then because the Pomeron is an SU(3) singlet, the $\pi P A_1$ coupling and the KPQ coupling are equal. Therefore the corresponding high-energy production cross sections for $\pi N \to A_1 N$ and for $KN \to QN$ will be equal[46]. Consequently we are entitled to take the observed equality of pion- and kaon-induced cross sections as a success for SU(3) in the resonance region, and not for abnormal duality. To some extent, this discussion has been diabolical advocacy. However, I do think one should be conservative in drawing inferences about duality from cross section ratios when the missing mass is small. In my view, the implications of the data remain to be understood[47].

Another way to probe the connection between s- and t-channel descriptions is provided by the finite missing-mass sum rules (FMSR)[48]. These relations, which are entirely analogous to the t=0 FESRs for the imaginary parts of two-body elastic amplitudes [see eq. (2) and the discussion of Fig. 5], take the form

$$\int_{\underline{M}^2}^{\overline{M}^2} d(M^2)\,(M^2)^N \left\{ \frac{d\sigma\,(a+b\to c+X)}{d^3p/E} + (-1)^{N+1} \frac{d\sigma\,(c+b\to a+X)}{d^3p/E} \right\}$$

$$= [1+(-1)^{N+1}\tau(\overline{\alpha})]\,(s/\overline{M}^2)^{2\alpha_i(t)-1}(\overline{M}^2)^{\overline{\alpha}(0)+N}$$

$$\times \frac{|\beta\xi|^2 \cdot \beta_{bb}^{\overline{\alpha}}(0)\ g_{ii}^{\overline{\alpha}}(t)}{\overline{\alpha}(0) + N + 1 - 2\alpha_i(t)}. \tag{20}$$

[Here the quantity $g_{ii}^{\overline{\alpha}}(t)$ is the "familiar" triple-Reggeon coupling.] There are tentative indications from FMSR analyses of abnormal duality for the diffractively-produced states[49,34] in that a value $\overline{\alpha}(0) \simeq 1$ is favored. These conclusions of course hinge on the identification of such states as resonances.

CHEW-LOW REDUX

Inclusive analysis reminds us that the total cross section is not the only property of $\pi\pi$ scattering which is accessible in πN scattering. Figure 21 shows the Mueller diagram for the two-particle inclusive reaction

$$p + \pi \to n + c + \text{anything}, \tag{21}$$

which can be related to the single-particle inclusive reaction

$$\pi + \pi_V \to c + \text{anything}, \tag{22}$$

where π_V is a virtual pion of invariant mass-squared t, as follows.

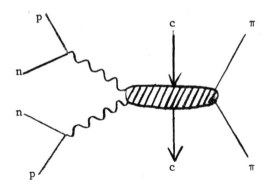

Fig. 21: Mueller–Regge diagram for reaction (21), which is appropriate in the triple-Regge limit for the subreaction p + π → n + anything.

In the right kinematical region [small t, large M^2] the connection is

$$\frac{1}{\sigma_{\pi\pi}(M^2,t)} \left. \frac{d\sigma(\pi + \pi_V \to c + \text{anything})}{dy_c} \right|_{M^2}$$

$$= \frac{d\sigma(p + \pi \to n + c + \text{anything})/dtd(M^2/s)dy_c}{d\sigma(p + \pi \to n + \text{anything})/dtd(M^2/s)} ,$$

(23)

where M is the missing mass from the detected neutron, $\sqrt{s}$ is the cm energy of the πp collision, and t is measured between proton and neutron. The importance of information on reaction (22) for the development of a microscopic theory of particle production cannot be overstated. I emphasize again that the measurements as they appear in (23) are of considerable interest, and that in this context the pole extrapolation can be regarded as a secondary goal.

ACKNOWLEDGMENTS

It is a pleasure to thank Dr. Paul Hoyer for many discussions of the topics presented here. For comments on triple-Regge analyses and on finite missing-mass sum rules I am grateful to Dr. R. D. Field, to Professor G. C. Fox, and to Dr. D. P. Sidhu. The Organizers are to be complimented for their efforts to make the Conference educational and pleasant.

FOOTNOTES AND REFERENCES

1. I thank Janos Kirz for help with data acquisition.

2. This argument is summarized by C. Lovelace, in *Phenomenology in Particle Physics - 1971*, ed. C. B. Chiu, G. C. Fox, and A. J. G. Hey (Pasadena: Caltech), p. 668.

3. M. G. Bowler, these proceedings.

4. D. Z. Freedman and J.-M. Wang, Phys. Rev. Lett. **17**, 569 (1966); Phys. Rev. **153**, 1596 (1967).

5. G. Veneziano, Nuovo Cimento $\underline{57A}$, 190 (1968). For a discussion specific to the $\pi\pi$ system, see J. Shapiro, Phys. Rev. $\underline{179}$, 1345 (1969).

6. See J. L. Rosner, SLAC-PUB-1391 for a recent review.

7. In the discussion period, S. J. Lindenbaum asked me to confess that I do believe weakly-coupled exotic states will be found at masses greater than about 2 GeV/c^2.

8. S. L. Glashow, these proceedings.

9. G. C. Fox, in *Experimental Meson Spectroscopy - 1972*, ed. A. H. Rosenfeld and K.-W. Lai (New York: AIP), p. 271; and Argonne Symposium on the Production of Heavy Excited Hadrons, March, 1974.

10. G. L. Kane, in $\pi\pi$ *Scattering - 1973*, ed. P. K. Williams and V. Hagopian (New York: AIP), p. 247; and Michigan preprint UM-HE-74-3.

11. P. Hoyer, Stony Brook preprint ITP-SB-74-11.

12. R. Dolen, D. Horn, and C. Schmid, Phys. Rev. $\underline{166}$, 1768 (1968).

13. P. Johnson, K. E. Lassila, P. Koehler, R. Miller, and A. Yokosawa, Phys. Rev. Lett. $\underline{30}$, 242 (1973).

14. F. Halzen and C. Michael, Phys. Lett. $\underline{36B}$, 367 (1971).

15. R. L. Kelly, Phys. Lett. $\underline{39B}$, 635 (1972).

16. G. Cozzika, *et al.*, Phys. Lett. $\underline{40B}$, 281 (1972).

17. G. C. Fox and C. Quigg, Ann. Rev. Nucl. Sci. $\underline{23}$, 281 (1973).

18. Strictly speaking, it is the behavior of the meson Regge pole (as distinct from Pomeron) exchange which should be anticipated by regularities in the resonance spectrum. According to the two-component theory of duality discussed below, low-energy resonances drive the high-energy Reggeon exchange, whereas low-energy background generates the Pomeranchuk singularity.

19. K. Igi and S. Matsuda, Phys. Rev. Lett. $\underline{18}$, 625, 822E (1967).

20. B. Hyams, C. Jones, P. Weilhammer, W. Blum, H. Dietl, G. Grayer, W. Koch, E. Lorenz, G. Lütjens, W. Männer, J. Meissburger, W. Ochs, U. Stierlin, and F. Wagner, in $\pi\pi$ *Scattering - 1973*, ed. P. K. Williams and V. Hagopian (New York: AIP), p. 206.

21. P. G. O. Freund, Phys. Rev. Lett. $\underline{20}$, 235 (1968); H. Harari, *ibid.*, p. 1395.

22. See C. Lovelace, Ref. 2, for more details and original references.

23. In a very extensive analysis of FESR calculations, M. R. Pennington has arrived at an asymptotic cross-section of 7^{+}_{-} mb.

24. E..L. Berger and G. C. Fox, Phys. Rev. **188**, 2120 (1969). I show model "A" of their Table III.

25. W. D. Walker, in $\pi\pi$ *Scattering - 1973*, ed. P. K. Williams and V. Hagopian (New York: AIP), p. 80.

26. C. Caso, F. Conte, G. Tomasini, P. Benz, P. Schilling, S. Ratti, D. Teodoro, G. Vegni, L. Mosca, and C. Lewin, Nuovo Cimento **3A**, 287 (1971).

27. A. Ukawa, Y. Oyanagi, and M. Fukugita, Tokyo preprint UT-206 (1973, unpublished).

28. C. Schmid, Phys. Rev. Lett. **20**, 628 (1968).

29. H. Harari and Y. Zarmi, Phys. Rev. **187**, 2230 (1969).

30. The odd partial waves, being uniquely $I_s=1$, will be similar for all values of I_t.

31. It receives the smallest $I_s=2$ contribution.

32. Here, in pursuit of the Chew-Low analogy, we discuss only the case in which a single trajectory is exchanged. The generalization is elementary.

33. G. F. Chew and F. E. Low, Phys. Rev. **113**, 1640 (1959).

34. R. D. Field and G. C. Fox, Caltech preprint CALT-68-434.

35. K. Abe, T. DeLillo, B. Robinson, F. Sannes, J. Carr, J. Keyne, and I. Siotis, contribution to the II International Conference on Elementary Particles, Aix-en-Provence, 1973.

36. P. Schlein, private communication to G. C. Fox.

37. F.-T. Dao, D. Gordon, J. Lach, E. Malamud, T. Meyer, R. Poster, P. E. Schlein, and W. E. Slater, Phys. Rev. Lett. **30**, 34 (1973).

38. P. Schlein is attempting to measure the πN total cross section in the reaction $p+p \to \Delta^{++}$ + anything at the ISR (private communication).

39. The first discussion of this point was given by G. F. Chew and A. Pignotti, Phys. Rev. Lett. **20**, 1078 (1968).

40. J.-M. Wang and L.-L. Wang, Phys. Rev. Lett. **26**, 1287 (1971); P. D. Ting and H. J. Yesian, Phys. Lett. **35B**, 321 (1971).

41. For a review of the Pomeron decoupling theorems, see R. C. Brower, C. E. DeTar, and J. H. Weis, CERN preprint TH-1817, to appear in Physics Reports.

42. A. B. Kaidalov, V. A. Khoze, Yu. F. Pirogov, and N. L. Ter-Isaakyan, Phys. Lett. 45B, 493 (1973); D. P. Roy and R. G. Roberts, RHEL preprint RL-74-022-T79; D. P. Sidhu, et al., to appear; and many more.

43. M. B. Einhorn, M. B. Green, and M. A. Virasoro, Phys. Lett. 37B, 292 (1971); Phys. Rev. D6, 1675 (1972).

44. R. G. Roberts and D. P. Roy, Phys. Lett. 47B, 247 (1973), and contribution to this Conference.

45. Yu. M. Antipov, et al., Nucl. Phys. B63, 141, 153, 175, 182, 189, 194, 202 (1973).

46. The same argument can be made in the Pomeron-hadron t-channel: PP can couple only to unitary singlet exchange.

47. It is interesting that the cross sections for the reactions $\pi N \to A_2 N$ and $KN \to K^{**}N$, which have Pomeron-like energy dependences at high energies, are approximately equal at 40 GeV/c, as required by the argument given here. [See Ref. 45.] It will be important to learn whether the energy-dependence and the equality of cross sections persists to higher energies.

48. M. B. Einhorn, J. Ellis, and J. Finkelstein, Phys. Rev. D5, 2063 (1972).

49. Chan Hong-Mo, H. I. Miettinen, and R. G. Roberts, Nucl. Phys. B54, 411 (1973).

Comment on Session

T. Ferbel

University of Rochester, Rochester, New York   14627

As session chairman I am obligated, at the very least, to try to offer words of wisdom concerning the subject of radiative widths of unstable elementary particles, and to remark on production processes which occur in the Coulomb field of nuclei, commonly referred to as Primakoff-production processes. Theoretical interest in the field has been growing lately, and judging from the fine presentations at this conference given by Gittelman and Rosen, the experimental work is fast approaching a high level of sophistication. Precision measurements of radiative widths are becoming available to test symmetry schemes and to confront predictions from quark models of elementary particles. I expect that within several years, experiments, particularly from NAL and CERN-II, will provide a great variety of measurements from which definitive radiative widths of particles, as well as new and exciting information on meson-photon and photon-photon scattering cross sections will be extracted. Although there has been a revival of theoretical interest in these processes, more detailed calculations concerning, for example, nuclear effects and non-Coulomb background amplitudes are still in order. This might also be an excellent time for enterprising theorists to calculate predictions for two-body decays of $J^P = 2^+$ mesons into photons and pseudoscaler mesons.

RADIATIVE WIDTHS AND PRIMAKOFF PRODUCTION

Jerome L. Rosen
Northwestern University, Evanston, Ill. 60201

I. TECHNIQUES

We begin our discussion of radiative widths by summarizing the techniques that have been employed to measure them.

1. $\gamma + x \to x^*$

Protons and neutrons are the only stable hadrons available as targets for photoproduction experiments. It would appear that the subject of baryonic radiative transitions is technically out of order at this conference. However, two very comprehensive treatments of such data have recently become available.[1,2] Photoproduction data comprises the largest portion of information concerning hadronic radiative transitions. The simple quark model seems to be fairly successful in predicting the signs of the various amplitudes.

2. $x^* \to x + \gamma$

A particular resonant state may be prepared and its radiative decay may be isolated by brute force, i.e. the radiative transition must be extracted from the potential background generated by strong decays which are generally 10^{2-3} times more copious. For example, consider the process $\rho^- \to \pi^- + \gamma$. The compiled upper limit to the radiative width is 0.5 mev.[3] The background is generated by $\rho^- \to \pi^- + \pi^0$ which is ~ 250 times more copious.

The above reasonable judgements are refuted by two notable exceptions. Nature has providently juxtaposed a large radiative decay with unusually small competing strong decay channels in the case of the ω^0 and φ^0 mesons. In fact, the processes $\omega^0 \to \pi^0 + \gamma$ and $\varphi^0 \to \eta^0 + \gamma$ are the only firmly established $V \to M + \gamma$ decays to date. Although we might have reasonably expected that some straightforward strong production channel would have served to produce the ω^0 and φ^0 states, these measurements were carried out by the storage ring reaction $e^- + e^+ \to V \to M + \gamma$.

3. $x + \gamma_c \to x^*$

In this approach, the coulomb field of a nucleus serves as an effective (zero frequency) photon target. The fact that photons are slightly off the mass shell is a minor and negligible technicality at high energies. Coherent x^* production by strong interaction with the nucleus is the principal technical difficulty. The coulombic process was first discussed by Primakoff[4] for the special case $\gamma + \gamma_c \to \pi^0$. The applicability to the η and η' was obvious once their existence was established. Professor Gittleman will discuss the pseudoscalar meson radiative width measurements in some detail at this conference.

A neglected point is the fact that 0^+ states can also be produced.[5] The reaction $\gamma + \gamma_c \rightarrow \varepsilon \rightarrow \pi + \pi$ has not been measured. The cross sections for this process is $\sim 10^3$ times that for $\gamma + A \rightarrow A + \rho^0$. The latter is a one photon process and has been extensively studied. Coherent production of a $\pi^0 + \pi^0$ final state could proceed only through the former channel. In principal the interference between the two channels in the $\pi^+ + \pi^-$ mode is measurable and some very clean information on the scaler dipion system (ε) could be obtained.

Good and Walker are responsible for the first qualitative discussion of the hadronic dissociation process in the nuclear coulomb field.[6] Although the experimental work has been meager in the intervening years, I hope to convince you by the end of this presentation that work in this field is about to go critical. The key is the availability of significantly higher energy beams. The coulomb field reaction cross section increases with energy while the nuclear coherent background channels decrease with energy.

4. For completeness we observe that 2 proton processes can also be observed in colliding beams.

$$e^- + e^+ \rightarrow e^- + e^+ + (\gamma+\gamma)_{virtual}$$
$$\downarrow$$
$$neutral\ mesons$$

Orito et al[7] have reported the first experiment of this type $(\gamma + \gamma \rightarrow \pi^+ + \pi^-)$. They find $\Gamma_{\varepsilon\gamma\gamma} = (9.6\ ^{+13.3}_{-8.0})$ kev.

II. VECTOR MESON-PSEUDOSCALER MESON RADIATIVE TRANSITIONS

We briefly summarize three of the theoretical ideas that have been employed to make predictions concerning $V \rightarrow M + \gamma$ radiative processes.[8]

1. <u>Pole Diagrams</u>. As an example, the radiative process $\rho \rightarrow \pi^+\gamma$ can be visualized as proceeding via $\rho \rightarrow \pi + \omega^0$ (virtual), followed by $\omega^0 \rightarrow \gamma$. The decay amplitude can be expressed in terms of the vector dominance model coupling constants.

2. <u>SU3</u>. The various radiative amplitudes can be related using SU3 symmetry exact and otherwise. Since the dynamics depend explicitly on the meson masses and in some cases mixing angles, the appelation "exact SU3" may be deceptive. These predictions are tabulated in Table I. The addition of symmetry breaking terms provides a large number of possibilities and auxiliary principals or prejudices must be invoked.

Table I

Exact SU(3) Predictions for Radiative Amplitudes

$$\langle\rho^\pm|\pi^\pm\gamma\rangle = \langle\rho^\circ|\pi^\circ\gamma\rangle$$

$$= \langle K*^\pm|K^\pm\gamma\rangle$$

$$= -\tfrac{1}{2}\langle K*^\circ|K^\circ\gamma\rangle$$

$$= \frac{1}{\sqrt{3}}\langle\omega_8|\pi^\circ\gamma\rangle$$

$$= -\langle\omega_8|\eta_8\gamma\rangle$$

$$= \frac{1}{\sqrt{3}}\langle\rho^\circ|\eta_8\gamma\rangle$$

$$\langle\omega_1|\pi^\circ\gamma\rangle = \sqrt{3}\,\langle\omega_1|\eta_8\gamma\rangle$$

$$\langle\omega_1|\eta_1\gamma\rangle = 0$$

$$\langle\rho_\circ|\eta_1\gamma\rangle = \sqrt{3}\,\langle\omega_8|\eta_1\gamma\rangle$$

3. <u>Naive Quark Model</u>. The SU3 predictions regarding relative decay amplitudes can be supplemented by a specific dynamical model such as the quark model. In this picture the radiative transitions are regarded as quark spin flip processes, i.e. magnetic dipole transitions. For example,

$$\langle\rho^-|\pi^-\gamma\rangle = \sqrt{\frac{2}{27}}\,\mu_p\,q$$

where $\mu_p = 2.79\ e/2M_p$. This gives the prediction $\Gamma(\rho^- \to \pi^- + \gamma) = 120$ kev.

Table II summarizes the published measurements.

Table II

	Γ_{exp} (kev)	Γ_{Quark} (kev)
$\omega^\circ \to \pi^\circ + \gamma$ [a]	890 ± 70	1160
$\omega^\circ \to \eta + \gamma$ [a]	130 ± 50	240
$K^{*+} \to K^+ + \gamma$ [b]	< 80 (95% C.L.)	80

[a]Reference 3, [b]Reference 9

III. THE GENERAL PRIMAKOFF PROCESS

The differential cross section for the process $x + \gamma_c \rightarrow x^*$ can be related to the corresponding photoproduction cross section $\sigma_\gamma(m^*)$ by the formula[10]

$$\frac{\partial^2 \sigma}{\partial t \partial m^{*2}} = \frac{Z^2 \alpha}{\eta \pi} \frac{\sigma_\gamma(m^*)}{(m^{*2}-m^2)} \frac{t'}{t^2} \, |F(t)|^2 \tag{1}$$

$$\eta = 1 \qquad \text{if } x \neq \gamma$$
$$\eta = \tfrac{1}{2} \qquad \text{if } x = \gamma$$

t = momentum transfer squared = $q^2 = t_o + t'$

$t' = q'^2 = (p\theta)^2$ = transverse momentum squared

$t_o = q_o^2 = [m^{*2}-m^2/2p]^2$

The quantity $F(t)$ is the "form factor" and is normalized such that $F(o) = 1$. In a sense, nuclear physics is concealed in this factor. There is a prevalent tendency to assume that form factors occuring in different types of coherent reactions are more or less similar. This is not precisely true. The structural form is quite different for electron scattering, **diffraction scattering**, **diffrac**tive dissociation, coulomb dissociation and ω^o exchange. The standard electromagnetic form factor is, to optical model, eikonal approximation,

$$F_{EM}(q) = \frac{1}{Ze} \int d^3r \; e^{i\vec{q}\cdot\vec{r}} \; \rho(\vec{r}) \tag{2}$$

For the Primakoff process however,

$$[\frac{t'}{t^2}]^{\frac{1}{2}} F(q) = \vec{q}' \cdot \vec{\mathcal{F}}(q)$$

where

$$\vec{\mathcal{F}} = \frac{1}{4\pi i} \int d^3r \; e^{-\frac{\sigma}{2}T(b)} \; e^{i\vec{q}\cdot\vec{r}} \; e^{i\chi_c(b)} \vec{E}(\vec{r})$$

We will not give the detailed definitions of the factors in this equation but rather only indicate the nature of the four successive terms in the integrand. The first factor takes into account the absorption of the incident (x) and final (x^*) hadronic systems by nuclear matter. For a heavy nucleus the factor is ≈ 0 for impact parameter $b \leqslant$ the nuclear radius. The second term is the usual coherent phase factor. The third term is the coulomb phase shift appropriate to charged hadrons. Finally, $E(\vec{r})$ is the electric field produced by $\rho(\vec{r})$.

If the system x^* is a resonance, formula (1) can be written in a second form

$$\frac{d\sigma}{dt} = \frac{8\pi Z\alpha^2}{\eta}\frac{(2J^*+1)}{(2J+1)}\Gamma(x^*\to x+\gamma)\left[\frac{m^*}{m^{*2}-m^2}\right]^3\frac{t'}{t^2}\,|F(t)|^2 \tag{3}$$

Thus the radiative width $\Gamma(x^*\to x+\gamma)$ is explicitly exhibited. In the practical world we must fit the coherent cross section for x production with an expression of the form

$$\frac{d\sigma}{dt} = |F_N + F_C|^2 \tag{4}$$

Here, F_N describes the coherent production amplitude for x^* formation by strong interaction with nucleus. The process $M + A \to A + V$ by ω^0 exchange is possible. Then F_N has the form

$$F_N = iC_{STR}^{\frac{1}{2}}\,A\,e^{i\varphi}\!\int d^3r\,e^{i\vec{q}\cdot\vec{r}}e^{-\frac{\sigma'}{2}T(b)}e^{i\chi_c(b)}\cdot(\vec{q}'\cdot\nabla)\rho(\vec{r}) \tag{5}$$

We note that if the transverse gradient factor $(\vec{q}'\cdot\nabla)$ were deleted the formula would describe diffractive dissociation. The quantity φ accounts for the intrinsic phase difference between the two amplitudes and A is the baryon number of the nucleus.

IV. THE MEASUREMENT OF $\pi^- + \gamma_c \to \rho^-$

We now describe a specific new measurement[11] which serves to illustrate some of the physics ideas previously discussed.[12] This experiment was carried out by a Northwestern-Rochester-C.M.U. group using the Lindenbaum-Ozaki spectrometer facility and an incident beam of 23 Gev/c π^-.

Figure 1 shows in highly schematic form, the detection geometry for measuring $\pi^- + A \to A + \pi^- + \pi^0$. Incident and final state π^- were measured with counters, hodoscopes and wire spark chambers. Detection of the final π^- was restricted to the upper half plane while the lower half plane was reserved for π^0 detection in an optical shower spark chamber. The target region was surrounded by an annulus of veto counters made of interleaved sheets of Pb and scintillator. This arrangement suppressed triggers with superfluous hadron production and/or incoherent nuclear excitation resulting in nuclear fragmentation.

Events were reconstructed assuming that the nuclear recoil energy was negligible as expected for the case of coherent reactions. The first signature of coherent reactions observed was the characteristic sharp Jacobion peak in the π^0 opening angle distribution. Afficionodo's of π^0 detection will recognize the invariant opening angle distribution plotted in figure 2. The sharp peak in effect, confirms longitudinal momentum balance. Note that there is background for R > 1 and in the unphysical region R < 1. This background results from the reaction $\pi^- + A \to A + \pi^- + \pi^0 + \pi^0$ which is predominantly coherent and is intrinsically two orders of magnitude larger than the coulomb field process. This channel provides background

when 2 of the 4 γ rays escape detection. Events with R < 1 correspond to the 2 detected γ rays originating from separate π^0's. The
yield of R < 1 events as a function of A agrees well with the yield
of coherent A_1 production as separately measured in the channel
$\pi^- + A \rightarrow A + (\pi^- + \pi^+ + \pi^-)$.

Figure 3 shows the cosmeticized R distribution after a background
subtraction using a Monte Carlo simulation based on the coherent A_1
production model for background generation. Similar curves exist for
other nuclear targets (C, Aℓ, Cu, Ag and Pb). A cut on R was ulti-
mately applied to the data (.95 < R < 1.125).

We digress to observe that two ancillary measurements were performed with the apparatus which served as control measurements
checking efficiency, acceptance and resolution. The first
of these, the decay in flight K^-(14 Gev/c) $\rightarrow \pi^- + \pi^0$ is topologically
similar to ρ^- decay and the decay rate is accurately known. The
second control was provided by the measurement of p(23 Gev/c) + $\gamma_c \rightarrow$
Δ^+.[13] The cross section $\sigma(\gamma + p \rightarrow \Delta^+)$ or equivalently $\Gamma(\Delta^+ \rightarrow p +^c \gamma)$
is well known. The Coulomb production of Δ^+ was found to agree with
the photoproduction data to an accuracy of ± 10%.

Figure 4 shows the mass spectrum of the small t' data for U and
Pb data combined. The yield is well fit by a ρ^- Breit-Wigner. The
fit analysis indicates that the final state is pure ρ^- with no more
than 10% intensity of other coulomb states. It should be noted that
coulomb production of states with masses higher than the ρ^- are
suppressed by the t_0 dependence built into the Primakoff formula.

Figures 5 and 6 show the measured distributions of decay angles
in the CM of the $\pi^- \pi^0$ system. The Gottfried-Jackson angles θ and ω
are displayed. The measurements are in good agreement with the ex-
pected $\sin^2\theta\sin^2\varphi$ distribution (for both Coulomb and ω^0 exchange).
The φ distribution for U, Pb data is a tour de force in that the
production plane is determined by the recoil momentum of the nucleus
($\vec{q}'$) which is typically ~ 50 mev/c. This momentum vector is extracted from measurements on the 23 Gev/c incident and final systems! Thus the azimuth angle distribution is a convincing demonstration of the experimental $\vec{q}'$ resolution.

The t' cross sections are displayed in Figure 7 in unormalized
form. Note the very small t' scale. The previously noted R cut has
been applied but the background has not been subtracted in this
figure. These yields are fitted with the aid of equations 3, 4 and
5 using Γ, C_{STR} and φ as free fitting parameters. Time does not
permit a detailed discussion of the fitting analysis. We limit ourselves to a few brief comments and summarized conclusions.

A typical shape fit is shown in figure 8. Considerable coupling
of the parameters is evidenced and only the destructive interference
solution φ = 180° ± 45° is clearly ruled out on this basis alone.

Figure 9 shows a fit to the total cross section as a function of
A. By total cross section we mean the integral of dσ/dt' from t'=0
to t'= (constant)$(R_{nucleus})^2$. It is seen that both the coulomb
process which scales as Z^2 and the saturating ω^0 process are in evidence. The saturation of the latter is the result of absorption.
A ρ^-N total cross section of 26 mb has been assumed in the optical

model along with other standard parameters for radius and surface diffuseness.

After weighing the consistency of the data, range of validity of the optical model etc. we conclude: the phase factor φ is in the vicinity of $270°$. $C_{STR} = (4.0 \pm 1.0)$ mb/(Gev/c)^4. This value is in reasonable agreement with a value of (4.9 ± 1.1)mb/(Gev/c)^4 obtained by extrapolating lower energy (16 Gev) hydrogen data. The radiative width $\Gamma(\rho^- \rightarrow \pi^- + \gamma) = (35 \pm 10)$ kev. This strongly disagrees with a simple SU3 prediction of 120 kev.

As noted earlier, the cross section for non resonant dipion production in the mass range $(0.5 - 1.0)$ Gev/c^2 is $(0 - 10)\%$.

V. CONCLUSION

We have observed that $V \rightarrow M + \gamma$ radiative transitions appear to be fundamental quantities directly related to the magnetic properties of the quarks or quark like structures which are believed to underly hadronic structure. The measurements to date are sparse. The newest measurement, $\Gamma(\rho^- \rightarrow \pi^- + \gamma)$ seems to contradict **naive predictions.**

Given the new availability of meson beams with energies greater than 50 Gev, present techniques are adequate for measuring a variety of radiative transitions by the coulomb dissociation technique to better than 10% accuracy. These transitions include

$$\pi^\pm + \gamma_c \rightarrow \rho^\pm$$
$$K^\pm + \gamma_c \rightarrow K^{*\pm}$$
$$K^0 + \gamma_c \rightarrow K^{*0}$$

Improved methods of photon detection will permit photon beam measurements of

$$\gamma + A \rightarrow A + V^0 \rightarrow \begin{array}{c} \pi^0 + \gamma \\ \eta^0 + \gamma \end{array}$$

where $V^0 = \rho^0, \omega^0, \varphi^0$. Some phase information could be gleaned from interference effects involving these amplitudes.

REFERENCES

1. R. Moorhouse, H. Oberlack, A. Rosenfeld, Phys. Rev. D, 1 (1974).
2. W. Metcalf, R. Walker, Cal. Tech. Report CALT-68-425 (1974).
3. Particle Data Group, Reviews of Modern Physics, April 1973.
4. H. Primakoff, Phys. Rev 81, 899 (1951).
5. J. Rosen, Proceedings of the Argonne Conference on Meson Spectroscopy (1969) unpublished.
6. M. Good and W. Walker, Phys. Rev. 120, 1855 (1960).
7. S. Orito, M.Ferrer, L. Paoluzi, R. Santanico, IV Experimental Meson Spectroscopy Conference.
8. W. Thirring, Physics Letters 16, 335 (1965); C. Becchi and G. Morpurgo, Physical Review 140B, 687 (1965); L. Soloviev, Physics Letters 16, 345 (1965); Y. Anisovitch et al., Physics Letters 16, 194 (1965); R. van Royer and V. F. Weisskopf, Nuovo Cimento 50A 617 (1967); A. Dar and V. F. Weisskopf, Physics Letters 26B, 670 (1968); L. M. Brown, H. Munczek, and P. Singer, Physics Review Letters 21, 707 (1968). For a review of the theoretical problems see G. Morpurgo, in Properties of the Fundamental Interactions (ed. A. Zichichi, 1972 Edelrice Composition, Bologna, Italy).
9. C. Bemporad et al., Nuclear Physics B51, 1 (1973).
10. A. Halprin, C. M. Anderson and H. Primakoff, Phys. Rev. 152, 1295 (1966).
11. B Gobbi, C. Meltzer, J. Rosen, H. Scott, S. Shapiro, L. Strawczynski (to be published).
12. This experiment has many physics and technical similarities with that described in reference 9.
13. R. Edelstein, B. Gobbi, E. Makuchowski, C. Meltzer, E. L. Miller, J. Rosen, H. Scott, S. Shapiro, L. Strawczynski (to be published).

330

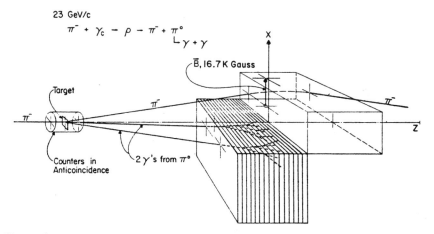

Fig. 1. Schematic diagram illustrating the basic geometry of the experimental arrangement.

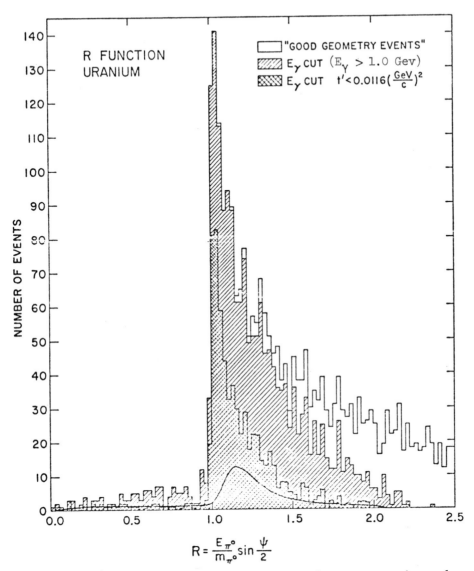

Fig. 2. R function for $\pi^- U$ data. The smooth curve superimposed on the lowermost spectrum is the result of a Monte Carlo simulation of the A_1 background.

332

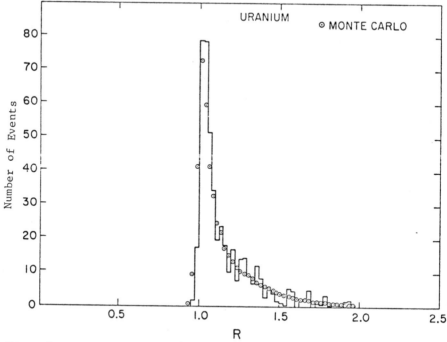

Fig. 3. R distrib. for the $\pi^- U$ data after applying cuts and subtracting the A_1 background. Superimposed are the Monte Carlo predictions.

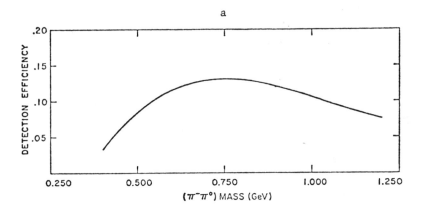

a

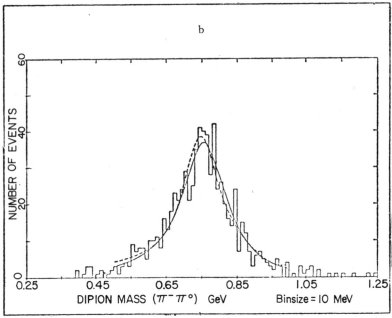

b

Fig. 4. a. Mass acceptance of the spectrometer.
b. $\pi^-\pi^0$ mass distribution of the low t' events on the Pb and U targets. The dashed curve is the best fit to the data for the hadronic model. The fit values are $m_0=0.754\pm0.004$ GeV and $\Gamma_0=0.130\pm0.009$ GeV. The solid curve is the best fit to the data for the Coulomb model. The fit values are $m_0=0.768\pm0.003$ GeV, $\Gamma_0=0.161\pm0.011$ GeV.

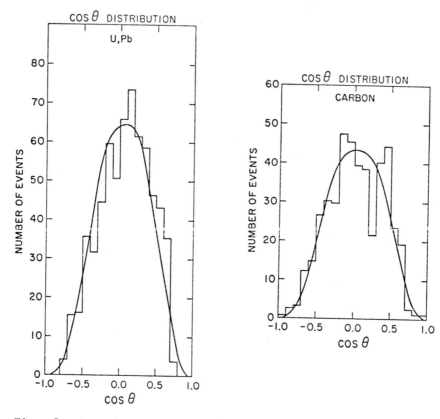

Fig. 5. Angular distribution in cosθ for the decay products π⁻ and π⁰ from ρ⁻.

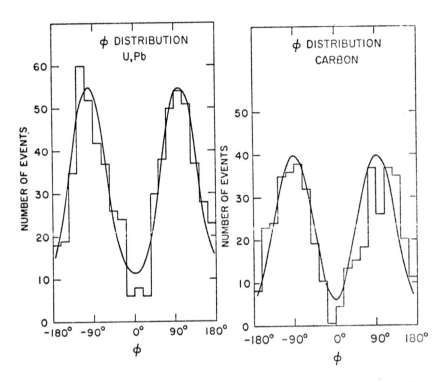

Fig. 6. Angular distribution in ϕ for the decay products π^- and π^0 from ρ^-.

336

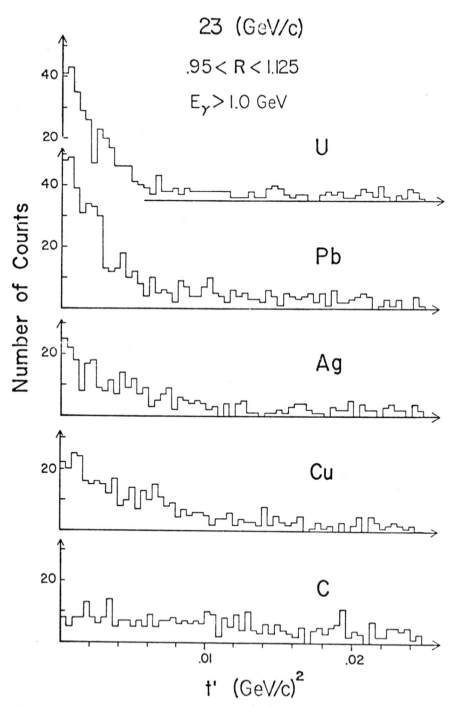

Fig. 7. Experimental t' spectrum for the elements U, Pb, Ag, Cu and C.

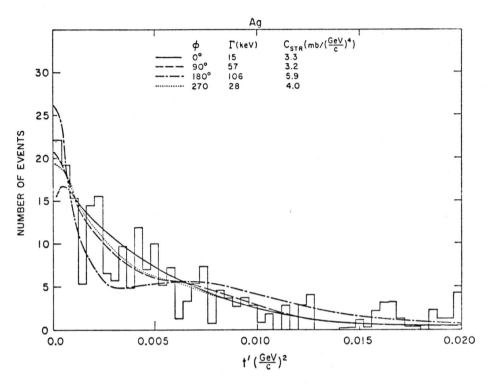

Fig. 8. Subtracted angular distribution for the Ag data. The curves are the result of fits to the Ag spectrum with the relative phase of the Coulomb and strong production amplitude fixed (ϕ = 0°, 90°, 180°, 270°).

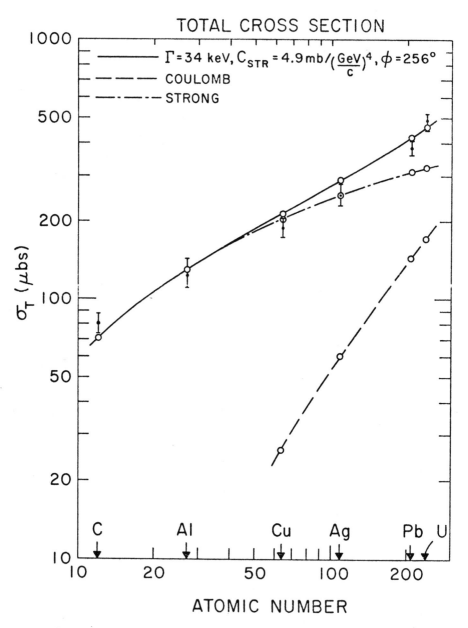

Fig. 9. Integrated total cross sections (t' < t'$_{max}$) together with the best fit obtained from the angular distributions.

$$t'_{max} = 0.44 \; A^{-2/3} \; (GeV/C)^2.$$

THE WIDTH OF THE NEUTRAL PSEUDOSCALAR MESONS[*]

Bernard Gittelman
Laboratory of Nuclear Studies, Cornell University
Ithaca, New York 14850

ABSTRACT

A discussion of the status of our knowledge of the π^0, η^0, and $X^0(960)$ width is given together with a brief description of recent measurements of these widths by a group at Cornell.

Our group at Cornell has measured the Primakoff effect for the π^0 and η^0 meson. An unsuccessful attempt was also made to measure the $X^0(960)$ by the same technique. I have been asked to describe this work and to summarize how it changes the current experimental situation on the width of these particles. The Cornell people participating in this are A. Browman, J. DeWire, K. Hanson, D. Larson, E. Loh, R. Lewis, and myself. The experiments consist of measuring the small angle photoproduction cross section of the meson on several nuclei at two or three energies in order to isolate the Primakoff amplitude. A schematic of the setup is shown in Figure 1. A collimated bremsstrahlung beam passes through a target of about 0.05 radiation length thickness, continues through a sweep magnet, between a pair of shower counter hodoscopes, and is stopped in a quantameter. Charged particles produced in the target are swept clear of the counters. The photon pairs from meson decays (for example, $\pi^0 \rightarrow 2\gamma$) are detected in the shower counters. The counters have a useful aperture of 36.6 cm. horizontally by 27.4 cm vertically. An event trigger is generated if more than 1 GeV of energy is deposited in each counter. The energy of each photon ($\delta E/E = \pm 0.11/\sqrt{E}$ (E in GeV), rms) and the position at which it entered the counter ($\delta s = \pm (3.6+7.0/E)$ mm, rms) are recorded for each event. Assuming that the photons came from the target, one constructs their momentum vectors. From these, the mass and momentum vector of the two photon system may be calculated. In Figure 2 several mass squared spectra from the π^0 experiment with a machine energy of 6.6 eV are shown. A π^0 peak sitting on top of a small background is seen. Events in the peak are used to establish the experimental angular distribution.

A detailed theory for the small angle photoproduction cross section on complex nuclei has been given by Morpurgo[1] and Faldt[2]. The coherent cross section is described by the sum of an amplitude for production in the Coulomb field and an amplitude for production in the nuclear field.

$$d\sigma/d\Omega = |T_c+T_n|^2$$

[*]Supported by the National Science Foundation.

This may be written as the sum of a term for Coulomb production, $d\sigma/d\Omega)_c$, a term for nuclear production, $d\sigma/d\Omega)_n$, and an interference term, $d\sigma/d\Omega)_{cn} = 2 \text{ Re}(T_c T_n^*)$. The determination of the magnitude of each of these amplitudes by measuring the cross section for several targets at two or three energies is possible because of their distinct dependence on production angle, atomic number, and energy. Angular momentum conservation requires both amplitudes to vanish in the forward direction. The Coulomb amplitude grows rapidly with angle, reaching a maximum at $\theta \simeq m^2/(2k^2)$ and then falls to zero. The nuclear amplitude has a similar angular dependence. However, the limited range of the nuclear field leads to a broader distribution. The nuclear amplitude reaches its maximum near $2/(KR)$, where R is the nuclear radius. Thus the angular distribution of Coulomb production is almost independent of atomic number, whereas the nuclear production is spread over a range proportional to $A^{-1/3}$. The distributions are illustrated in Figure 3 where $1/z^2 (d\sigma/d\Omega)_c$ and $1/z^2 (d\sigma/d\Omega)_n$ are plotted for several nuclides. The energy dependence of the nuclear production is not specified without making an assumption about the energy dependence of the spin non-flip nucleon amplitude. We have found it adequate to assume that the square of this amplitude has the same energy dependence as the hydrogen cross section (i.e. $d\sigma/d\Omega$(hydrogen) $= f(t-t_{min})$). The energy dependence of Coulomb and nuclear production are illustrated in Figure 4 where the integrated cross sections divided by Z^2 are plotted vs. Z.

A description of the forward cross section must also include a term to describe incoherent production. The angular variation of this term is gentler than that of the coherent processes. We have taken it to be isotropic. The incoherent cross section should be approximately proportional to the number of nucleons, A. Including reabsorption effects reduces this somewhat. We have found $A^{0.75}$ gives a satisfactory description of large angle data. Including an appropriate normalization constant for each component of the cross section and introducing an angle for the relative phase between the Coulomb and nuclear amplitudes, the cross section becomes

$$d\sigma(\theta)/d\Omega = C_c (d\sigma(\theta)/d\Omega)_c + C_n (d\sigma(\theta)/d\Omega)_n$$
$$+ \sqrt{C_c C_n} \ (d\sigma(\theta,\phi)/d\sigma)_{cn} + C_b (d\sigma(\theta)/d\Omega)_b$$

THE WIDTH OF THE π^0 MESON

Knowledge of the π^0 width is based on the half dozen experiments listed in Table I. The experiments cover three independent techniques. Von Dardel was able to deduce the π^0 lifetime by measuring the number of positrons emerging from a thin foil inside the CERN proton synchrotron as a function of the foil thickness. Stamer measured the distance between birth ($K^+ \rightarrow \pi^+ + \pi^0$) and death ($\pi^0 \rightarrow e^- + e^+ + \gamma$) of a sample of 232 events in an emulsion stack. The remaining values come from measurements of the Primakoff effect. In an abstract submitted to this conference, the Cornell group has reported a preliminary value of 7.3±0.8 eV for this width. In this experiment data were recorded for bremsstrahlung endpoint energies

of 4.4 and 6.6 GeV. At 4.4 GeV, in addition to running with the
detectors straddling the beam line at 0°, we also recorded data with
the counters displaced by 15 milliradians from the beam line. This
enabled us to pick up the maximum in the coherent nuclear signal,
thus providing a more definitive measurement of the magnitude of
this contribution. Our procedure for fitting the angular distributions consisted of integrating the differential cross section over
the bremsstrahlung spectrum, folding in the detector acceptance and
angular resolution. The resulting expression for the number of
events in each angular bin can be expressed as

$$N(\theta) = C_c N_c(\theta) + C_n N_n(\theta) + \sqrt{C_c C_n}\, N_{cn}(\theta,\phi) + C_b N_b(\theta)$$

The coefficients C_c, C_n, C_b, and ϕ are the parameters that are
varied in order to fit the data. To extract unique answers the data
from five targets in a given data set are fitted simultaneously. All
of the data has also been fitted with a single set of values for C_c
and C_n, while allowing ϕ and C_b to take on different values at the
two energies. The results of the fits are given in Table II. The
fitted angular distributions along with the individual contributions
to $N(\theta)$ are shown in Figures 5, 6, and 7.

THE WIDTH OF THE η^0 MESON

There have been two measurements of the width of the η^0 meson,
both based on the Primakoff effect. In 1967, Bemporad et al.
reported a value for the partial width $\Gamma(\eta \to 2\gamma) = 1.0 \pm 0.2$ KeV.[9] A
recently reported Cornell measurement[10] gave $\Gamma(\eta^0 \to 2\gamma) = 0.324 \pm 0.046$
KeV. The experiment of Bemporad et al. was carried out at
bremsstrahlung energies of 4.35 and 6.0 GeV with targets of Zn, Cu,
and Pb. The Cornell experiment was made at bremsstrahlung energies
of 5.8, 9.0, and 11.5 GeV with targets of Be, Al, Cu, Ag, and U. A
sample of the angular distributions at 9.0 GeV is shown in Figure 8.
During the past year we have tried to understand the difference
between the two results.[11] Using the Cornell data sample at 5.8
GeV, events were reselected to closely imitate the geometry of the
earlier experiment. An analysis, which paralleled that made by
Bemporad et al., was made on these events. The number of events in
the forward peak per incident equivalent quanta was found to be
approximately the same in the two experiments, although there was an
uncertainty about how to treat the background in the Cornell data
sample. On the basis of this study we concluded that the discrepancy between the two results could not be explained by some gross
experimental error (such as beam monitor, detector efficiency, dead
time, etc.).
Next we calculated the number of events expected in the several
counter pairs of the Bemporad et al. experiment. The calculation
was done both for the published cross section values of their experiment, and for the values found in the Cornell experiment. The
results are illustrated in Figure 9. The data points are those
published by Bemporad et al. The straight line labeled "bkgnd"
is an arbitrarily drawn background level. The dashed curve is the

Table I. Experiments Measuring the π^0 Width

Author	Value (eV)	Year	Reference
Von Dardel	6.3±1.1	1963	3
Bellettini	9.0±1.1	1965	4
Stamer	6.6±3.3	1966	5
Bellettini	11.8±1.3	1970	6
Kryshkin	7.2±0.6	1970	7
Browman	7.3±0.8	(Preliminary)	8
Weighted Average	7.7±0.4		

Table II. Fits to the π^0 Angular Distribution

Data Set	$\Gamma(\pi^0 \to 2\gamma)$ (eV)	C_n	$\|\Phi\|$ (radians)	C_b	χ^2/DF
4.4 GeV on Axis	7.1±0.1	1.30±0.27	1.03±0.17	3.1±0.5	203/190
4.4 GeV off Axis	7.4±0.3	1.40±0.05	1.18±0.08	4.0±0.4	236/215
6.6 GeV	7.5±0.1	1.25±0.14	0.88±0.14	7.9±0.8	219/210
All Data Combined	7.32±0.07	1.40±0.04	1.16±0.04[+] 0.84±0.07[†]	3.9±0.6[+] 8.2±1.5[†]	670/623

[+] 4.4 GeV
[†] 6.6 GeV

result of the calculation using the values of Bemporad et al.
($\Gamma(\eta \rightarrow 2\gamma)$ = 1.0 KeV, C_n = 0.). The solid curve is the result
using the Cornell values ($\Gamma(\eta^0 \rightarrow 2\gamma)$ = .33 KeV, C_n = 1., ϕ = 0°).
The dashed curve is a better fit to the data, but we believe that
the solid curve is not ruled out by 3.5 standard deviations. One
might ask how the two calculated curves can look so much alike in
view of the distinct differences illustrated in Figure 2. To answer
this, one must appreciate the effect introduced by the interference
term and the smearing produced by the detector angular resolution.
Our conclusion is that the quoted error in the paper of Bemporad et
al. is too small.

THE WIDTH OF THE X(960) MESON

Yesterday morning, Dr. Binnie described a measurement of the
width of the X(960) by their group at Rutherford.[12] They observe
the reaction $\pi^- + P \rightarrow X^0 + N$ near threshold. They detect the neutron
with a time-of-flight system to identify the reaction. The fine
mass resolution of the apparatus is pegged on the momentum resolution of the incident beam. They obtain a 2.5 std. deviation upper
limit on the X^0 width of 0.8 MeV. In the same experiment they detect the $X^0 \rightarrow 2\gamma$ decay mode in a pair of shower counters and obtain
a value of (2.5 ± 0.7)% for the 2γ branching ratio.
The Cornell group has tried to measure the X^0 width with the
same apparatus used for the π^0 and η^0 experiments. They found themselves limited by a target related background in the X^0 mass region.
No X^0 mass peak was seen above the background. The sensitivity of
the experiment was not sufficient to provide an interesting upper
limit on the X^0 width (using 2.5% for the two photon branching ratio,
the experimental upper limit on the total width was $\approx$10 MeV). We
intend to make another attempt at measuring the X^0 width using the
Primakoff technique by detecting the ρ-γ decay mode.

THEORETICAL CONSEQUENCES

Our measurement of the π^0 width has no new theoretical implications in the sense that it does not change the commonly accepted
value of this quantity.[13] Those theorists who were happy with 7.8
eV may remain so. Perhaps our experiment has made this value somewhat more compelling. (Two recent papers on the theory of the π^0
width are given in references 14 and 15.)
The Cornell value for the η^0 width has some impact on the
SU(3) mixing model of the pseudoscalar mesons. Recent reviews of
this subject are listed in references 16, 17, and 18. Within the
framework of this model, the partial width of the neutral pseudoscalar mesons into photon pairs is given by

$$\Gamma(\pi^0 \rightarrow 2\gamma) = 3m_\pi^3 g^2$$

$$\Gamma(\eta^0 \rightarrow 2\gamma) = m_\eta^3 (+g \cos(\alpha) + f \sin(\alpha))^2$$

$$\Gamma(X^0 \rightarrow 2\gamma) = m_X^3 (-g \sin(\alpha) + f \cos(\alpha))^2$$

Here α is the octet-singlet mixing angle, and g (f) is the coupling of the SU(3) octet (singlet) to the photons. The factor of 3 in the π^0 rate comes from U spin conservation. A measurement of all three decay rates would provide enough information to independently determine f, g, and α. Presumably, the value α so obtained would be the same as derived from the mass formula and one could resolve the famous problem of quadratic vs. linear masses. Lacking the $X^0 \rightarrow 2\gamma$ decay rate, we can turn the problem around and use the mixing angle obtained from the mass formula to predict this width. In Table III values for f/g, $\Gamma(X^0 \rightarrow 2\gamma)$, and $\Gamma(X^0)$ are given for each case. In working out these numbers, an alternative solution to the above quadratic equations was dismissed as having been ruled out by the upper limit of Binnie et al.

Table III. $\eta-\eta'$ Mixing Model Prediction for the X^0 Width

Using $\Gamma(\pi^0 \rightarrow 2\gamma) = 7.7 \pm 0.4$ eV
$\Gamma(\eta^0 \rightarrow 2\gamma) = 324 \pm 46$ eV

	Linear Mass Formula	Quadratic Mass Formula
α	23.8°	10.6°
f/g	1.1 ± 0.3	2.1 ± 0.7
$\Gamma(X^0 \rightarrow 2\gamma)$, KeV	0.39 ± 0.26	3.4 ± 1.8
$\Gamma(X^0)$, KeV	16 ± 10	136 ± 72

REFERENCES

1. G. Morpurgo, Nuovo Cimento 31, 569 (1964).
2. G. Faldt, Nucl. Phys. 43, 591 (1972).
3. G. Von Dardel et al., Phys. Letters 4, 51 (1963).
4. G. Bellettini et al., Nuovo Cimento 40, 1139 (1965).
5. P. Stamer et al., Phys. Rev. 151, 1108 (1966).
6. G. Bellettini et al., Nuovo Cimento 66A, 243 (1970).
7. V. I. Kryshkin et al., Soviet JETP 30, 1037 (1970).
8. A. Browman et al., Phys. Rev. Lett. in preparation. Since this conference, further corrections have raised the value of the width to 7.6 eV. There is a small reduction in the quoted error.
9. C. Bemporad et al., Phys. Letters 25B, 380 (1967).
10. A. Browman et al., Phys. Rev. Lett. 32, 1067 (1974).
11. Almost all of the information we required to compare the two experiments is contained in Ref. 9. The two exceptions are the quantameter calibration constant, 3.35×10^{15} GeV/Coulomb, and the target thicknesses, Zn = 1.02 mm; Ag = 0.51 mm; Pb = 0.52 mm. We wish to thank the various members of the Bonn-Pisa collaboration for supplying this information and for openly discussing their experiment with us (5 years after publication!!).
12. A. Duane et al., Phys. Rev. Lett. 32, 425 (1974).
13. Particle Data Group, "Review of Particle Properties", Rev. Mod. Phys. 45, No. 2, Part II, April 1973.
14. S. D. Drell, Phys. Rev. D7, 2190 (1973).
15. P.G.O. Freund and S. Nandi, Phys. Rev. Lett. 32, 181 (1974).
16. F. D. Gault et al., "Review of the η-η' Mixing Problem", Univ. of Durham preprint, Durham, England.
17. A. Kotlewski et al., Phys. Rev. D8, 348 (1973).
18. A. Bramon and M. Greco, Phys. Letters 48B, 137 (1974).

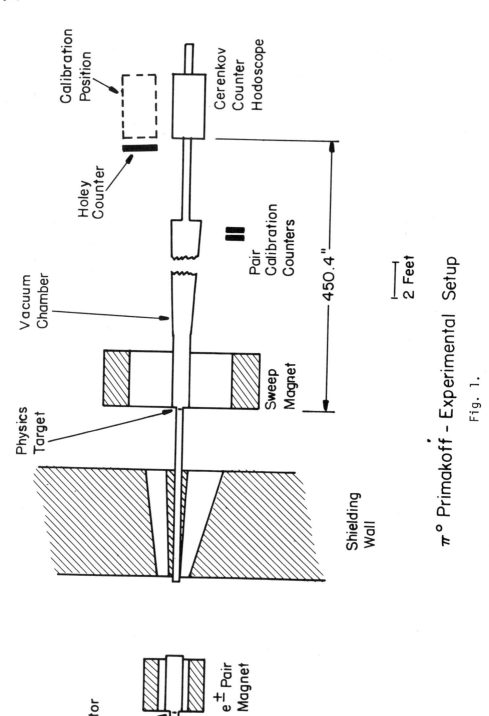

$\pi°$ Primakoff - Experimental Setup

Fig. 1.

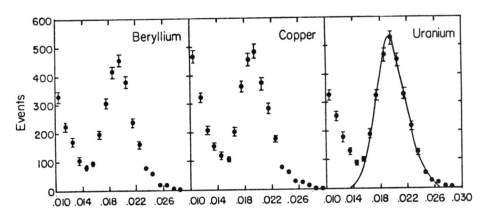

Two Photon Mass Squared (GeV^2)

Fig. 2. Two photon mass squared spectra from the π^0 experiment. The solid curve shown on the uranium spectrum was calculated using the measured energy and spatial resolution of the counters.

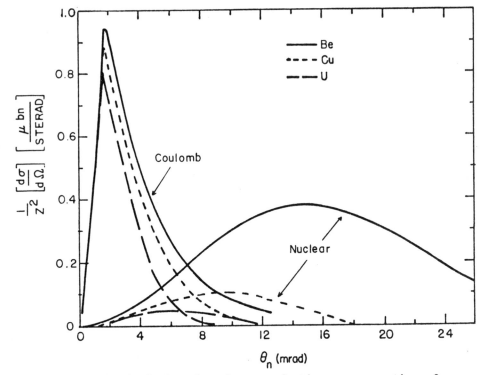

Fig. 3. The Coulomb and nuclear production cross sections for mesons with 8.7 GeV incident photons. The Coulomb cross section has been plotted for $\Gamma(\eta^0 \to 2\gamma) = 0.1$ KeV.

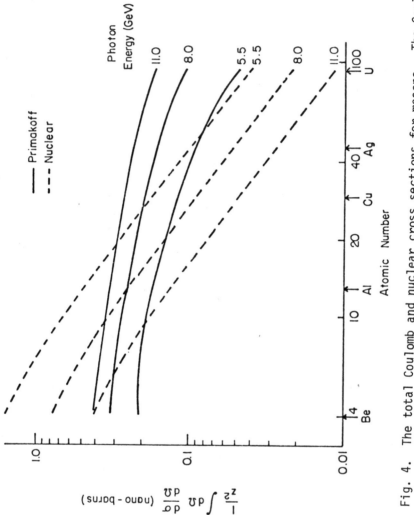

Fig. 4. The total Coulomb and nuclear cross sections for mesons. The Coulomb cross section has been plotted for $\Gamma(\eta \rightarrow 2\gamma) = 1.0$ KeV.

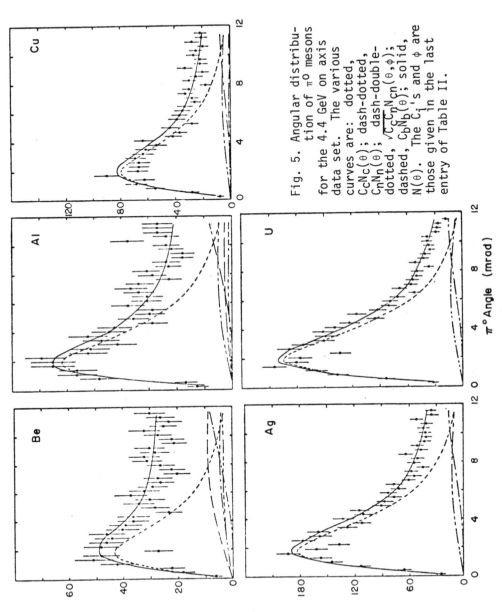

Fig. 5. Angular distribution of π^0 mesons for the 4.4 GeV on axis data set. The various curves are: dotted, $C_c N_c(\theta)$; dash-dotted, $C_n N_n(\theta)$; dash-double-dotted, $\sqrt{C_c C_n} N_{cn}(\theta,\phi)$; dashed, $C_b N_b(\theta)$; solid, $N(\theta)$. The C_i's and ϕ are those given in the last entry of Table II.

Events

π° Angle (mrad)

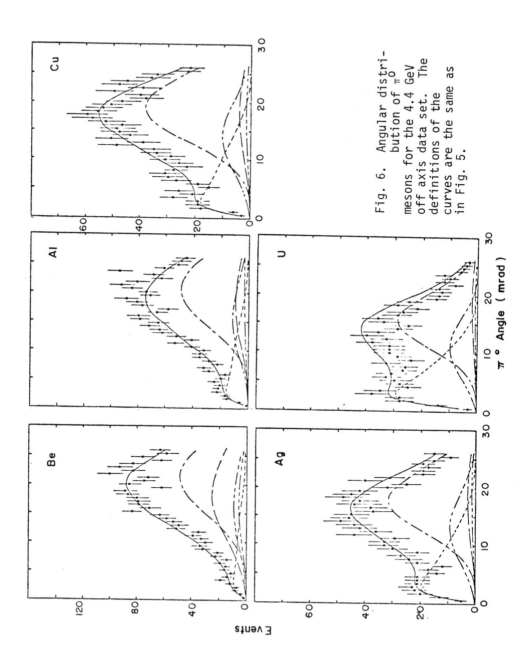

Fig. 6. Angular distribution of π^0 mesons for the 4.4 GeV off axis data set. The definitions of the curves are the same as in Fig. 5.

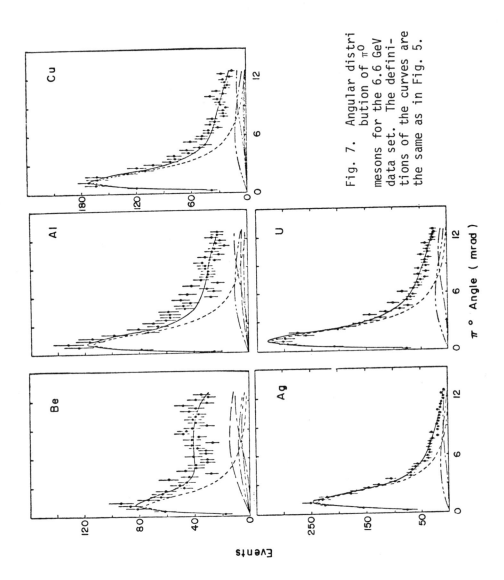

Fig. 7. Angular distribution of π^0 mesons for the 6.6 GeV data set. The definitions of the curves are the same as in Fig. 5.

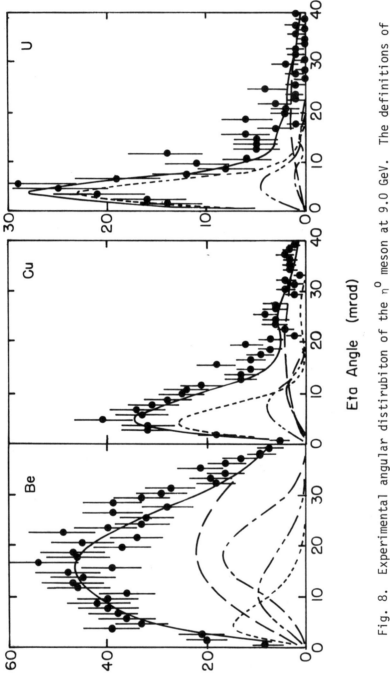

Fig. 8. Experimental angular distribuiton of the η^0 meson at 9.0 GeV. The definitions of the curves are the same as in Fig. 5 with $C_c = 0.31$; $C_n = 1.25$; $\phi = -0.63$; and $C_b = 3.1$ (see Ref. 10).

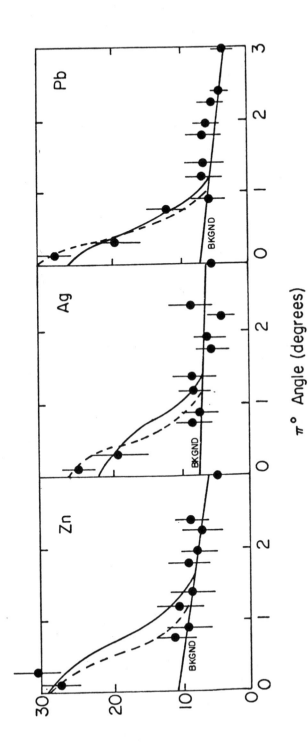

$\pi°$ Angle (degrees)

Fig. 9. Comparison of the 6.0 GeV data of Bemporad et al. with the rate expected for pure Primakoff production (dashed), and for a sum of Primakoff and nuclear (solid). See the text.

REVIEW OF SPIN-PARITY ANALYSES OF $\pi\omega$ AND Kω SYSTEMS[*]

S.U. Chung
Brookhaven National Laboratory, Upton, New York 11973

The purpose of this talk is to give a review of the progress in spin-parity (J^P) analyses on the $\pi\omega$ system.[1] In addition, first attempts at a similar analyses on the Kω system near the threshold are presented and its production and decay properties compared with those of the $\pi\omega$ system.

There are five new works on the $\pi\omega$ J^P analysis, one[2] a re-analysis of the work presented at the NAL Conference and the other four,[3-6] submitted to this conference, representing new endeavors on this subject. All these works find that the most likely J^P for the B is 1^+. In particular, Chung, et al.,[3] Chaloupka,[4] and Morrison, et al.,[6] have independently performed mass-independent partial-wave analyses in which the possibility of interference among different J^P states is explicitly taken into account. They find that the B region is dominated by a 1^+ wave, produced mainly by natural-parity exchange. The same conclusion is also reached by Karshon, et al.,[5] who have performed a more conventional background-subtracted moment analysis.

One may ask the question: Does a $\pi\omega$ system around 1250 MeV harbor just one resonant 1^+ wave? Although a preponderance of evidence coming from πp interactions indicates that the 1^+ wave is the only one with a "bump" structure in this region, one cannot as yet give a definitive answer due mainly to the fact that the partial-wave solutions are not unique if interference effects are allowed in the analysis. Chung, et al.,[3] find that an adequate description of their data can be achieved with the waves 2^-, 1^- and 0^- in and below the B region (N.B. no 1^+ wave in the solution!). There are unpleasant aspects to this solution, such as a large ρ_{22} component for the 2^- state and the possibility of two resonances 2^- and 1^- in the B region. Also the goodness of fit to the data for this solution is not as good as that for the 1^+ solution. However, they feel that the solution cannot be ruled out with mathematical certainty at this time.

There are hints that a 1^- object, so-called $\rho'(1250)$, might exist in the $\pi\omega$ system in the vicinity of the B. Frankiel, et al.,[7] from study of the reaction $\bar{p}p$(at rest) $\rightarrow \pi^+\pi^-\omega$, find that the data require both 1^+ and 1^- objects in the B region. It would be of some interest to see if the situation persists when the assumption of S-wave capture of the $\bar{p}p$ system is relaxed (this is what they assumed in the analysis). On the other hand, Ballam, et al.,[8] observe an enhancement at 1.24 GeV in the reaction $\gamma p \rightarrow p\pi^+\pi^-$ MM at 2.8, 4.7 and 9.3 GeV (see Fig. 1). They argue that the

---

* Work performed under the auspices of the U.S. Atomic Energy Commission.

most likely decay mode of the object "B" is $\pi^0\omega$ and, in addition, they claim that the production cross-section is nearly constant as a function of the photon energy, suggesting Pomeron exchange. Then the simplest explanation of their data would be the presence of a diffractively produced 1^- object, namely the $\rho'(1250)$. However, one cannot rule out the possibility that the genuine $J^P=1^+$ B-meson is produced diffractively in this reaction, violating the Morrison-Gribov rule. An additional hint for the $\rho'(1250)$ comes from the e^+e^- experiments. Conversi, et al.,[9] in a paper submitted to this Conference, argue that they see evidence of the $\rho'(1250) \to \pi^0\omega$ in the reaction $e^+e^- \to \pi^+\pi^-\pi^0\pi^0$. The statistics, as is seen in Fig. 2, is too meager to be taken seriously; more data are clearly needed to clear up the situation here.

For completeness, a few relevant details are given from the papers on the B analysis submitted to the Conference. The $\pi\omega$ mass spectra from these papers are shown in Figs. 3a-c, 4a, 5a, and 7a. The mass of the B is typically found to be in the range 1220 to 1240 MeV and the width in the range 130 to 160 MeV. An important parameter from the theoretical point of view is the decay amplitudes of the $J^P=1^+$ B-meson. They can be expressed in terms of either the helicity decay amplitude F_λ (λ is the ω helicity) or the orbital angular momentum for the $\pi\omega$ decay (D and/or S wave for the $J^P=1^+$ object). The measured values of these are given in Table I; it is seen that $|F_0|^2$ is relatively small. However, according to Chung, et al.,[3] $|F_0|^2$ is definitely non-zero. The best evidence comes from a plot of the unnormalized moment which is proportional to $|F_0|^2$ (see Fig. 5c); this plot shows a bump structure in the B region, demonstrating that $|F_0|^2$ cannot be zero. From a partial-wave analysis of the $\pi\omega$ system, they quote $|F_0|^2 = 0.16 \pm 0.04$ for the $J^P=1^+$ B-meson.

Rosner,[10] in discussing the $SU(6)_W$ classification scheme, asserts that the $|F_0|^2$ is the only free parameter describing nearly all decays of the 0^+, 1^+ and 2^+ families. He finds that a satisfactory description of these decays emerges using the value 0.13 for $|F_0|^2$. In addition, there is the work of P. Shen and K. Kang,[11] who did a N/D dynamical calculation of the $\pi\omega$ scattering process. They find a resonant state in the 1^+ wave and predict 0.18 for the ratio $|D/S|$ for this state, which is compatible with the experimental measurements (see Table I).

Turning now to some further details of the J^P analyses, one can see that the $\pi\omega$ region below and in the B is nearly all in the 1^+ state, as seen in Figs. 6 and 7a (the points with error bars). Chung, et al.,[3] quote for the density-matrix elements $\rho_{mm'}$ of the 1^+ state: $\rho_{11} = 0.23 \pm 0.02$, $\rho_{1-1} = -0.13 \pm 0.03$, and $Re\rho_{10} = 0.09 \pm 0.02$, evaluated in the s-channel helicity frame with $t' < 1$ GeV^2. Note that both ρ_{00} and $\rho_{11} - 1-1$ are large and significantly non-zero; this indicates the importance of the natural-parity exchange in the B production. The same conclusion is reached by Chaloupka,[4] as is seen in his plots of $N\rho_{00}$, $N(\rho_{11}-\rho_{1-1})$ and $N(\rho_{11}+\rho_{1-1})$ for the 1^+ state (Figs. 7b, 7c and 7d). In addition, Karshon, et al.,[5] reach the same conclusion from their analysis of the background-subtracted moments. Finally, it should be pointed out that all the analyses find a sharp forward peak in the $d\sigma/dt'$ distribution for the 1^+ state (see Fig. 8) and that the slope of the t' distribution is roughly $3\sim4$ for $t' < 1$ GeV^2.

No extensive partial-wave analysis of the $K\omega$ system has yet been performed due mainly to paucity of statistics. First attempts at the $K\omega$ analysis, as submitted to this Conference, have been made by Protopopescu, et al.,[12] at BNL and by the ABBCHLV collaboration.[6] Their analyses of the threshold enhancement (see Fig. 9) find that the 1^+ wave is dominant. In addition, Protopopescu, et al., quote $|D/S| = 0.08 \pm 0.08$ and $\rho_{00}(1^+) = 1.0 \pm 0.1$ for $M(K\omega)$ in the range 1.24 to 1.40 GeV. Thus, it is seen that a predominantly S-wave 1^+ state is produced (presumably) by Pomeron exchange. Further evidence of Pomeron exchange is afforded by the p_{LAB}^{-n} behavior of the $K\omega$ cross-section. As shown in Fig. 10 (provided by Morrison[6]), $n \approx 0$ for the $K\omega$ threshold region, indicating Pomeron exchange for the $K\omega$ system. In contrast, the $\pi\omega$ threshold region has a markedly different cross-section dependence ($n \approx 2.3$), indicating importance of exchange other than the Pomeron.

The situation regarding the $\pi\omega$ system may be summarized as follows. The B-meson has $J^P=1^+$ with the $|D/S|$ ratio between 0.2 and 0.3, and the B production from πp interactions is mostly via natural-parity exchange with a sharp forward peak in the t' distribution. The spin-parity for the B is most unlikely to be anything other than 1^+, although it is not impossible to fit the B region with a solution (rather unsatisfactory in some respects) containing little 1^+ wave. The situation regarding the B region from the $\bar{p}p$(at rest), γp and e^+e^- interactions is less satisfactory. Here a real possibility exists for a 1^- object in the B region; however, the evidence is not yet compelling, and a more sophisticated analysis on more abundant statistics (e^+e^- in particular) is sorely needed.

The $K\omega$ threshold region is found to be mainly in the $J^P=1^+$ state. However, unlike that of the $\pi\omega$ system, the $K\omega$ production from K^-p interactions is almost purely via Pomeron exchange. From the generalized Morrison-Gribov rule, then, the $K\omega$ threshold region has its C parity positive, the same as that of the K^- incident particle. Therefore, even if the $K\omega$ threshold enhancement turns out to be dominated by a genuine resonance, it is not likely to belong to the same octet to which the B belongs, for C is negative for the B meson.

So, for the moment, the B "stands" by itself, and there is as yet no clear indication where the strange member or the isoscalar partners of the B are to be found.

REFERENCES AND FOOTNOTES

1. Review of Spin-Parity Analysis of the B Meson, S.U. Chung, Proc. of the XVI Int'l. Conf. on High Energy Physics, I, 96 (1972). For ease of reference, published references on the J^P analyses through 1973, not explicitly referenced in this review, are given: G. Ascoli, et al., Phys. Rev. Lett. 20, 1411 (1968); A. Werbrouck, et al., Lett. Nuovo Cimento 4, 1267 (1970); I.A. Erofeev, et al., Soviet J. of Nucl. Phys. 11, 450 (1970); M. Afzal, et al., Nuovo Cimento 15A, 61 (1973). For a complete listing on the B meson, see Review of Particle Properties, Phys. Lett. 50B, April 1974.

2. N. Armenise, et al., Lett. Nuovo Cimento 8, 425 (1973).

3. S.U. Chung, et al., Phys. Lett. 47B, 526 (1973) (paper #19).

4. V. Chaloupka, A. Ferrando, M.J. Losty, L. Montanet, J. Alitti, B. Gandois, J. Louie, Spin Parity Analysis of the $\omega\pi$ System in the B Meson Mass Region, submitted to Phys. Lett. B.

5. U. Karshon, G. Mikenberg, Y. Eisenberg, S. Pitluck, E.E. Ronat, A. Shapira, G. Yekutieli, Production and Decay Mechanism of the B.Meson and the $\omega\pi$ System in π^+p Interactions at 5 GeV/c (paper #67).

6. Aachen/Berlin/Bonn/CERN/Heidelberg/London/Vienna Collaboration, Comparison of Low Mass $(\pi^+\omega)$ and $(K^-\omega)$ Systems and Evidence that the Reaction $K^-p \rightarrow (K^-\omega)p$ is Mainly Diffractive (paper #82).

7. Frankiel, et al., Nucl. Phys. B47, 61 (1972).

8. J. Ballam, G.B. Chadwick, Y. Eisenberg, E. Kogan, K.C. Moffeit, I.O. Skillicorn, H. Spitzer, G. Wolf, H.H. Bingham, W.B. Fretter, W.J. Podolsky, M.S. Rabin, A.H. Rosenfeld, G. Smadja, G.P. Yost, P. Seyboth, A Search for B and $\rho'(1250)$ Production in the Reaction $\gamma p \rightarrow p\pi^+\pi^-$ + Neutrals at 2.8, 4.7 and 9.3 GeV, SLAC-PUB-1364, submitted to Nucl. Phys. B.

9. M. Conversi, L. Paoluzi, F. Ceradini, S. d'Angelo, M.L. Ferrer, R. Santonico, M. Grilli, P. Spillantini, V. Valente, On the Possible Existence of a Vector Meson $\rho'(1250)$.

10. J.L. Rosner, The Classification and Decays of Resonant Particles, SLAC-PUB-1391, submitted to Physics Reports.

11. P. Shen and K. Kang, Phys. Rev. D7, 269 (1973).

12. S.D. Protopopescu, S.U. Chung, R.L. Eisner, N.P. Samios, R.C. Strand, Partial Wave Analysis of the $K\omega$ Spectrum (paper #21).

TABLE I

Reference	B Events[a]	$\vert F_o \vert^2$	$\vert D/S \vert$ [c]
N. Armenise, et al.[2] $\pi^- p$ at 9.1 GeV/c	78	0.27 ± 0.13	–
S. Chung, et al. [3] $\pi^+ p$ at 7.1 GeV/c	1092[b]	0.16 ± 0.04	0.21 ± 0.07 [d]
V. Chaloupka [4] $\pi^- p$ at 3.9, 5, and 7.5 GeV/c	1950[b]	0.10 ± 0.06	0.30 ± 0.10 [d]
U. Karshon, et al. [5] $\pi^+ p$ at 5 GeV/c	584	0.08 ± 0.08	0.35 ± 0.25 [d]

(a) Estimated B events in the interval $1.1 < M(\pi\omega) < 1.4$ GeV.

(b) These are the 1^+ events obtained from the partial-wave analysis of the $\pi\omega$ system in the B region with $t' < 1$ GeV^2 (0.6) for Chung (Chaloupka).

(c) $D = \sqrt{\frac{2}{3}} (-F_o + F_1)$, $S = \sqrt{\frac{1}{3}} (F_o + 2F_1)$.

(d) These analyses find F_o and F_1 relatively real and $F_o F_1 > 0$.

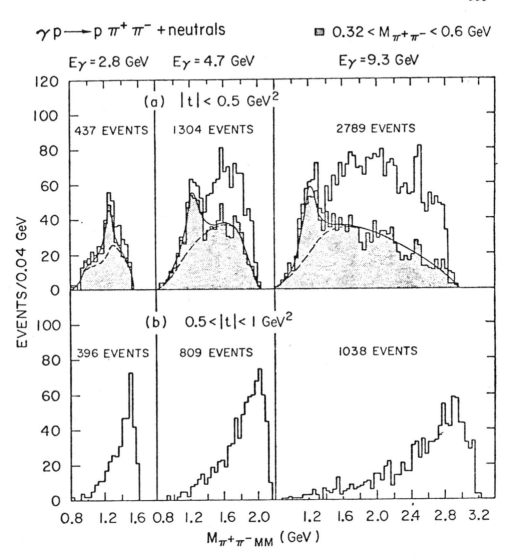

Fig. 1.   M($\pi^+\pi^-$MM) from the reaction $\gamma p \rightarrow p\pi^+\pi^-$ + neutrals at E_γ = 2.8, 4.7 and 9.3 GeV (Ref. 8).   The shaded histograms result from a selection which tends to favor the $\pi^0\omega$ decay mode of the "B" over others.

360

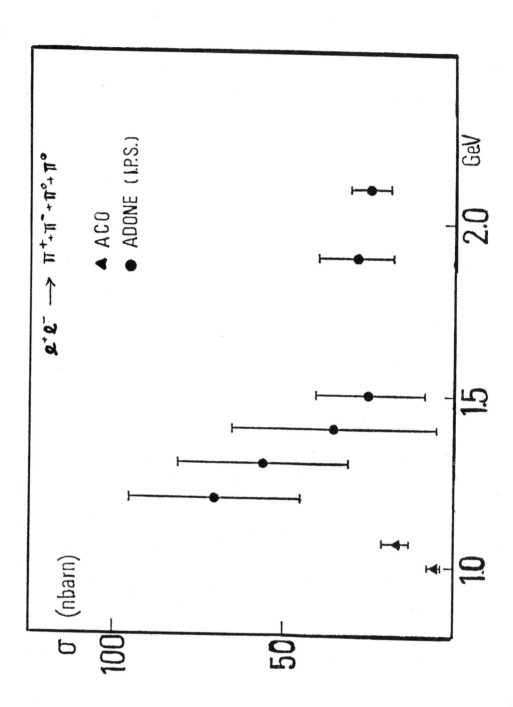

Fig. 2.   Cross-section vs. center-of-mass energy for the reaction e^+e^-
$\rightarrow \pi^+\pi^-\pi^0\pi^0$ (Ref. 9).

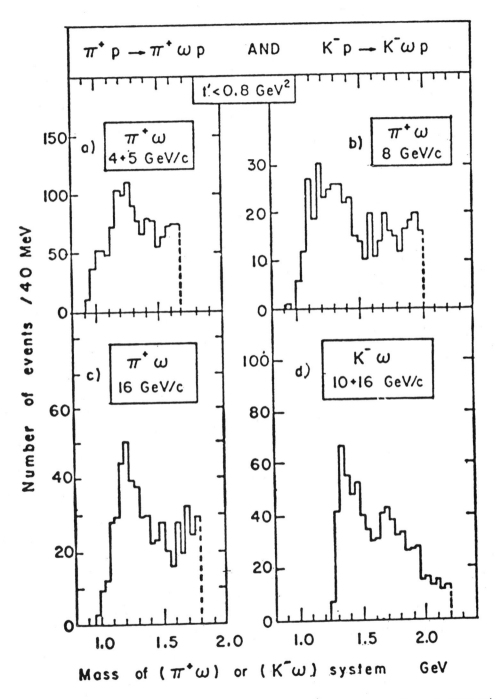

Fig. 3(a-c) M($\pi^+\omega$) from the reaction $\pi^+p \rightarrow \pi^+\omega p$ at 4, 5, 8, and 16 GeV/c.
(d) M($K^-\omega$) from the reaction $K^-p \rightarrow K^-\omega p$ at 10 and 16 GeV/c (Ref. 6).

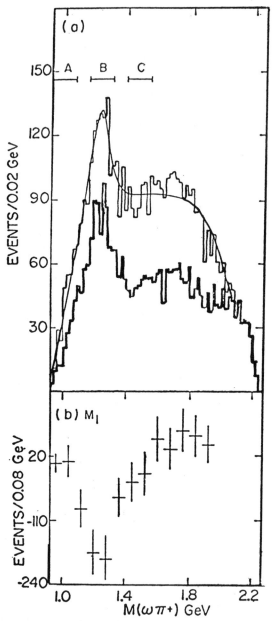

Fig. 4(a) M($\pi^+\omega$) from the reaction $\pi^+p \to \pi^+\omega p$ at 5 GeV/c (Ref. 5); the lower histogram results after the elimination of the $\Delta^{++}\omega$ events. (b) Unnormalized moments proportional to $3|F_0|^2-1$.

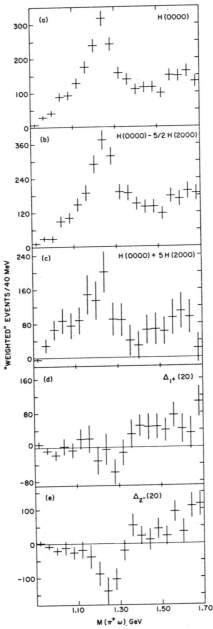

Fig. 5(a) M($\pi^+\omega$) from the reaction $\pi^+p \rightarrow \pi^+\omega p$ at 7.1 GeV/c with $\Delta^{++}\omega$ events out and $t' < 1$ GeV^2 (Ref. 3); the events from the ω control region have been included with appropriate negative weights to "eliminate" the non-ω background events.
(b)[c] Unnormalized moments proportional to $|F_1|^2$ $[|F_0|^2]$.
(d)[e] Unnormalized moments which should be zero if the $\pi\omega$ system is in the 1^+ $[2^-]$ state.

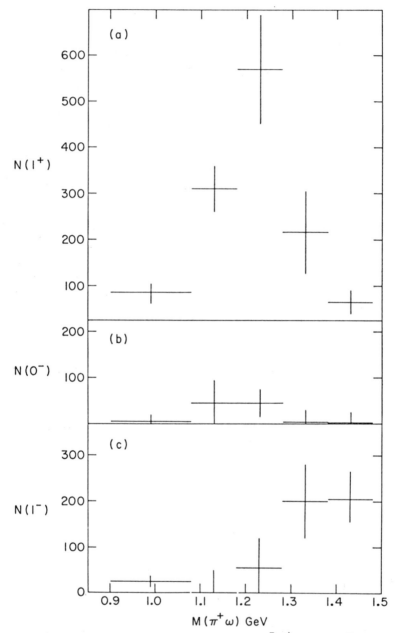

Fig. 6. Fitted numbers of events in states of $J^P = 1^+$, 0^- and 1^- as a function of $M(\pi^+\omega)$ (Ref. 3).

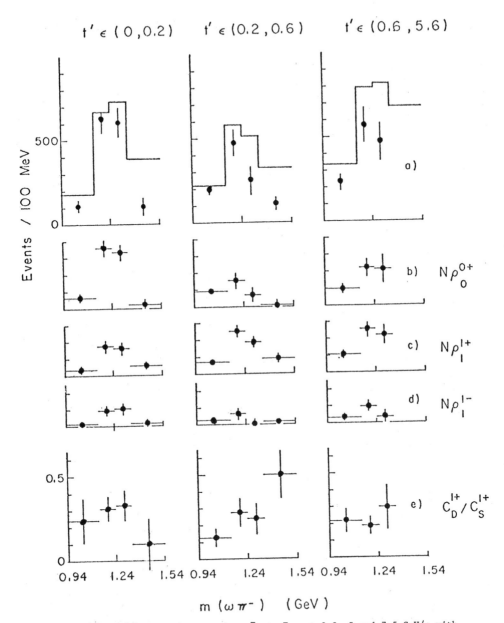

Fig. 7(a) $M(\pi^-\omega)$ from the reaction $\pi^-p \to \pi^-\omega p$ at 3.9, 5 and 7.5 GeV/c with $N^*\omega$ events out (Ref. 4); the points with error bars correspond to fitted events with $J^P=1^+$.

(b-d) $N\rho_{oo}$, $N(\rho_{11}-\rho_{1-1})$ and $N(\rho_{11}+\rho_{1-1})$ for the 1^+ wave where N stands for the 1^+ events; $\rho_{mm'}$ are evaluated in the Jackson frame.

(e) $|D/S|$ for the 1^+ wave as a function of $M(\pi^-\omega)$.

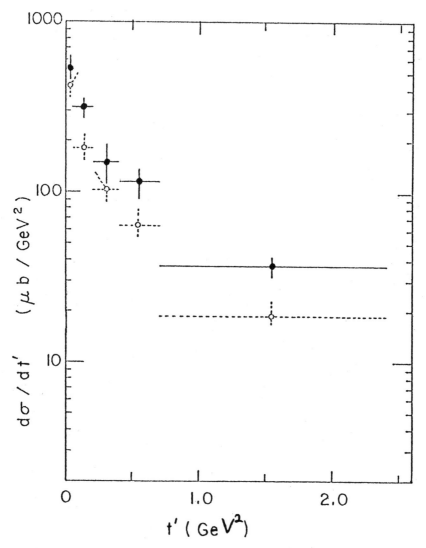

Fig. 8.   dσ/dt' distribution for the 1⁺ wave at 3.9 GeV/c (Ref. 4).   Solid
bars correspond to fitted events from the partial-wave analysis.
The dashed bars correspond to the events found in a mass fit to
the B region.

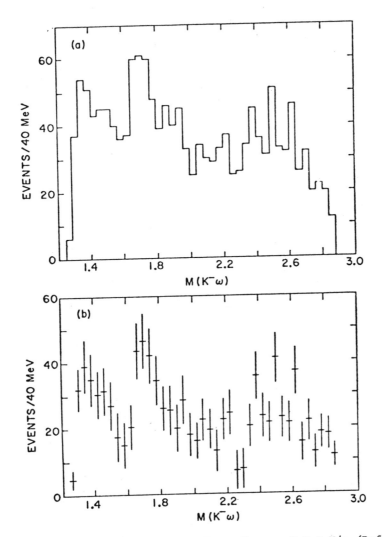

Fig. 9 (a) M(K⁻ω) for the reaction K⁻p → K⁻ωp at 7.3 GeV/c (Ref. 12).
(b) The same spectrum with non-ω background events subtracted out
(cf. Fig. 5a).

368

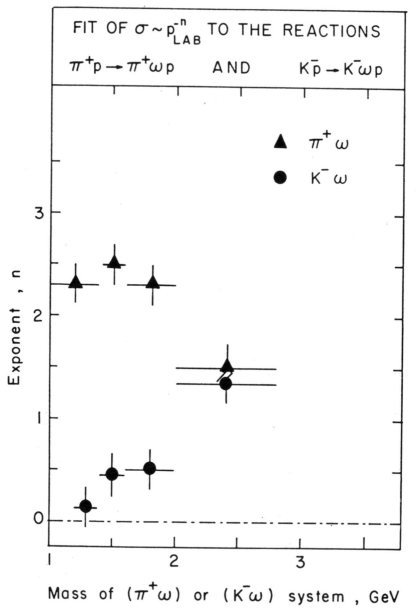

Fig. 10. Fit of σ ∼ p_{LAB}^{-n} to the reactions $\pi^+p \to \pi^+\omega p$ and $K^-p \to K^-\omega p$ (Ref. 6); fitted exponents n vs. $M(\pi^+\omega)$ and $M(K^-\omega)$ are plotted.

SYMMETRIES OF CURRENTS AS SEEN BY HADRONS*

Frederick J. Gilman

Stanford Linear Accelerator Center, Stanford University, Stanford, Ca. 94305

ABSTRACT

The transformation properties of hadron states and current operators under the $SU(6)_W$ algebra of currents are reviewed. A transformation from current to constituent quark bases is introduced, and the algebraic properties of certain transformed current operators are abstracted from the free quark model. The resulting theory yields selection rules, relations among widths, and relative signs of amplitudes for both pion and real photon transitions among hadrons. The agreement with experiment found, especially for amplitude signs, lends strong support both to the proposed theory of current-induced-transition amplitudes and to the assignment of hadrons to constituent quark model multiplets. The theory may then be used to classify states and to predict properties of yet unseen decays, thereby providing a new tool in hadron spectroscopy.

INTRODUCTION

Someday, when we have a real theory of hadrons and their interactions, we will be able to calculate all the current-induced-transitions among them. That is, we will be able to calculate the matrix elements for the vector and axial-vector current induced processes: $V_\mu(x)$ + Hadron $\rightarrow$ Hadron' and $A_\mu(x)$ + Hadron $\rightarrow$ Hadron'. At the particular point $q^2 = 0$, if we were then willing to invoke the vector dominance hypothesis or the PCAC hypothesis, such amplitudes could be related to those for the purely hadronic processes ρ + Hadron $\rightarrow$ Hadron' and π + Hadron $\rightarrow$ Hadron', respectively.

To find ourselves in this happy situation, even at $q^2 = 0$, we must actually solve two problems at once. First, we must know the properties of currents — what symmetry properties do they possess, what are their commutation relations? Second, we need to understand the structure of hadrons — what are their relations to one another, how are the currents flowing within them distributed?

It is progress on these questions which has been most dramatic since the last conference in this series and which forms the principal subject of this talk. These two problems have been attacked by relating them — through a so-called "transformation between current and constituent quarks". The end result is a well-defined theory of vector and axial-vector transition matrix elements within the context of the quark model. We shall review the formulation of the theory and its application to pion and real photon decays of hadrons.

The agreement of the theory with experiment that we shall find lends support both to the theory of current-induced-transitions and to the quark model for hadron spectroscopy. Particularly the success in predicting amplitude

---

*Work supported by the U.S. Atomic Energy Commission.

signs indicates that in a sense, not only do hadrons fall into recognizable quark model multiplets, but the relation between their wave functions is at least roughly given by the quark model as well.

CURRENTS AND QUARKS

In order to formulate a theory of current-induced-transitions among hadrons composed of quarks we need a group theoretic framework for labeling the states and operators involved. For this purpose it is natural to turn to an algebra of charges formed by integrating weak and electromagnetic current densities over all space. We use currents because: (a) it is plausible to work to lowest order in the weak or electromagnetic interaction but to all orders in strong interactions; (b) the symmetries and commutation relations of such currents are relatively well understood; and (c) matrix elements of currents are measured in the laboratory, or if not, in cases of relevance to us they are related by the Partially Conserved Axial-Vector Current Hypothesis (PCAC) to pion amplitudes which are measured.

To start with, consider vector and axial-vector charges:

$$Q^\alpha(t) = \int d^3x \, V_0^\alpha(\vec{x}, t) \tag{1a}$$

$$Q_5^\alpha(t) = \int d^3x \, A_0^\alpha(\vec{x}, t) , \tag{1b}$$

where α is an SU(3) index which runs from 1 to 8 and $V_\mu^\alpha(\vec{x}, t)$ and $A_\mu^\alpha(\vec{x}, t)$ are the local vector and axial-vector current densities with measurable matrix elements. The vector charges are just the generators of SU(3). These integrals over the time components of the current densities are assumed to satisfy the equal-time commutation relations proposed by Gell-Mann[1]

$$\left[Q^\alpha(t), Q^\beta(t)\right] = i f^{\alpha\beta\gamma} Q^\gamma(t)$$

$$\left[Q^\alpha(t), Q_5^\beta(t)\right] = i f^{\alpha\beta\gamma} Q_5^\gamma(t) \tag{2}$$

$$\left[Q_5^\alpha(t), Q_5^\beta(t)\right] = i f^{\alpha\beta\gamma} Q^\gamma(t) ,$$

where $f^{\alpha\beta\gamma}$ are the structure constants of SU(3). Sandwiched between nucleon states at infinite momentum, the last of Eqs. (2) gives rise to the Adler-Weisberger sum rule.[2] From this point on, we shall always be considering matrix elements to be taken between hadron states[3] with $p_z \to \infty$.

For the purposes at hand we need a somewhat larger algebraic system then that provided by the measurable vector and axial-vector charges in Eqs. (1), which form the algebra of SU(3) × SU(3) according to Eqs. (2). To obtain the larger algebra we adjoin to the integrals over all space of[4] $V_0^\alpha(\vec{x}, t)$ and $A_z^\alpha(\vec{x}, t)$, those of the tensor current densities $T_{yz}^\alpha(\vec{x}, t)$ and

$T_{zx}^{\alpha}(\vec{x}, t)$. In the free quark model these charges have the form:

$$\int d^3x \, V_0^{\alpha}(\vec{x}, t) \quad \sim \quad \int d^3x \, \psi^+(x) \left(\frac{\lambda^{\alpha}}{2}\right) \mathbb{1} \, \psi(x)$$

$$\int d^3x \, A_z^{\alpha}(\vec{x}, t) \quad \sim \quad \int d^3x \, \psi^+(x) \left(\frac{\lambda^{\alpha}}{2}\right) \sigma_z \, \psi(x)$$

$$\int d^3x \, T_{yz}^{\alpha}(\vec{x}, t) \quad \sim \quad \int d^3x \, \psi^+(x) \left(\frac{\lambda^{\alpha}}{2}\right) \beta \, \sigma_x \, \psi(x) \tag{3}$$

$$\int d^3x \, T_{zx}^{\alpha}(\vec{x}, t) \quad \sim \quad \int d^3x \, \psi^+(x) \left(\frac{\lambda^{\alpha}}{2}\right) \beta \, \sigma_y \, \psi(x)$$

where $\psi(x)$ is the Dirac (and SU(3)) spinor representing the quark field. When commuted using the free quark field commutation relations, these charges act algebraically like the product of SU(3) and Dirac matrices $(\lambda^{\alpha}/2) \mathbb{1}$, $(\lambda^{\alpha}/2) \sigma_z$, $(\lambda^{\alpha}/2) \beta \sigma_x$, and $(\lambda^{\alpha}/2) \beta \sigma_y$ respectively.[5] The Dirac matrices $\beta \sigma_x$, $\beta \sigma_y$, and σ_z form the so-called W-spin.[6] They are invariant under boosts in the z direction and the corresponding charges are "good", in the sense that they have finite (generally non-vanishing) matrix elements between states as $p_z \rightarrow \infty$. This makes them the correct set of charges to use to label states in terms of their internal quark spin components. If we let $\alpha = 0$ correspond to the SU(3) singlet representation (and λ^0 be a multiple of the unit matrix), then Eqs. (3) consist of 36 charges which close under commutation. They act like an identity operator plus 35 other generators of an SU(6) algebra. We call this algebra the SU(6)$_W$ of currents[5] because of its origin. Q^{α} and Q_5^{α} then essentially[4] form a chiral SU(3) × SU(3) subalgebra of this larger algebra.

Given such an algebra, we define the smallest representations of it (other than the singlet), the 6 and $\overline{6}$ representations, as the current quark (q) and current anti-quark ($\overline{q}$) respectively. We may build up all the larger representations of SU(6)$_W$ out of these basic ones.

Can then real baryons be written as three current quarks, qqq, and real mesons as current quark and anti-quark, $q\overline{q}$, with internal angular momentum L, as in the constituent quark model[7] used for hadron spectroscopy? While possible in principle, it is a disaster when compared with experiment. For it leads to $g_A = 5/3$, zero anomalous magnetic moment of the nucleon, no electromagnetic transition from the nucleon to the 3-3 resonance (Δ), no decay of ω to $\gamma\pi$, etc. It would also yield results for masses like $M_N = M_{\Delta}$, $M_{\pi} = M_{\rho}$, etc. The hadron states we see can not be simple in terms of current quarks. They must lie in mixed representations of the SU(6)$_W$ of currents. Work in past years has shown directly that hadron states are quite complicated when viewed in terms of current algebra.[8]

We may restate this complication in terms of the definition of an operator V for any hadron:

|Hadron> ≡ V|simple qqq or $q\overline{q}$ state of current quarks >

$$= \text{|simple qqq or } q\overline{q} \text{ state of constituent quarks} > \tag{4}$$

All the complication of real hadrons under the $SU(6)_W$ of currents (i.e., in terms of current quarks) has been swept into the operator V. On the other hand, real hadrons are supposed to be simple in terms of the "constituent quarks" used for spectroscopy purposes, as indicated by the second equality in Eq. (4). In other words, the transformation V connects the two simple descriptions in terms of current quarks and constituent quarks.[9] It is for this reason that it is sometimes called the "transformation from current to constituent quarks".[10,11]

Up to this point we have only managed to restate the problem via Eq. (4). But as often happens, phrasing the problem right is a major way toward the solution. For what we are after in the end are matrix elements of various current operators, $\mathcal{O}$. Using Eq. (4) and assuming V is unitary we may write

$$< \text{Hadron}' | \mathcal{O} | \text{Hadron} >$$

$$= < (\text{simple current quark state})' | V^{-1} \mathcal{O} V$$

$$| (\text{simple current quark state}) > \quad . \tag{5}$$

This has two important advantages. First, we may study the properties of $V^{-1}\mathcal{O}V$ in isolation, and then apply what we learn to the matrix elements of $\mathcal{O}$ between any two hadron states. Second, even though V itself is very complicated and contains (by definition) all information on the current quark composition of each hadron, it is possible that the object $V^{-1}\mathcal{O}V$ for some operators $\mathcal{O}$ may be relatively simple in its algebraic transformation properties.

This last possibility is of course exactly what we shall assume on the basis of calculations done in the free quark model. In that model, Melosh[12] and others[13,14,15] have been able to formulate and explicitly calculate the transformation V. While one would not take the details of the transformation found there as correctly reflecting the real world, one might try to abstract the algebraic properties of some transformed operators $V^{-1}\mathcal{O}V$, from such a calculation. In cases of interest, this turns out to be equivalent to assuming that the transformed operators $V^{-1}\mathcal{O}V$ have the algebraic properties of the most general combination of single quark operators consistent with SU(3) and Lorentz invariance.

Thus, while Eq. (3) shows that Q_5^α itself behaves under the $SU(6)_W$ of currents as simply

$$\int d^3x \, \psi^+(x) \left(\frac{\lambda^\alpha}{2} \right) \sigma_z \, \psi(x) \quad ,$$

a direct calculation in the free quark model shows that algebraically $V^{-1}Q_5^\alpha V$ behaves as a sum of two terms:[16]

$$V^{-1}Q_5^\alpha V \sim \left(\frac{\lambda^\alpha}{2} \right) \sigma_z$$

$$+ \left(\frac{\lambda^\alpha}{2} \right) \left[(\beta\sigma_x + i\beta\sigma_y)(v_x - iv_y) - (\beta\sigma_x - i\beta_y)(v_x + iv_y) \right] \, , \tag{6}$$

where the products of Dirac and SU(3) matrices are understood to be taken between quark spinors (and integrated over all space). Here v_μ is a vector in configuration space, so that $v_x \pm iv_y$ raises (lowers) the z component of angular momentum (L_z) by one unit. The particular combination of Dirac matrices and vector indices in the two terms in Eq. (6) is dictated by the demands that the total $J_z = 0$ and the parity be odd for the axial-vector charge, Q_5^α, and for $V^{-1}Q_5^\alpha V$.

For the vector charge, Q^α, we must have

$$V^{-1}Q^\alpha V = Q^\alpha \, , \tag{7}$$

since we want these charges to be the generators of SU(3), both before and after the transformation. However, the first moment of the charge density,[17]

$$D_+^\alpha = i \int d^3x \left(\frac{-x-iy}{\sqrt{2}}\right) V_0^\alpha(\vec{x}, t) \, , \tag{8}$$

is not a generator and is transformed non-trivially by V. One finds in the free quark model that in algebraic properties $V^{-1}D_+^\alpha V$ behaves as a sum of four terms under the SU(6)$_W$ of currents:[18]

$$V^{-1}D_+^\alpha V \sim \left(\frac{\lambda^\alpha}{2}\right) \mathbb{1} \, (v_x + iv_y)$$

$$+ \left(\frac{\lambda^\alpha}{2}\right) (\beta \sigma_x + i\beta \sigma_y)$$

$$+ \left(\frac{\lambda^\alpha}{2}\right) \sigma_z (v_x + iv_y)$$

$$+ \left(\frac{\lambda^\alpha}{2}\right) (\beta \sigma_x - i\beta \sigma_y)(v_x + iv_y)(v_x + iv_y) \, , \tag{9}$$

where again the Dirac and SU(3) matrices are understood to be taken between quark spinors.

We abstract the algebraic properties of $V^{-1}Q_5^\alpha V$ and $V^{-1}D_+^\alpha V$ given in Eqs. (6) and (9) from the free quark model and assume them to hold in the real world. We are then able to treat matrix elements of Q_5^α and D_+^α between hadron states as follows:

(1) We identify the hadrons with qqq or q$\bar{q}$ states of the constituent quark model where the total quark spin S is coupled to the internal angular momentum L to form the total J of the hadron. The states so constructed fall into SU(6)$_W \times$ 0(3) multiplets. Meson states formed in this simple manner are enumerated in Table I, where candidates are given for the isospin 1 and 0 "slots" for each J^{PC} value from among the observed mesons.[7] The sad state of meson spectroscopy is reflected in the lack of established states even at the L = 1 level. The situation for baryons is of course much better,[7] there being one or more established candidate for every J^P value in the SU(6)$_W \times$ 0(3) multiplets $\underline{56}$ L = 0, $\underline{70}$ L = 1, and $\underline{56}$ L = 2.

Table I Meson states of the constituent quark
model and possible $I = 1$ and 0 candidates. [7]

$SU(6)_W \times 0(3)$ Multiplet	$SU(3)$ Multiplet	Quark Spin S	J^{PC}	Candidates $I=1$ and 0
$\underline{35} + \underline{1}$	$\underline{8} + \underline{1}$	1	1^{--}	ρ, ω, ϕ .
$L = 0$	$\underline{8} + \underline{1}$	0	0^{-+}	π, η, X^0 ?
$\underline{35} + \underline{1}$	$\underline{8} + \underline{1}$	1	2^{++}	A_2, f, f'
$L = 1$	$\underline{8} + \underline{1}$	1	1^{++}	$A_1 ?, D, ?$
	$\underline{8} + \underline{1}$	1	0^{++}	$\delta, \epsilon ?, S^* ?, \epsilon' ?$
	$\underline{8} + \underline{1}$	0	1^{+-}	$B, ?, ?$
$\underline{35} + \underline{1}$	$\underline{8} + \underline{1}$	1	3^{--}	$g, \omega_3 ?, ?$
$L = 2$	$\underline{8} + \underline{1}$	1	2^{--}	$F_1 ?, ?, ?$
	$\underline{8} + \underline{1}$	1	1^{--}	$\rho' ?, ?, ?$
	$\underline{8} + \underline{1}$	0	2^{-+}	$A_3 ?, ?, ?$

(2) Since very few weak axial-vector transitions are measured, given a matrix element of Q_5^α, we use PCAC to relate it to a measured pion transition amplitude. Application of the golden rule then yields:

$$\Gamma(H' \to \pi^- H) = \frac{1}{4\pi f_\pi^2} \frac{p_\pi}{2J'+1} \cdot \frac{(M'^2 - M^2)^2}{M'^2} \sum_\lambda |<H', \lambda| (1/\sqrt{2}) (Q_5^1 - iQ_5^2) |H, \lambda>|^2,$$

(10)

where $f_\pi \simeq 135$ MeV. The factors in Eq. (10) are forced on us by PCAC and kinematics — there are no arbitrary phase space factors.

For real photon transitions, matrix elements of $D_+^3 + (1/\sqrt{3}) D_+^8$ are directly proportional to the corresponding Feynman amplitudes. The width for $H' \to \gamma H$ is given by[17]

$$\Gamma(H' \to \gamma H) = \frac{e^2}{\pi} \frac{p_\gamma^3}{2J'+1} \sum_\lambda |<H', \lambda|D_+^3 + (1/\sqrt{3}) D_+^8|H, \lambda - 1>|^2.$$

(11)

(3) Given a matrix element of Q_5^α or D_+^α between hadron states which is related to measurements by either Eq. (10) or (11), we transform using V from simple constituent to simple current quark states. The particular matrix element is thus rewritten in terms of $V^{-1}Q_5 V$ or $V^{-1}D_+ V$, and simple current quark states. We know the algebraic properties of all these quantities under the $SU(6)_W$ of currents via abstraction of Eqs. (6) and (9) from the free quark model and our identification of hadrons with quark model states. We may then apply the Wigner-Eckart theorem to each term to express it as a Clebsch-Gordan coefficient[19] (of $SU(6)_W$) times a reduced

matrix element. Since the same reduced-matrix element occurs in many different transitions, relations among the corresponding transition amplitudes follow.

CONSEQUENCES OF THE THEORY

The experimental consequences of the theory outlined in the last section have been considered by a number of authors. [12,20-29] These consequences fall into the following three categories:

(1) <u>Selection Rules</u> For transitions by pion or photon emission from states (either mesons or baryons) with internal angular momentum L' to those with L, one finds[22,23]

$$\left| |L' - L| - 1 \right| \leq \ell_\pi \leq L + L' + 1 \tag{12a}$$

$$\left| |L' - L| - 1 \right| \leq j_\gamma \leq L + L' + 1 , \tag{12b}$$

where ℓ_π and j_γ are the total angular momentum carried off by the pion and photon in the overall transition.

For example, ℓ_π can be 0 or 2 ($\ell_\pi = 1$ is forbidden by parity), but not 4 for a pion decay from L' = 1 to L = 0. Thus the decay of the $D_{15}(1670)$, the $J^P = 5/2^- N^*$ resonance with L' = 1, into $\pi\Delta$ is forbidden in g-wave ($\ell_\pi = 4$), although otherwise allowed by kinematical considerations. Similarly, only $j_\gamma = 1$ is allowed for L' = 0 to L = 0 photon transitions, although $j_\gamma = 2$ (and even $j_\gamma = 3$ for $\Delta \to \gamma\Delta$) is generally permitted by kinematics. This particular rule is well-known for $\Delta \to \gamma N$, where it is just the successful quark model result[30] that the transition is purely magnetic dipole in character, i.e. the possible electric quadrupole amplitude is forbidden. The inequalities in Eqs. (12) might be regarded as the generalization of these particular results to all L and L' in the present theoretical context.

Note that for $|L - L'| \geq 3$ the lower limit of the inequalities becomes operative in a non-trivial way, forbidding low values of ℓ_π or j_γ which would otherwise have been favored kinematically. Unfortunately, the relevant hadron states which would provide an interesting test of this have not yet been found.

Selection rules of a different sort govern the number of independent reduced matrix elements. For pion transitions from a hadron multiplet with internal angular momentum L' down to the ground state hadrons with L = 0, there are at most two independent reduced matrix elements, corresponding to the two terms in Eq. (6). For real photon transitions between the same two multiplets there are at most four independent reduced matrix elements, corresponding to Eq. (9).

In general structure, the theory described above includes various concrete quark model calculations, both non-relativistic[31] and relativistic.[32] In fact, a one-to-one correspondence exists between the quantities calculated in such models and the reduced matrix elements in the present theory. However, such models are usually much more specific, with parameters like the strength of the "potential", quark masses, etc. fixed. Since the quantities corresponding to reduced matrix elements are expressed explicitly in terms of such parameters, they are computable numerically and the scale of the reduced matrix elements is determined.

Also included in the general structure of the theory are the results following from assuming strong interaction $SU(6)_W$ conservation.[6] For pion transitions, this corresponds in the present theory to retaining only the first term in $V^{-1}Q_5^\alpha V$. Since this hypothesis fails experimentally, various ad hoc schemes for breaking $SU(6)_W$ have been proposed.[33] Such schemes still fall within the general structure of amplitudes presented above[34] and they are similar in giving relations between amplitudes while not setting their absolute scale.[35] However, as we shall see below, they are generally more restrictive in that they tie together pion and rho decay amplitudes.

(2) Decay Widths The simplest such set of relations are those for pion transitions from $L' = 0$ to $L = 0$ mesons. Here there is only one reduced matrix element (the second term in Eq. (6) has $\Delta L_z = \pm 1$ and so can not contribute when $L' = L = 0$), so that the amplitudes for $\rho \to \pi\pi, K^*(890) \to \pi K$, and $\omega \to \pi\rho$ are all proportional. The ratio of the amplitudes for the first two processes may be obtained from $\Gamma(\rho \to \pi\pi)/\Gamma(K^* \to \pi K)$, while the amplitude for the latter is obtainable from $\omega \to 3\pi$ and rho dominance. Within errors, the ratio of the three amplitudes is that predicted by the theory.[36]

For pion transitions from mesons with internal angular momentum $L' = 1$ to those with $L = 0$, both terms in Eq. (6) are possible and there are consequently two independent reduced matrix elements which describe all such decays. Rather than performing a fit to all the data, we choose two measured widths as input and thereby determine all the other decay rates. For this purpose we take $\Gamma(A_2 \to \pi\rho) = 71.5$ MeV, from the latest particle data tables,[37] and $\Gamma_{\lambda=0}(B \to \pi\omega) = 0$. This latter condition, the vanishing of the helicity zero (longitudinal) decay of $B \to \pi\omega$, is suggested by high statistics experiments[38] which find the transverse decay to be strongly dominant. While probably not exactly zero, we take this as a very reasonable first approximation to the data. Exact vanishing of $\Gamma_{\lambda=0}(B \to \pi\omega)$ corresponds to only the second term in $V^{-1}Q_5^\alpha V$, with the algebraic properties of $(\lambda^\alpha/2)$ $[(\beta\sigma_x + i\beta\sigma_y)(v_x - iv_y) - (\beta\sigma_x - i\beta\sigma_y)(v_x + iv_y)]$, having a non-zero reduced matrix element. This well illustrates the experimental necessity of a non-trivial transformation V; for if $V = \mathbb{1}$, only the term behaving as $(\lambda^\alpha/2)\sigma_z$ would be present and the predicted helicity structure for $B \to \pi\omega$ would be completely opposite that observed.

The results[39] can be seen in Table II. The correct values for $\Gamma(A_2 \to \pi\rho)/\Gamma(K^*(1420) \to \pi K^*)$ and $\Gamma(f \to \pi\pi)/\Gamma(K^*(1420) \to \pi K)$ may be regarded as testing the SU(3) component of the theory, while, for example, the value of $\Gamma(A_2 \to \pi\rho)$ or $\Gamma(K^*(1420) \to \pi K^*)$ relative to $\Gamma(f \to \pi\pi)$, $\Gamma(K^*(1420) \to \pi K)$ or $\Gamma(A_2 \to \pi\eta)$ tests the full theory, including the phase space factors in Eq. (10), since one is relating d-wave pion decays into pseudoscalar vs. vector mesons. As for the other decays in the Table, we note that: (a) other strong interaction decay modes of the B meson very likely exist, as we discuss later, although $\pi\omega$ is certainly dominant; (b) the "real" A_1 resonance still remains to be found for comparison with the theory; (c) the now established $I = 1$ scalar meson, δ, only has $\pi\eta$ as a possible strong decay channel, so the total width should almost coincide with that into $\pi\eta$; (d) we have chosen 1300 MeV, the mass where the s-wave πK phase shift[37] goes through 90°, as the mass of the strange, $J^P = 0^+$ meson[40]
The overall agreement found in Table II between theory and experiment is quite good, with the exception of $\Gamma(A_2 \to \pi X^\circ)$. While mixing of the pseudoscalar mesons is such as to alleviate this discrepancy, reasonable mixing angles do not change   the width appreciably from the value in Table II.

Table II Decays of L'=1 mesons to L=0 mesons by pion emission.[39]

Decay	Γ(predicted) (MeV)	Γ(experimental)[37] (MeV)
$A_2(1310) \to \pi\rho$	71.5 (input)	71.5 ± 8
$K^*(1420) \to \pi K^*$	27	29.5 ± 4
$f(1270) \to \pi\pi$	112	141 ± 26
$K^*(1420) \to \pi K$	55	55 ± 6
$A_2(1310) \to \pi\eta$	16	15 ± 2
$A_2(1310) \to \pi X^0$	5	< 1
$B(1235) \to \pi\omega, \lambda = 0$	0 (input)	$\Gamma_{total} = 120 \pm 20$ $\pi\omega$, with $\lambda=1$ strongly dominant,[38] only mode seen
$\lambda = 1$	75	
$A_1(1100) \to \pi\rho, \lambda = 0$	63	??
$\lambda = 1$	31	
$\delta(970) \to \pi\eta$	41	50 ± 20
$\kappa(1300) \to \pi K$	380	?, broad

A more likely source of trouble lies in the theoretical assignment of the X^0 to be dominantly that SU(3) singlet pseudoscalar meson associated with the octet containing the pion and eta. In any case, an actual measurement of the $A_2 \to \pi X^0$ decay width, rather than an upper limit, would be an interesting quantity to determine experimentally.

For L' = 2 mesons decaying by pion emission to the L = 0 states, there are again two independent reduced matrix elements. About the only decay width determined with any certainty is $g \to \pi\pi$. The meagre information available on other decays is consistent with the theory within the large experimental errors.[23]

For photon decays of mesons the data are even more sparse, although there are plenty of theoretical predictions.[28] In fact, only a few decays among L' = 0 mesons are actually measured, where there is just one possible reduced matrix element. Fixing this from $\Gamma(\omega \to \gamma\pi)$, the predictions[41] are collected in Table III. What widths have been measured are consistent with the predictions of the theory, although at the limits of the error bars in several cases.

There are a large number of pion and photon transitions among baryons which are predicted by the theory. They are compared with experiment elsewhere.[22,23,28] Overall there is fair agreement between theory and experiment, with a number of predicted pion widths "right on the nose", but others off by factors of 2 to 3. In many of these cases there are large experimental uncertainties, as well as the theoretical uncertainty inherent in using the narrow resonance approximation to compute decays of one broad resonance into another.

Table III Decays of L' = 0 mesons to other
L = 0 mesons by photon emission.

	Γ(predicted) no mixing (KeV)	Γ(predicted) $\theta_p = -10.5^\circ$ (KeV)	Γ(experimental)[37] (KeV)
$\omega \to \gamma\pi$	870 (input)	870 (input)	870 ± 60
$\rho \to \gamma\pi$	92	92	$30\pm10 < \Gamma < 80\pm10$ (Ref. 42)
$\phi \to \gamma\pi$	0	0	< 14
$\rho \to \gamma\eta$	36	56	< 160 (Ref. 43)
$\omega \to \gamma\eta$	5	7	< 50
$\phi \to \gamma\eta$	220	170	126 ± 46
$X^o \to \gamma\rho$	160	120	$0.27\ \Gamma(X^o \to \text{all})$
$X^o \to \gamma\omega$	15	11	
$\phi \to \gamma X^o$	0.5	0.6	

(3) <u>Relative Signs</u> In the process $\pi N \to N^* \to \pi\Delta$, the couplings to both πN and $\pi\Delta$ of all the N^*'s with a given value of L are related by $(SU(6)_W)$ Clebsch-Gordan coefficients to the same reduced matrix element(s). The signs of the amplitudes for passing through the various N^*'s in $\pi N \to \pi\Delta$ are then computable group theoretically. The correctness of these sign predictions is crucial, for while, for example, one may be willing to envisage a small amount of mixing of the constituent quark states, and corresponding corrections of say, 20%, to amplitudes (and 40% to widths), this will not change their signs. A wrong sign prediction could well spell the end of the theory!

This in fact seemed to be the case last year[44] when a comparison of the theoretical predictions[22,45] was made with the amplitude signs observed in an earlier phase shift solution of $\pi N \to \pi\Delta$ by the LBL-SLAC collaboration.[46] Since then a newer solution[47,48] with much better χ^2 has been found — in fact, the new solution is the only one left once additional data in the previous energy "gap" between 1540 and 1650 MeV is used as a constraint.[49]

The present situation with regard to amplitude signs for intermediate N^*'s with L = 1 in $\pi N \to N^* \to \pi\Delta$ is shown[50] in Table IV. Aside from an overall phase (chosen so as to give agreement with the sign of the $DD_{15}(1670)$ amplitude), there is one other free quantity. This is the relative size of the reduced matrix elements of the two terms in $V^{-1}Q_5^\alpha V$ or, what turns out to be equivalent, the sign of an s-wave relative to a d-wave transition amplitude. In Table IV we have fixed this by using the sign of the $SD_{31}(1640)$ amplitude. All other signs for N^*'s in the $\underline{70}$ L = 1 multiplet are then predicted theoretically. The seven other signs determined experimentally agree with these predictions. The sign of the s-wave relative to d-wave amplitude is such as to show that the reduced matrix element of the second term in $V^{-1}Q_5^\alpha V$,

Table IV Signs of resonant amplitudes[50] in $\pi N \to N^* \to \pi\Delta$ for N^*'s in the $\underline{70}$ L=1 multiplet of $SU(6)_W \times 0(3)$. Amplitudes are labeled by $(\ell_{\pi N}\ell_{\pi\Delta})_{2I,2J}$ and the resonance mass in MeV.

Resonant Amplitude	Theoretical Sign	Experimental Sign[48]
$DS_{13}(1520)$	−	−
$DD_{13}(1520)$	−	−
$SD_{11}(1550)$	+	?
$SD_{31}(1640)$	+ (input)	+
$DS_{33}(1690)$	−	−
$DD_{33}(1690)$	−	−
$DD_{15}(1670)$	+ (input)	+
$DS_{13}(1700)$	−	−
$DD_{13}(1700)$	+	+
$SD_{11}(1715)$	+	+

Table V Signs of resonant amplitudes[50] in $\pi N \to N^* \to \pi\Delta$ for N^*'s in the $\underline{56}$ L=2 multiplet of $SU(6)_W \times 0(3)$. Amplitudes are labeled as in Table IV.

Resonant Amplitude	Theoretical Sign	Experimental Sign[48]
$FP_{15}(1688)$	− (input)	−
$FF_{15}(1688)$	+	+
$PP_{13}(1860)$	−	?
$PF_{13}(1860)$	+	?
$FF_{37}(1950)$	−	−
$FP_{35}(1880)$	−	?
$FF_{35}(1880)$	−	−
$PP_{33}(\quad)$	+	?
$PF_{33}(\quad)$	+	?
$PP_{31}(1860)$	+	?

with the algebraic properties of $(\lambda^\alpha/2) [(\beta\sigma_x + i\beta\sigma_y)(v_x - iv_y) - (\beta\sigma_x - i\beta\sigma_y)(v_x + iv_y)]$, is dominant for $L'=1$ to $L=0$ pion transitions of baryons, just as it is for $L'=1$ to $L=0$ pion transitions of mesons.

For N^*'s with $L=2$, many of the amplitudes have not been seen experimentally. As the overall phase is already fixed, there is just one parameter free. Again this is the relative size of the two possible reduced matrix elements, only now it corresponds to the sign of a p-wave relative to an f-wave pion decay amplitude. We use the $FP_{15}(1688)$ amplitude in Table V to fix this sign[50]— it corresponds to the reduced matrix element of the first term in $V^{-1}Q_5^\alpha V$, behaving algebraically as $(\lambda^\alpha/2)\sigma_z$, being dominant. All other signs (3) which are measured in Table V agree with the theory.

Another reaction where relative signs are predicted is $\gamma N \to N^* \to \pi N$. This involves the theory at both the γN N^* and πN N^* vertices. Although the situation is more complicated, there are also more amplitudes determined experimentally. An analysis[26,28] of the situation shows that not only are there 15 or so signs correctly predicted, but the information on the πN N^* vertex so obtained agrees with that from $\pi N \to N^* \to \pi\Delta$ as to which term in $V^{-1}Q_5^\alpha V$ has the dominant reduced matrix element.

With our confidence in the theory for giving correct amplitude signs thus established, we may use the theory as a tool to classify new resonances. For example, a $P_{33}(1700)$ state is seen[48] in $\pi N \to N^* \to \pi\Delta$ and other reactions.[37] Does it belong to a state of three constituent quarks with

$L=0$ or with $L=2$? Both such "slots" are open in the constituent quark model, the former being the partner of the Roper resonance, $P_{11}(1470)$, and the latter a relative of the third resonance, $F_{15}(1688)$. Fortunately the amplitude sign in $\pi N \to N^* \to \pi\Delta$ corresponding to these two choices is opposite. Experiment then allows a determination of the correct assignment: the $P_{33}(1700)$ belongs with the $P_{11}(1470)$ and has $L = 0$, as shown in Table VI. We have thus established both non-strange members of a new (although long suspected) quark model multiplet.

Table VI Signs of resonant amplitudes[50] in $\pi N \to N^* \to \pi\Delta$ for N^*'s in a radially excited $\underline{56}$ $L = 0$ multiplet of $SU(6)_W \times O(3)$. Amplitudes are labeled as in Table IV.

Resonant Amplitude	Theoretical Sign	Experimental Sign[48]
$PP_{11}(1470)$	+	+
$PP_{33}(1700)$	−	−

Finally note the inelastic reaction $\pi N \to N^* \to \rho N$. If strong interaction $SU(6)_W$ conservation is assumed, the ρNN^* and πNN^* (or $\pi\Delta N^*$) vertices are related since the π and ρ are in the same strong interaction or constituent $SU(6)_W$ multiplet. The same result holds in broken $SU(6)_W$ schemes.[33] As far as the transformation from current to constituent quarks is concerned, there is no such relation, for only by using PCAC and vector dominance, respectively, are pion and rho vertices obtainable from axial-vector and vector current amplitudes — amplitudes which are themselves totally unrelated. An examination of the Argand diagrams from the LBL-SLAC analysis[48] shows that the π and ρ couple differently to the N^*'s with $L = 1$. This particularly spells trouble for the so-called "ℓ-broken $SU(6)_W$" scheme,[33] as emphasized by Faiman[51] recently.

SOME APPLICATIONS AND OPEN QUESTIONS

Let us then consider some further problems in meson spectroscopy which can be treated using the theory of pion and photon transitions we have been discussing:

(1) Is the $\rho'(1600)$ a $q\bar{q}$ state with $L = 0$ or $L = 2$? In the first case we would have a radial excitation (of the ρ), while in the second we would be filling out the $L = 2$ multiplet (see Table I). Just as we were able to classify the $P_{33}(1700)$ using amplitude signs, a similar application of the theory permits an unambiguous classification here also. In particular, it turns out that the relative signs of the amplitudes for $\pi\pi \to \rho' \to \pi\omega$ and $\pi\pi \to g \to \pi\omega$ (or $\pi\pi \to \rho' \to \overline{K}K^*$ and $\pi\pi \to g \to \overline{K}K^*$) are the same (opposite) for $L = 0$ ($L = 2$). Amplitude analysis of this kind should be possible given the new generation of spectrometers discussed by Leith[52] at this conference.

(2) Can we have a ρ' state which decays to $\pi\omega$ and not $\pi\pi$? This possibility, which is sometimes invoked[53] for a $\rho'(1250)$ state, is difficult to understand in the theory of pion transitions discussed above. The Clebsch-Gordan coefficients yield a factor of 2 (1/2) for $\Gamma(\rho' \to \pi\omega)/\Gamma(\rho \to \pi\pi)$ if $L = 0$ ($L = 2$), while phase space always favors the $\pi\pi$ mode. Thus, without

invoking a very particular mixture of L = 0 and 2, ρ' states should have comparable $\pi\pi$ and $\pi\omega$ decay modes.

(3) <u>Where are the isoscalar $J^P = 1^+$ mesons ?</u> A direct calculation of the width of an isoscalar partner, H, of the B meson to decay into $\pi\rho$ shows that

$$\tilde{\Gamma}(H \to \pi\rho) = \tilde{\Gamma}(B \to \pi\omega) \tag{13a}$$

if H is the eighth component of an octet,

$$\tilde{\Gamma}(H \to \pi\rho) = 2\tilde{\Gamma}(B \to \pi\omega) \tag{13b}$$

if H is the appropriate SU(3) singlet, and

$$\tilde{\Gamma}(H \to \pi\rho) = 3\tilde{\Gamma}(B \to \pi\omega) \tag{13c}$$

if H is an ideally mixed combination of singlet and octet.[54] Here $\tilde{\Gamma}$ denotes the reduced width, with phase space taken out. As the H is presumably heavier than the B, one should be looking for a broad object — at least 100 MeV wide, and more likely 200 to 300 MeV wide! No wonder it's been hard to find.

On the other hand, the isoscalar partner of the A_1, the D, has no decay by pion or kaon emission into the ground state L = 0 mesons. It can only decay to other L = 1 mesons by pion emission, and should be relatively narrow, as indeed is seen experimentally[37] for the state at 1285 MeV.

(4) <u>Pion decays among L = 1 mesons.</u> The decay[37] $D \to \pi\delta$ seems to be the dominant decay of the D(1285), and the recently discovered $A_2 \to \pi\pi\omega$ mode[55] may well proceed (virtually) via $A_2 \to \pi B \to \pi\pi\omega$. The existence of these pionic transitions among L = 1 mesons suggests that decays like $B \to \pi\delta \to \pi\pi\eta$, $B \to \pi A_1 \to \pi\pi\rho$, and $D \to \pi A_1 \to \pi\pi\rho$ should also occur. While there is insufficient data on other decays to make a definite prediction for the latter three, one expects widths of roughly 10 to 20 MeV. Until these possible modes are investigated one should use caution in assigning the B decay width entirely to $\pi\omega$.

A similar situation holds for pion transitions from L = 2 to L = 1. The decay $\omega(1675) \to \pi B \to \pi\pi\omega$ seems to have been recently detected,[56] where the ω (1675) is presumed to be the isoscalar companion of the $g(J^{PC} = 3^{--})$. Decays like $F_1(J^{PC} = 2^{--}) \to \pi\delta$, $g \to \pi H$, etc. should occur with comparable rates.

(5) <u>Bounds on widths.</u> One would like to go beyond symmetry relations and obtain information on the absolute magnitude of some amplitudes. One method of attack is to use the last of the commutation relations in Eq. (2) in the form

$$[Q_5^+, Q_5^-] = Q^3 \tag{14}$$

where $Q_5^\pm = (1/\sqrt{2})(Q_5^1 \pm i Q_5^2)$. Sandwiching this between I = 1, $I_z = 1$ meson states, H^+, with helicity λ, and assuming no I = 2 mesons yields

$$\sum_{H'} |<H'^0, \lambda|Q_5^-|H^+, \lambda >|^2 = 1. \tag{15}$$

Therefore

$$|<H'^{0}, \lambda | Q_5^- | H^+, \lambda >| \leq 1 , \qquad (16)$$

and PCAC then implies a bound on $\Gamma(H' \to \pi H)$. Unfortunately this is not very useful in practice, for it only tells us that $\Gamma(D \to \pi \delta) \leq 310$ MeV — about a factor of 10 too large; $\Gamma(B \to \pi \delta) \leq 135$ MeV — roughly the total width and probably also a factor 10 too big; and $\Gamma(\rho \to \pi \pi) \leq 300$ MeV — which is closer to the true width but still not very useful. Equation (16) only assures us that things can't be really wild.

(6) Masses. Information on masses can be obtained[57,58] by using the commutator

$$\left[Q_5^+(t), \frac{d \, Q_5^+(t)}{dt} \right] = 0 , \qquad (17)$$

where the right hand side is zero under the assumption that there is no $I = 2$ sigma-term. It is clear that masses now enter, since the time derivative is proportional to the commutator with the Hamiltonian. One then probes the structure at a deeper level than when one just uses commutators of charges.

If the transformation V was simply the identity, then it is possible to show that the solution to Eq. (17) is

$$M^2 = M_0^2(L) + M_1^2(L) \, \vec{S} \cdot \vec{L} , \qquad (18)$$

i.e., states with internal angular momentum L are only split in mass by a spin-orbit term of arbitrary magnitude.[59] When $V \neq \mathbb{1}$, the situation becomes very complicated. It is clear that $V^{-1}M^2V$ can not be like a single quark operator in algebraic properties, for this would result in $M_\pi^2 = M_\rho^2$, $M_N^2 = M_\Delta^2$, as Eq. (18) would also have given. Thus we can not abstract some quantities from the free quark model — we do not want its mass spectrum, in particular.[60] While Eq. (17) has been used to derive interesting results for masses in terms of the complicated mixing of representations of current algebra realized by real hadrons,[57,58] it has so far proven difficult to extract much useful information directly from it[61] using the transformation V. This is an important area of further research.

CONCLUSION

The theory of pion and photon transitions which we have outlined has had great success in predicting the signs of amplitudes — more than 25 relative signs are correctly predicted in the reactions $\pi N \to N^* \to \pi \Delta$ and $\gamma N \to N^* \to \pi N$. There is also at least fair success in predicting the relative magnitude of decay amplitudes, particularly for mesons.

This success lends support both to the theory of current-induced-transitions we have presented and to the assignment of hadron states to constituent quark model multiplets. In particular, the amplitude signs found to be in agreement with experiment mean that, at least in a rough sense, the relationship between the wave functions of different hadrons is that of the quark model. At $q^2 = 0$ one sees evidence for a quark picture of hadrons which is just as compelling as that obtained in a very different way as $q^2 \to \infty$ in deep inelastic scattering.

Aside from pushing further on questions like masses, the extension[29] to $q^2 \neq 0$ current induced transitions, the relationship[62] of V and PCAC, etc., what is most needed is a deeper understanding of why we can get away with such simple assumptions — why can we abstract anything relevant about transformed current operators from the free quark model? Even given that, why can we recognize so clearly the hadrons corresponding to the constituent quark model states? Why aren't the multiplets more badly split in mass and mixed? Most of all, to answer these and other questions we need at least part of the dynamics, at which point we might be able to calculate magnitudes of the matrix elements as well.

REFERENCES

1. M. Gell-Mann, Phys. Rev. 125, 1067 (1962).
2. S. L. Adler, Phys. Rev. Letters 14, 1051 (1965); W. I. Weisberger, Phys. Rev. Letters 14, 1047 (1965).
3. The theory may alternately be formulated in terms of light-like charges and hadron states at rest. This more modern language makes clearer some of the theoretical questions which arise, but for the charges and operators we consider here the two formulations give equivalent results.
4. Between states with $p_z \to \infty$, the z and t components of the vector and axial-vector currents have equal matrix elements. Therefore, the integral over all space of $A_z^\alpha(\vec{x}, t)$ has the same matrix elements as $Q_5^\alpha(t)$.
5. R. Dashen and M. Gell-Mann, Phys. Letters 17, 142 (1965).
6. H. J. Lipkin and S. Meshkov, Phys. Rev. Letters 14, 670 (1965), and Phys. Rev. 143, 1269 (1966); K. J. Barnes, P. Carruthers, and F. von Hippel, Phys. Rev. Letters 14, 82 (1965).
7. For a discussion of the constituent quark model states and the assignment of observed hadrons see F. J. Gilman in Experimental Meson Spectroscopy — 1972, AIP Conference Proceedings No. 8, edited by A. H. Rosenfeld and K. W. Lai (American Institute of Physics, New York, 1972); p. 460; R. H. Dalitz, invited paper presented at the Triangle Meeting on Low Energy Hadron Physics, Nov. 5-7, 1973 and Oxford preprint, 1974 (unpublished); J. L. Rosner, Phys. Reports (to be published) and Stanford Linear Accelerator Center Report No. SLAC-PUB-1391, 1974 (unpublished); M. Bowler, invited talk at this conference.
8. Older work on current algebra representation mixing is reviewed by H. Harari in Spectroscopic and Group Theoretical Methods in Physics (North Holland, Amsterdam, 1968), p. 363, particularly for baryons, and in F. J. Gilman and H. Harari, Phys. Rev. 165, 1803 (1968) for mesons. For some recent results see A. Casher and L. Susskind, Phys. Rev. D9, 436 (1974); M. Ida, Kyoto preprint RIFP-182, 1973 (unpublished); and T. Kuroiwa, K. Yamawaki, and T. Kugo, Kyoto preprint KUNS-277, 1974 (unpublished).
9. M. Gell-Mann, in Elementary Particle Physics, P. Urban, ed. (Springer Verlag, New York, 1972), p. 733.
10. Such a transformation was suggested by R. F. Dashen and M. Gell-Mann, Phys. Rev. Letters 17, 340 (1966) in connection with saturating local current algebra. The free quark model transformation given there is just that finally settled on by Melosh, ref. 2.

384

11. A phenomenological scheme for transforming charges is discussed by F. Buccella et al., Nuovo Cimento 69A, 133 (1970) and 9A, 120 (1972).

12. H. J. Melosh IV, Caltech thesis (1973) (unpublished); and Phys. Rev. D9, 1095 (1974). See also ref. 10.

13. S. P. de Alwis, Nucl. Phys. B55, 427 (1973); S. P. de Alwis and J. Stern, CERN preprint TH. 1679, 1973 (unpublished); E. Eichten, J. Willemsen and F. Feinberg, Phys. Rev. D8, 1204 (1973).

14. W. F. Palmer and V. Rabl, Ohio State University preprint, 1974 (unpublished). See also E. Celeghini and E. Sorace, Firenze preprint, 1974 (unpublished); and T. Kobayashi, Tokyo preprint TUETP-73-29, 1973 (unpublished).

15. H. Osborn, Caltech preprint CALT-68-435, 1974 (unpublished).

16. Under the chiral SU(3) × SU(3) subalgebra of the SU(6)$_W$ of currents, the two terms in Eq. (6) transform as $(8,1) - (1,8)$ and $(3,\bar{3}) - (\bar{3},3)$, respectively.

17. The operators D^α_+ have $J_z = +1$. The corresponding operators D^α_- with $J_z = -1$, and all their matrix elements, are related (up to a sign) by a parity transformation. Hence we need only consider the properties of D^α_+.

18. Under the chiral SU(3) × SU(3) subalgebra of the SU(6)$_W$ of currents the four terms in Eq. (9) transform as $(8,1) + (1,8)$, $(3,\bar{3})$, $(8,1) - (1,8)$, and $(\bar{3},3)$ respectively.

19. J. C. Carter, J. J. Coyne, and S. Meshkov, Phys. Rev. Letters 14, 523 (1965) and S. Meshkov, private communication; C. L. Cook and G. Murtaza, Nuovo Cimento 39, 531 (1965). W-spin Clebsch-Gordan coefficients are just those of SU(2).

20. F. J. Gilman and M. Kugler, Phys. Rev. Letters 30, 518 (1973).

21. A. J. G. Hey and J. Weyers, Phys. Letters 48B, 263 (1973).

22. F. J. Gilman, M. Kugler and S. Meshkov, Phys. Letters 45B, 481 (1973).

23. F. J. Gilman, M. Kugler and S. Meshkov, Phys. Rev. D9, 715 (1974).

24. A. J. G. Hey, J. L. Rosner, and J. Weyers, Nucl. Phys. B61, 205 (1973).

25. A. Love and D. V. Nanopoulos, Phys. Letters 45B, 507 (1973).

26. F. J. Gilman and I. Karliner, Phys. Letters 46B, 426 (1973).

27. A. J. G. Hey and J. Weyers, Phys. Letters 48B, 69 (1974).

28. F. J. Gilman and I. Karliner, Stanford Linear Accelerator Center Report No. SLAC-PUB-1382, 1974 (unpublished).

29. F. E. Close, H. Osborn, and A. M. Thomson, CERN preprint TH. 1818, 1974 (unpublished).

30. C. Becchi and G. Morpurgo, Phys. Letters 17, 352 (1965).

31. D. Faiman and A. W. Hendry, Phys. Rev. 173, 1720 (1968); ibid. 180, 1572 (1969); L. A. Copley, G. Karl, and E. Obryk, Phys. Letters 29B, 117 (1969) and Nucl. Phys. B13, 303 (1969).

32. See, for example, R. P. Feynman, M. Kislinger and F. Ravndal, Phys. Rev. D3, 2706 (1971).

33. Such schemes center around adding an $L = 1$ "spurion" in a 35: see J. C. Carter and M. E. M. Head, Phys. Rev. 176, 1808 (1968); D. Horn and Y. Ne'eman, Phys. Rev. D1, 2710 (1970); R. Carlitz and M. Kislinger, Phys. Rev. D2, 336 (1970). Specific broken SU(6)$_W$ calculational schemes with the same general algebraic structure as the theory considered here have been developed by L. Micu, Nucl. Phys. B10, 521 (1969); E. W. Colglazier and J. L. Rosner, Nucl. Phys. B27, 349 (1971); W. P. Petersen and J. L. Rosner, Phys. Rev. D6, 820 (1972);

W. P. Petersen and J. L. Rosner, Phys. Rev. D7, 747 (1963). See also D. Faiman and D. Plane, Nucl. Phys. B50, 379 (1972).

34. The algebraic structure of broken SU(6)$_W$ schemes and their relation to the present theory are discussed in Ref. 24.

35. The relation between various quark model, SU(6)$_W$, broken SU(6)$_W$, and constituent to current quark transformation calculations of pion decay amplitudes is discussed by H. J. Lipkin, NAL preprint NAL-PUB-73/62-THY, 1973 (unpublished).

36. See the discussion in Ref. 20, particularly footnote 13.

37. N. Barash-Schmidt et al., "Review of Particle Properties," Phys. Letters 50B, No. 1 (1974).

38. S. U. Chung et al., Brookhaven preprint BNL 18340, 1973 (unpublished); V. Chaloupka et al., CERN preprint, 1974 (unpublished); U. Karshon et al., Weizmann Institute preprint WIS-73/44-Ph, 1973 (unpublished) and references to previous work therein.

39. This is essentially an updated version of results found in F. J. Gilman, M. Kugler, and S. Meshkov, Refs. 22 and 23, and also in J. L. Rosner, Ref. 7. The decay widths are calculated from the expressions in Table I of ref. 23, with quark model assignments as described there.

40. We have not treated the decay of the I = 0 scalar mesons into $\pi\pi$ in Table III because of the unclear situation in assigning the observed states to the quark model multiplet. For recent assessments see J. L. Rosner, Ref. 7 and D. Morgan, Rutherford preprint RL-7-063, 1974 (unpublished).

41. This is an updated version of Table II of F. J. Gilman and I. Karliner, Ref. 28. The decay widths can be calculated from Table I of Ref. 28, with the quark model assignments as described there.

42. J. Rosen, invited talk presented at this conference.

43. M. E. Nordberg et al., Cornell preprint CLNS-239, 1973 (unpublished).

44. F. J. Gilman, in Baryon Resonances - 73 (Purdue University, West Lafayette, 1973), p. 441.

45. The relative signs of these resonant amplitudes in theories with the same general algebraic structure have been considered by R. G. Moorhouse and N. H. Parsons, Nucl. Phys. B62, 109 (1973) in the context of the quark model and by D. Faiman and J. L. Rosner, Phys. Letters 45B, 357 (1973) in the context of broken SU(6)$_W$.

46. D. Herndon et al., LBL Report No. LBL-1065, 1972 (unpublished). See also U. Mehtani et al., Phys. Rev. Letters 29, 1634 (1973).

47. R. J. Cashmore, lectures presented at the Scottish Universities Summer School in Physics, Stanford Linear Accelerator Center Report No. SLAC-PUB-1316, 1973 (unpublished).

48. A. H. Rosenfeld et al., Stanford Linear Accelerator Center/LBL preprint SLAC-PUB-1386/LBL 2633, 1974 (unpublished); R. S. Longacre et al., Stanford Linear Accelerator Center/LBL preprint SLAC-PUB-1390/LBL 2637, 1974 (unpublished).

49. R. J. Cashmore et al., Stanford Linear Accelerator Center/LBL preprint SLAC-PUB-1387/LBL 2634, 1974 (unpublished).

50. Tables IV and V update the comparison of theory and experiment contained in Table VIII of Ref. 23 (see also Ref. 45). The ordering in angular momentum and isospin Clebsch-Gordan coefficients is the same as in Ref. 23 (see particularly footnote 59) even though there is a changed isospin convention in the new experimental papers, Ref. 48.

51. D. Faiman, Weizmann Institute preprint WIS-74/16 Ph, 1974 (unpublished).

52. D. W. G. S. Leith, invited talk at this conference.

53. A. Bramon, Nuovo Cimento Letters $\underline{8}$, 659 (1973) and references to previous work therein.

54. In calculating Eqs. (13) we are assuming that the "ϕ-like" state decouples from $\pi\rho$ in order to relate the SU(3) singlet and octet couplings. The results follow directly from Table I of Ref. 23. See also, J. L. Rosner, Ref. 7.

55. U. Karshon et al., Weizmann Institute preprint WIS-74/2-Ph, 1974 (unpublished) and references to earlier work therein.

56. J. Diaz et al., Phys. Rev. Letters $\underline{32}$, 262 (1974).

57. F. J. Gilman and H. Harari, Phys. Rev. Letters $\underline{19}$, 723 (1967) and Ref. 7.

58. S. Weinberg, Phys. Rev. $\underline{177}$, 2604 (1969).

59. This result has been discussed in a somewhat different context by C. Boldrighini et al., Nucl. Phys. 22B, 651 (1970).

60. Note that the properties of the states and operators under $J_x \pm i J_y$ are at the same level as those under M^2, i.e., both depend not just on kinematics but on dynamics. One should not expect to have correct properties with respect to $J_x \pm i J_y$ (i.e., the angular conditions) unless one also has a reasonable mass spectrum. This presumably applies to the free quark model and any conclusions to be drawn from it (see Ref. 15). I thank R. Carlitz and W. K. Tung for emphasizing this to me. A similar philosophy is expressed by S. P. de Alwis and J. Stern, CERN preprint TH. 1783, 1974 (unpublished).

61. For a recent attempt in this direction, see F. Buccella, F. Nicolo, and A. Pugliese, Nuovo Cimento Letters $\underline{8}$, 244 (1973).

62. The connection of chiral representation mixing and the Nambu-Goldstone mechanism of chiral symmetry realization is explored by R. Carlitz and W. K. Tung, University of Chicago preprint EFI 74/19, 1974 (unpublished).

CHARM: AN INVENTION AWAITS DISCOVERY*

Sheldon Lee Glashow
Harvard University, Cambridge, Massachusetts 02138

A most important question in experimental meson spectroscopy is to determine what are the hadronic quantum numbers. Charm, a conjectured strong interaction quantum number for which the theoretical raison d'etre is is all but compelling, has not yet been found in the laboratory. I would bet on charm's existence and discovery, but I am not so sure it will be the hadron spectroscopist who first finds it. Not unless he puts aside for a time his fascination with such bumps, resonances, and Deck-effects as have been discussed at length at this meeting. Charm will not come so easily as strangeness, yet no concerted, deliberate search has been launched.

WHAT IS CHARM?

It is a new hadronic quantum number suggested in '64 and not yet found which should lead to a new level of associated production. Well above threshold for the production of two oppositely charmed hadrons, the production cross sections for charmed hadrons should be comparable to those for strange hadrons: at high energies charm is not a small effect.

In other words, charm is a fourth kind of quark. Nucleons and pions are made up of p and n quarks and antiquarks. Strange particles contain one or more λ quarks which each carry the same electric charge as the n quark and one unit of strangeness. Analogously, charmed hadrons contain one or more p' quarks which each carry the same electric charge as the p quark and one unit of charm. Similarity of the masses of p and n quarks shows itself as approximate isospin invariance; a higher λ quark mass leads to broken but still observable SU(3) symmetry; an even higher p' quark mass leads to an SU(4) symmetry so badly broken as not yet to have been seen in nature.

The lowest-lying charmed hadrons are those containing just one p' or $\bar{p}$' quark. The lightest charmed baryons (pnp', etc.) are a $Q = 1$ isosinglet and a $Q = 2,1,0$ isotriplet, and may be produced in association with

*Work supported in part by the NSF under grant GP40397X.

oppositely charmed mesons ($p\bar{p}'$, $n\bar{p}'$), comprising an iso-doublet with Q = 0, -1.

When first invented, the justification for charm was just symmetry: Because there are four leptons, there also should be four kinds of quarks. Then, the weak couplings of leptons and quarks are identical. Just an extrapolation of antique ideas of lepton-hadron symmetry. Today, better reasons have been found for the existence of charm.

GLEEK: THREE OF ANYTHING
(The Oxford Universal Dictionary)

Consider these three logically independent propositions involving the word three:
1. Baryons are made of three quarks.
2. There are three kinds of quark: p n, λ.
3. Each quark comes in three colors.
The first statement seems to be true: we cannot sensibly construct baryons from other numbers of quarks. We allege the second statement to be false: we need four kinds of quarks The last statement is optional: there can be three colors, or ten colors, or various other larger numbers of colors. But, the simplest possibility is surely three. This irrelevant digression is offered only to allay possible confusions.

THE NEED FOR CHARM

Imagine a conventional (pre-gauge-theory) model of weak interactions mediated by a charged W. Cabibbo's hadronic weak current, $\bar{p}\gamma_\mu (1 + \gamma_5)(n \cos\theta + \lambda \cos\theta)$,

offers an entirely adequate description of weak phenomena in lowest order perturbation theory. But look what happens in higher order: we obtain $\Delta S = 2$ non-leptonic amplitudes and $\Delta S = 1$ couplings of neutral lepton currents of order $G(G\Lambda^2)$ where Λ is the cutoff momentum). Since these amplitudes are measured to be very small, Λ cannot exceed several GeV. So low a cutoff is distressing to the theorist. This is the problem solved by the introduction of charm, and of a new term in the weak hadronic current, $\bar{p}'\gamma_\mu (1 + \gamma_5)(\lambda \cos\theta - n \sin\theta)$.

Unwanted amplitudes cancel in the SU(4) limit, and otherwise are of order $G(G\Delta^2)$ where Δ is a measure of the mass
difference between p and p' quarks. The problem is
solved to all orders if only the p' mass is not too
large: <u>charmed hadrons cannot weigh more than several
GeV</u>.

What is a problem in conventional models is even
more a problem in currently fashionable and probably
relevant gauge theories. It is promoted to a difficulty
in first-order perturbation theory. Gauge theories
necessarily involve the neutral current got by commuting
the charged current with its adjoint: this is the source
of the famous neutral current effects recently discovered
at CERN and NAL. But, this procedure applied to Cabibbo's
current yields a neutral current carrying strangeness,
which causes the appearance of the unwanted phenomena in
lowest order. Charm, of course, saves the day. With its
charmed addition, the modified weak current yields a
truly neutral current carrying neither charge nor
strangeness.

Not only does charm solve the problem of unwanted
amplitudes neatly and completely, and to all orders in
perturbation theory, but no alternative solution seems
to work. The vector dominance methodology of Bars,
Halpern, and Yoshimura generally leads to parity violation at order α; the pseudo-Cabibbo alternative of Georgi
and Glashow is probably ruled out by experiment. Charm
seems the only way.

WHAT CHARMED HADRONS DO

We have said that they weigh several GeV, that they
cannot be too heavy lest the unwanted phenomena reappear.
Surely made in associated production at NAL, they may
even appear in pictures already taken. They must have
short lifetimes compared to hyperons and kaons. Because
their decay amplitudes depend on $\cos\theta$ and not $\sin\theta$, their
decay rates are enhanced by a factor of twenty. Another
factor of at least five and more probably fifty comes
from their high masses. We expect lifetimes shorter
than 10^{-12} sec. and probably of order 10^{-13} sec. They
should decay both semi-leptonically (producing a charged
lepton pair and hadrons) and non-leptonically, probably
with comparable amplitudes. Although they will have a

wide variety of decay modes, the decay products will almost always include a strange particle. This is because the p' quark prefers to become a λ quark (via $\cos\theta$) than an n quark (via $\sin\theta$) in its weak couplings.

HOW THE HADRON SPECTROSCOPIST CAN FIND CHARM

He can search for V's--a particle with lifetime 10^{-13} sec. and $\gamma = 100$ travels 0.3 cm. before decaying. He can look for <u>very</u> sharp bumps in $K\pi$, $K\pi\pi$ etc. in complex-topology events at high energy. He can look for prompt leptons in events producing strange particles. He can probably think of more ingenious experiments, but because the charmed particles are produced in pairs, the typical missing-mass experiment--so important for the spectroscopy of normal hadrons--is quite useless.

HOW THE NEUTRINO PHYSICIST CAN FIND CHARM

Here we produce single charmed particles, not associated pairs. The charmed current, $\overline{p}'\gamma_\mu(1 + \gamma_5)\lambda$ is inoperative below charm threshold. It is also inoperative above charm threshold, unless there are some p' or λ quarks or antiquarks in the nucleon. Assuming this is the case, we expect truly remarkable things to happen when the neutrinos are sufficiently energetic.

A word about charm threshold: it is <u>not</u> simply the mass of the lightest charmed baryon, M_B. For $\overline{\nu}$ scattering, charm threshold is $M_M + m(\Lambda)$ where M_M is the mass of the lightest charmed meson. For ν scattering, charm threshold is the smaller of $M_M + m(p) + m(K)$ and $M_B + m(K)$. To see this, it is only necessary to remember that the weak current changes p' to λ (or vice versa) and to keep track of the quarks.

When the neutrino energy somewhat exceeds charm threshold, charm production will take place in the region of y near one and x near zero, with W at or above charm threshold. One anticipates a characteristic anomaly in this kinematic region. At larger energies, the disease spreads to smaller y and larger x. It should be easiest to detect these effects in $\overline{\nu}$ scattering which at low energies displays a $(1 - y)^2$ dependence in accord with the naive quark picture. A sudden change in this distribution, in the total cross sections, or in their ratio

$\sigma(\overline{\nu})/\sigma(\nu)$ would signal charm production.

Looked at closely, the charmed events should be truly anomalous. They must generally involve two strange particles--one from the production of the charmed state, and one from its decay. A significant fraction of the charmed events will involve two oppositely-charged leptons, one coming from the semi-leptonic decay of the charmed state.

HOW THE e^+e^- ANNIHILIST CAN FIND CHARM

We bravely apply the naive quark model to e^+e^- annihilation even though simple scaling results seem to fail. The virtual photon produces a pair of "free" quarks which then rearranges itself into hadrons. In this rearrangement, many pairs of quarks must be produced as fill between the back-to-back primary quark-anti-quark. We assume that the fill is composed predominantly of (light) p and n quarks and anti-quarks. This implies the following properties of the final states. When the available energy is insufficient to produce a pair of charmed particles, only the couplings of the pnoton to p, n, and λ quarks are relevant:

A) Seventy-Five percent of the events will be I = 1, the rest I = 0.

B) One sixth of the events (i.e., when the primary pair is $\overline{\lambda}\lambda$) will contain a $\overline{K}K$ pair.

These properties do seem to describe what is seen at SPEAR and CEA. When enough energy is available the p' quark will also participate, and will cause an abrupt increase in the I = 0 cross section. The total hadronic cross-section should increase by 5/3, and the preceding properties are modified to become:

A') Forty-Five percent will be I = 1, the rest I = 0.

B') One-tenth of the events will contain a $\overline{K}K$ pair from a primary $\overline{\lambda}\lambda$ pair. An additional four-tenths of the events will involve a pair of charmed mesons. Because each charmed meson produces a kaon in its decay, fully half the events will contain a $\overline{K}K$ pair.

Thus, the signals for charm production are: a sudden increase in the yield of kaons and of the overall multiplicity, and the appearance of prompt leptons coming from the semi-leptonic decays of charmed mesons.

WHAT TO EXPECT AT EMS-76

There are just three possibilities:

1. Charm is not found, and I eat my hat.

2. Charm is found by hadron spectroscopers, and we celebrate.

3. Charm is found by outlanders, and you eat your hats.

UPPER LIMITS ON THE PRODUCTION OF PARTICLES WITH CHARM QUANTUM NUMBERS IN pp COLLISIONS AT 400 GeV/c*[#]

T. Ferbel

University of Rochester, Rochester, New York 14627

(On Behalf of the Michigan-Rochester Collaboration)[†]

ABSTRACT

Upper limits are discussed for the production of particles with charm quantum numbers in pp collisions at 400 GeV/c.

In his presentation at the recent conference on Experimental Meson Spectroscopy, Glashow discussed the consequences of introducing an additional SU(3) singlet quark, having "charm" quantum number, for providing an aesthetically pleasing lepton-hadron symmetry and a natural suppression mechanism for neutral strangeness-changing weak currents. A particularly interesting consequence of Glashow's scheme is the predicted existence of new charmed particles which decay weakly having lifetimes in the range of 10^{-13} to 10^{-11} sec.[1] These particles could be produced in hadronic collisions, in pairs having a net charm of zero. The masses of these charmed objects would be expected to be ≤ 2 GeV (even for baryons), and their production cross sections at high energies would be comparable to yields observed for K-meson production in pp collisions.

In this note we wish to point out that rather small upper limits for the production of charmed particles, having the previously described properties, are already available, and, consequently, if the suggested model is to be useful in the area of weak neutral currents, modifications might be needed to account for the lack of observation of charmed particles in hadronic collisions.

The 30-inch NAL bubble chamber is well suited for detecting the production of short-lived objects which have macroscopic decay mean-free paths (ℓ). In particular, the scanning efficiency for neutral particles which

*This note was presented as a comment by the author at the conference on Experimental Meson Spectroscopy (Held at Northeastern Univ. 1974).

#Research supported by the U.S. Atomic Energy Commission.

†C. Bromberg, T. Ferbel, P. Slattery (Rochester), J. Cooper, R. Diamond, A. Seidl and J. C. Vander Velde (Michigan).

decay into two or more particles (even number of charged tracks) for 3 cm $\leq \ell \leq$ 30 cm is almost 100%. Similarly, the efficiency for observing charged particles which decay into three or more particles (odd number of charges) is also excellent for 3 cm < ℓ < 30 cm. If charmed particles are produced in pp collisions with momenta in excess of 100 GeV/c, their mean-free paths would lie, typically, in this sensitive range.

We have scanned ~ 20,000 frames of film from an exposure of the 30-inch NAL bubble chamber to 400 GeV/c protons. A search was made for anomalous decays of neutral and charged particles produced in primary pp collisions. In particular, the film was scanned for topologies which involved a materialization of 2, 4, 6 or more tracks (even in number), or the decay of a charged track into 3, 5, 7 or more odd-charged prongs, on any frame which contained a primary pp interaction.

A total of ~ 700 V's (two charged particles which suddenly materialize in the bubble chamber) were observed. In addition, one 8-pronged neutral decay candidate, and one 3-pronged charged-particle decay candidate were also observed. The measurable V events were consistent with being interpretable as $K^O_s \to \pi^+\pi^-$ (~ 200 events), $\Lambda^O \to p\pi^-$ (~ 80 events), or $\bar{\Lambda}^O \to \bar{p}\pi^+$ (~ 10 events) decays, or with being e^+e^- pairs resulting from materializations of photons (~ 400 events). The 8-pronged "V" can be interpreted as resulting from an interaction near the secondary vertex, caused by one of the prongs from a neutron interaction (odd-pronged star). The 3-pronged event is not a K-decay, but could be interpreted as a secondary interaction yielding a 4-pronged event having a low-momentum proton recoil which is too short to observe in the bubble chamber. We consider the two anomalous events as representing upper limits for the production of charmed particles.

The cross sections for K^O_s, Λ^O, and $\bar{\Lambda}^O$ production are 6 ± 1 mb, 3.4 ± 0.6 mb and 0.40 ± 0.15 mb, respectively.[2] Assuming charmed particle masses (M) between 0.5 GeV and 4 GeV, having average momenta of 160 GeV/c (i.e., produced forward in the pp center of mass), and, further, supposing charged-particle decay modes for the charmed objects, we calculate and display in Fig. 1 approximate upper limits for the production of charmed particles as a function of their expected mean lives. For comparison we provide the value of the total cross section for inclusive production of antiprotons at 400 GeV/c.[3] We wish to note that the detection efficiency for observing possible charmed particles remains substantial for mean lives $\geq 10^{-11}$ because of the additional low-momentum source (production in the backward hemisphere of the center of mass) which we have ignored in obtaining Fig. 1.

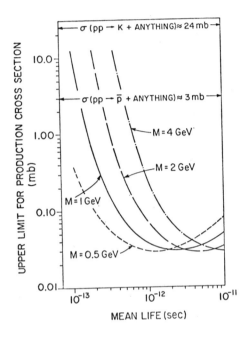

Fig. 1. Upper limits for the production of charmed particles in pp collisions at 400 GeV/c. The results are displayed as a function of the masses and mean lives of the hypothesized objects.

From the graph presented in Fig. 1 we conclude that if charmed particles are produced in pp collisions at high energies as copiously as K mesons ($\sigma^K_{TOT} \sim 24$ mb) then either their mean lives must lie below the expected range of 10^{-13} to 10^{-11} sec, or their masses must exceed 4 GeV. From a theoretical viewpoint, these alternatives do not appear to be appealing ones.

I wish to thank S. Prasad for helpful comments.

REFERENCES

1. For a previous discussion of these SU(4) ideas see Bjorken and Glashow, Phys. Letts. 11, 255 (1964).
2. A. Seidl et al., Bull. Am. Phys. Soc. 19, 467 (1974). A similar study at 102 GeV/c also provided no evidence for charmed particles.
3. M. Antinucci et al., Lett. Nuovo Cimento 6, 121 (1973).

396

CHAIRMAN'S COMMENTS
A SURVEY OF THE THEORY CONTRIBUTIONS

R. L. Arnowitt
Department of Physics, Northeastern University
Boston, Massachusetts 02115

ABSTRACT

A brief survey is given of some of the highlights of
the theory contributions to the Conference.

The Program Committee, in its infinite wisdom, has chosen a
theorist to chair this final session of EMS '74 on Prospects in Meson
Spectroscopy. While in certain ways it might be amusing to have a
theorist's comments on an experimentalist's comments on the future of
meson spectroscopy, discretion suggests that it might be more appropriate for this chairman to restrict his discussion to the theoretical
contributions to the Conference. It is somewhat heartening to see
the large amount of theory presented at a conference, which by its
very name, is mainly experimental in nature. The theoretical proposals have ranged from concrete efforts to deduce $\pi\pi$ and $K\pi$ phase
shifts to the **proposals** of Glashow, and utilized most of the current
ideas of high energy theory. We will say here a few words about these
theoretical suggestions. [1]
The quark model has been highly useful in understanding the
systematics of hadron spectroscopy, and one cannot help but be
impressed by its successes in describing the baryon states. The case
for the boson states is not so clear-cut. Even leaving aside the
missing and questionable entries of Fig. 9 in Dr. Leith's talk [2]
(which hopefully more diligent experimental investigation will
uncover [3]) a number of difficulties exist. Experimentally,
there may be as many as four $J^P = 0^+$ mesons below 2 GeV [4] and there
is clearly a large amount of non-resonant background in this channel.
Also, the Q region remains in a state of darkness. Thus the existence
or non-existence of the A_1 (which is, of course, an intrinsically
interesting issue) is not the most serious question facing the quark
picture, and one must deal with both background and resonances in some
channels.
The above problems suggest the need for dynamical rather than
just group theoretical input for the quark model. An interesting
dynamical proposal by Melosh and others was described in the talk by
Gilman. [5] Melosh distinguishes between two types of quarks
"constituent" quarks and "current" quarks. The former are the usual
quarks of the observed hadron states, while the latter make the
various currents appear simple. By postulating a relation between
the two types of quarks, information on the particle matrix elements
of the currents can be obtained. Thus predictions of electromagnetic
processes and pion processes (via PCAC) can be made. At present, the
theory predicts a number of signs of resonant amplitudes and a number

of decay widths. The former are in rather good agreement with present data analyses, [5] while there are sizable discrepancies for a large number of the decay widths. [6] Clearly better dynamical hypotheses are still needed. [7]

The talk by Quigg [8] was concerned mainly with the application of Regge theory to meson spectroscopy. The idea of duality [that the sum of direct channel resonances is equivalent to crossed channel Regge poles, while the Pomeron arises from direct channel background] has been implemented by the finite energy sum rules (FESR) and has had many successes in baryon phenomena. It is thus interesting to see these ideas also working well for the $\pi\pi$ system. This is particularly heartening since detailed models based on duality, such as the Lovelace-Veneziano model, give poor fits to the $\pi\pi$ phase shifts. [9]

A particularly exciting new possibility is the determination of very high energy $\pi\pi$ cross sections by means of inclusive reactions. As first pointed out by Lusignoli and Srivastava, [10] the inclusive process $\pi^- + p \rightarrow n + X$, for small p-n momentum transfer, should be governed by a peripheral pion. The inclusive cross section (summed over all X) then turns out to be proportional to $\sigma_{\pi\pi}$. (The attendent Chew-Low problems of extrapolating to the pion pole, of course, still exist.) At NAL energies, kinematics can allow $\sigma_{\pi\pi}$ to be determined in this fashion up to c.m. energies of about 10 GeV. Similarly, K induced and photo induced inclusive reactions determine the πK and γK cross sections. The high energy behavior of the $\pi\pi$ cross section has been considered by theorists for many years, and its experimental determination would be of very great interest.

A comprehensive discussion of different theoretical approaches to the problem of determining the $\pi\pi$ and Kπ phase shifts below 2 GeV were presented in the talk of Nath. [9] While many approaches can fit part of the data, two procedures appear to be in accord with the present data for a large number of phase shifts and over a wide range of energy: (1) Those based on the Roy equations (and the closely related fixed-t dispersion techniques) and (2) those based on current algebra methods. The Roy equations are rigorous consequences of crossing symmetry and analyticity. However, in their actual use in determining the lower partial waves one must insert in information concerning the higher partial waves, and the high energy behavior of all partial waves. Thus how to parameterize this input data may be debated. The current algebra procedure is based on the presumably valid ideas of chiral algebra, PCAC and CVC. However, in their implementation a smoothness and narrow resonance approximation is usually made. In addition, both methods have an arbitrariness as to how unitarity is imposed.

In spite of the above questions, both methods are generally in quite good agreement with the various measured $\pi\pi$ phase shifts. (One can see that this is non-trivial by looking at some of the theory that doesn't fit the data! [9]) In addition, the current algebra produces a reasonable fit to the present πK phase shift data. There is, however, one rather striking disagreement between theory and experiment, and this surprisingly at threshold. Both the Roy

equations and current algebra give a P-wave scattering length close
to the Weinberg value of [9] $a^1_1 \simeq 0.033 m_\pi^{-3}$. The experimental
phase shift analyses seem to imply a value of 2-3 times larger. [4]
Also, the experimental P-wave effective range appears to be small
while the theoretical one is large. What theoretical effects could
account for such a discrepancy? Perhaps the Roy equations require
some special parameterization near the P-wave threshold, and perhaps
there is a remarkably large hard meson current algebra threshold correction. [11] Further, the experimental a^1_1 comes close to violating
the Olsson sum rule for the quantity [12] $(2a^0_o - 5a^2_o) - 18a^1_1$.
Thus if the present data in the vicinity of the P-wave threshold is
substantiated, it would imply the existence of new effects in a
region that hitherto fore had been thought to be well understood.
Further determinations of the threshold parameters (both scattering
lengths and effective ranges) would be greatly welcomed by the
theorists.

FOOTNOTES

1. While Glashow's proposals undoubtedly possess CHARM, we will
   refrain from commenting on them here.
2. D. W. G. S. Leith, Proceedings of this Conference.
3. However, as pointed out by C. Quigg (Proceedings of this
   Conference) one may say that "to a good first approximation, no
   new mesons have been established since Meson Conferences began!"
4. W. Männer, Proceedings of this Conference.
5. F. J. Gilman, Proceedings of this Conference.
6. See, e.g., F. J. Gilman, M. Kugler and S. Meshkov, Phys. Letters
   48B, 481 (1973).
7. Thus the transformation between constituent and current quarks
   may depend on the quark interaction structure (in contrast to
   Melosh's simple assumption).
8. C. Quigg, Proceedings of this Conference.
9. P. Nath, Proceedings of this Conference.
10. M. Lusignoli and Y. Srivastava, Lett. Nuovo Cimento 4, 550
    (1972). See also C. Quigg, Proceedings of this Conference. It
    is not necessary to assume the triple-Regge expansion to arrive
    at the basic results.
11. Large hard meson corrections near threshold do exist in πN
    scattering (though they tend to accidentally self-cancel there).
    A similar mechanism doesn't seem to exist for the boson case,
    however.
12. See Ref. 9, Eqs. (3.2) and (3.3) and discussion following these
    equations.

PROSPECTS IN MESON SPECTROSCOPY*

David W. G. S. Leith

Stanford Linear Accelerator Center, Stanford University, Stanford, Ca., 94305

ABSTRACT

A review of the prospects for experimental meson spectroscopy is presented. The history of baryon spectroscopy is given and the lesson for meson studies is underlined. Several areas of new opportunity for meson experiments are described — namely, the new multiparticle spectrometers, their application to photoproduction investigations, and the new generation of e^+e^- colliding beam facilities. Accumulating evidence for the existence of meson towers is presented. Finally, the possible use of high energy and large momentum transfer to give a new (more democratic) look at the meson world, is discussed.

INTRODUCTION

Before taking our "perspective" on the future of meson spectroscopy, I feel that it is important to say something to justify the very large efforts in new experiments and data analysis which are the natural extension of what we have heard these last two days, and which is also the strong conclusion of this review. I am also partly prompted to make these pious remarks as a defensive reaction to the title the organizing committee were originally bandying about — "Is there any point doing experiments below 50 GeV?"

I think the answer to that question is "yes", and that, at least for the moment, the study of meson spectroscopy is both important and fundamental. The goal of our high energy physics endeavours is to find the basic building blocks of matter, to explain the forces between them and the dynamics of their interactions. In our attempt to achieve these ends we make models which supposedly describe nature. In turn, these models have something to say about how the basic constituents of matter "dress" themselves for the observable world and of their properties. One of the important ways we have of learning more, is by confronting the predictions of these theories with data on the spectrum of particles which exist in nature, and on their specific properties. Therefore, the study of meson spectroscopy is not merely a botanical collecting of numbers, not just "stamp collecting", but it is, in fact, one of the main testing grounds of our understanding of fundamental processes.

One of the other conclusions of this meeting must be that there is a great deal still to do in the field of meson spectroscopy and that there are very many unanswered questions close to the heart of our understanding of the classification of resonant states.

In the following sections we will review the history of baryon spectroscopy and the lessons that it holds for meson spectroscopy, (Section II); look at the classification of the known meson states and what needs to be done to clarify the situation, and consider the new experimental opportunities becoming available — (the multiparticle spectrometers (MPS, OMEGA, LASS); the new possibilities for photoproduction studies; and the new generation of e^+e^- storage

---

*Work supported by the U.S. Atomic Energy Commission.

ring facilities) — Section III; gather the evidence for the existence of towers of meson states, (Section IV); briefly exort new efforts in search of exotic states (Section V); and finally present the possibility that high energy and large P_T collisions may be a useful tool to look at the meson world, (Section VI).

BARYON HISTORY AND ITS LESSON FOR MESON SPECTROSCOPY

Baryon spectroscopy is in rather good shape and has, in recent years, been the major testing ground for our theoretical ideas on the classification of resonant states. The good progress in this field is in large proportion due to the intense and systematic programs of many groups over the last 10-15 years. An indicator of the quality of the data is given in Fig. 1. The total cross sections have been accurately measured every (5-10) MeV, and the $\pi^{\pm}p$ elastic and charge exchange cross sections measured every (10-20) MeV up to 2500 MeV. Systematic polarisation measurements for $\pi^{\pm}p$ elastic scattering complete the story. Detailed analysis of this data — the elastic phase shift analysis programs of Lovelace[1], and Bareyre[2] — have provided most of the information on N* resonances to date. It is interesting to note the scale of these analyses — the recent study of πN scattering by Lovelace[2] used some 25,000 data points. The situation for Λ's and Σ's is also in fair shape thanks to a comparably systematic study of the s-channel K^-p reactions,[3] ($K^-p \to K^-p$, $\overline{K}^0n$, $\pi\Lambda$, $\Sigma\pi$).

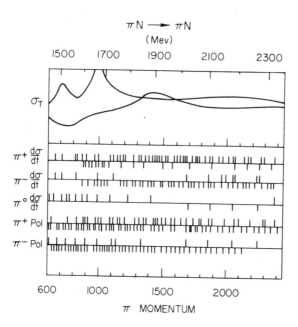

FIG. 1--Summary of data used in elastic phase shift analysis.

To see the sensitivity of the partial wave analysis technique, let us examine the total cross sections for π^-p as shown in Fig. 2. One sees clear indications of the bump structure which was initially identified as the 1st, 2nd, 3rd πN resonances. However, below the cross sections, shown as arrows, are the various πN resonances found in the phase shift analysis described above. The number of resonant states far exceeds the visible structure in the cross section. These analyses are approaching their limit of sensitivity — resonances with couplings to the πN channel $x_{\pi N} \gtrsim 0.1$ are being found, but states with weaker coupling will have to be investigated in production experiments where they may be formed in a more favorable channel (e.g. via ρ-exchange).

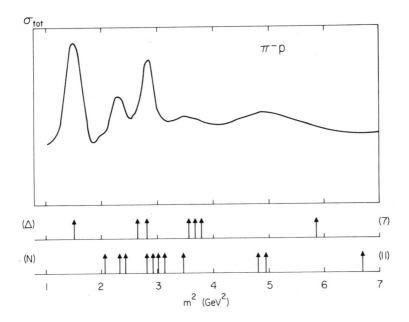

FIG. 2--π^-p total cross section structure and the resonances found in partial wave analyses.

We know a great deal about the resonances (mass, width, $x_{\pi N}$, J, P, I-spin) from these partial wave analyses, but we can never learn of the sign of the coupling constant (since that always appears squared in elastic processes),

$$\sigma(\pi N \to \pi N) \propto g_{\pi N}^2$$

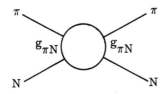

FIG. 3--Formation and decay of a resonance in elastic channel.

nor do we have any information on how the N* states couple to other channels. As Gilman[4] has pointed out, it is just these parameters — the coupling signs and the branching ratio into the various decay modes — which we need to know for the classification of these states into the various supermultiplets.

To investigate these parameters one must study inelastic scattering. Some attempts at study of inelastic πN scattering have been undertaken in which the Dalitz Plot moments have been analysed, or cuts on the data have been made to isolate a quasi-two body reaction, (e.g. $\pi N \to \Delta \pi$).[5] These studies were interesting and gave useful information on the dominant resonance structure, but were unable to make fullest use of the information inherent in the data. (Such treatment looses information on the interference between states of different J, or different P, or different magnetic substate J_z — such interferences are very sensitive probes of the resonant structure within multibody reactions).

Recently a SLAC-LBL collaboration[6,7] has studied inelastic πN scattering in the reaction $\pi N \to \pi\pi N$ at energies up to 2000 MeV using the isobar model.[8] There, the presence of overlapping resonances in the final state is taken into account, and an attempt is made to make use of the information contained in the interferences between the specific amplitudes. The transition amplitude for reaching a given final state is written as the coherent sum of the two body processes

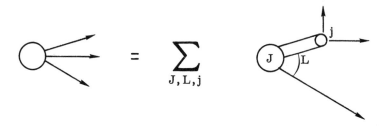

FIG. 4--Schematic description of the isobar model.

The amplitude is written

$$T(W, \omega_1, \omega_2, \alpha, \beta, \gamma)$$

$$= \sum_{J,L,S,I,\ell} A^{IJLS\ell}(w)\, C^I\, X^{JLS\ell}(W,\omega_1,\omega_2,\alpha,\beta,\gamma)\, B^L(\omega_1,\omega_2)$$

ω_1, ω_2 — energies, α, β, γ — Euler angles

C^I — C. G. coefficient products

X — ang. mom. decomposition factor

B — final state factor — (B.W. or Watson Theorem)

A — the partial wave amplitudes.

The data on all single pion production processes were then fit at each center of mass energy, to determine the partial wave amplitudes, $A^{IJLS\ell}$. For $\pi N \to \pi\pi N$, we have three possible two-body final states: $\Delta\pi$, ρN, ϵN.

Figure 5 shows the cross section for one of the single pion production reactions, $\pi^- p \to \pi^+\pi^- n$, from (1300-2000) MeV. Below, an indication is given of the resonances found in each angular momentum wave from the partial wave analysis (PWA). Clearly, the PWA has been able to dig out much more structure than we could see in the cross section.

This analysis of the inelastic scattering allows an almost complete description of πN scattering up through 2000 MeV. In Table I the N* resonances found in this energy region are listed, together with their various decay modes.

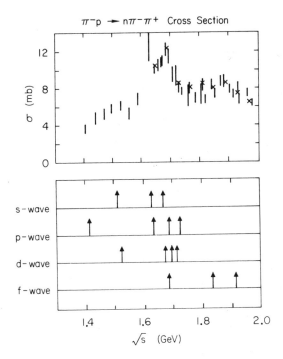

FIG. 5--The cross section for the reaction $\pi^-p \rightarrow \pi^+\pi^-$ in below 2000 MeV, and the resonant structure found from a partial wave analysis.

The final column shows the sum of all the measured branching ratios; it is interesting to observe that almost all the resonance coupling has been accounted for in this analysis.[7]

The inelastic partial wave analysis of the $\pi N \rightarrow \pi\pi N$ reaction also provides an Argand diagram for each wave. The results for one wave, the P11 wave,[9] are reproduced in Fig.6. The Argand diagrams for the reactions $\pi N \rightarrow \pi N$, $\pi\Delta, \rho N$ and ϵN are each displayed on the left, while on the right-hand side the energy dependence of the total inelastic cross section, and the square of the scattering amplitude for each of the inelastic reactions is plotted. Two resonances are found in this wave at 1415 MeV and 1730 MeV,[6] and from inspection of the energy dependence of the scattering amplitude one can see that the lower mass state couples to $\Delta\pi$ and ϵN, while the 1730 MeV state couples to $\Delta\pi$, ρN and ϵN. On the Argand plots on the left, one can see the circular anticlockwise motion of the amplitude, characteristic of resonant structure. The coupling strength to each decay mode is given by the amplitude of the circular motion, while the sign of the coupling is given by whether the amplitude points up (positive) or down (negative) on the Argand plot, at the resonance energy.[10]

Gilman[4] and others,[11] have emphasized that coupling signs and branching ratios are very important for the classification of resonant states. Since the nucleon and the delta, (and the rho-meson and the pion) belong to the same SU(6) super-multiplet, the reactions

$$\pi N \rightarrow \pi N , \qquad \pi N \rightarrow \pi\Delta$$

and
$$\pi N \rightarrow \pi N , \qquad \pi N \rightarrow \rho N$$

are related by SU(6) Clebsch-Gordan coefficients. This allows the intermediate isobar state to be well characterized by its decay properties; this method was very successfully applied to the classification of several new states found in the πN isobar analysis.[7,12]

In summary, the lesson from baryon spectroscopy is that high statistics, systematic experiments together with sophisticated partial wave analysis

Table I N* Couplings for LBL–SLAC Isobar Analysis.

Resonance	Mass	πN	$\pi\Delta$	ρN	ϵN	Σx_i
S_{11}	1520	✓		✓	✓	0.4
S_{11}	1675	✓	✓	✓	✓	0.8
S_{31}	1625	✓	✓	✓		1.0
P_{11}	1415	✓	✓		✓	1.0
P_{11}	1730	✓	✓	✓	✓	1.0
P_{13}	1695	✓		✓		1.0
P_{33}	1900	✓	✓			1.0
D_{13}	1525	✓	✓	✓		1.0
D_{13}	1710	✓	✓	✓	✓	1.0
D_{33}	1725	✓	✓	✓		1.0
D_{15}	1660	✓	✓			1.0
F_{15}	1680	✓	✓	✓	✓	1.0
F_{35}	1870	✓	✓	✓		1.0
F_{37}	1930	✓	✓	✓		0.8

studies (and not just bump hunting) are necessary for a complete probe of the full resonant structure. These studies must include both the elastic and the inelastic scattering channels; the former is probably the most sensitive probe of resonant structure (providing the coupling is large enough), but the latter is required for information on coupling signs and branching ratios, which are important in the classification of the states.

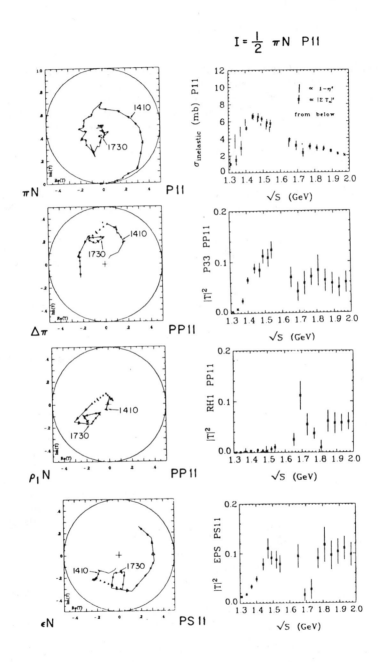

FIG. 6--The Argand plots and the energy dependence of the inelastic cross section and the scattering amplitudes for the $\pi\Delta$, ρN and ϵN sub-states, all for the P11 wave. These results are taken from the analysis of Ref. (6).

MESON SPECTROSCOPY

A. General.

The mesons are in a much sadder situation — or as Fox described it so poetically — "The Great Meson Scandal". The reasons for the discrepancy between the mesons and baryons are clear and well known — mesons may only be observed in production reactions, and therefore must couple strongly to a healthy exchange trajectory; form factors complicate production as a function of momentum transfer; the meson cloud round the proton (or the t-channel couplings) provide a very "low luminosity" target for meson experiments; — and we could go on.

An example of the complication in the meson world is found in considering the equivalent of the total cross section data discussed above. Figure 7 shows the missing mass spectrum in $\pi^- p$ collision at 11 GeV/c, where the recoil proton was detected.[13]

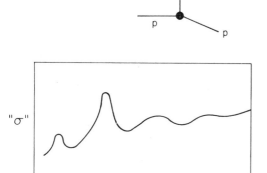

This may be considered as a "total cross section" in a meson-meson collision. However, it is really the sum of many total cross sections, with each term having its own s- and t-dependence:

$$"\sigma" = a\,(s,\,t) \cdot \sigma\,(\pi\,\mathbb{P})$$

$$+\,b\,(s,\,t) \cdot \sigma(\pi\pi)$$

$$+\,c\,(s,\,t) \cdot \sigma(\pi\,\rho)$$

$$+ -- \;\; -- \;\; --$$

Fortunately, the lesson from baryon spectroscopy is that a detailed study of this structure is not very useful, and that we must settle down to the long, systematic programs of study of exclusive two-body and multibody scattering.

FIG. 7--The missing mass spectrum in the reaction $\pi^- p \to p(MM)$, which may be interpreted as a "total cross section" in a meson-meson scattering problem.

One area in which meson studies have some advantage over the corresponding baryon experiments is that the selection rules governing particle decays are much more stringent. For example, the πN channel couples to all I, J^P isobar states, while statistics, J^P, C, G, I — all combine to limit the various meson couplings. Table II[14] displays the effect of the selection rules for a few meson channels, to illustrate the point, for each of the J^P states in the L = 1 and L = 2 quark model supermultiplets. We notice that some decay modes, $(\pi\rho,\ \pi\omega)$, are quite selective while others, like $K^*\overline{K}$ or $K^*\overline{K}^*$, couple to most spin-parity states. These properties may be very useful in different circumstances; for example, we might use one of the very selective channels to attempt to identify a very rare state, or use a decay mode like $\overline{K}K^*$ when measuring relative coupling signs through interferences of many overlapping resonances.

Table II Selection Rules for Meson Decays

DECAYS	0++		1++		1+-		1--		2++		2--		2-+		3--	
	I=0	I=1	I=0	I=1	I=0	I=1	I=0	I=1	I=0	I=1	I=0	I=1	I=0	I=1	I=0	I=1
$\pi\pi$	√							√	√							√
K$\bar{\text{K}}$	√	√					√	√	√	√					√	√
$\pi\omega$						√		√						√		√
$\eta\omega$					√		√				√		√			
$\pi\rho$			√	√			√				√	√	√	√		
K$\bar{\text{K}}$*			√	√	√	√	√	√	√	√	√	√	√	√	√	√
K*$\bar{\text{K}}$*	√	√	√	√	√	√	√	√	√	√	√	√	√	√	√	√

The emphasis for meson spectroscopy is then to concentrate on amplitude structure in elastic and inelastic exclusive processes. Some substantial progress has been made already:

— the π-π scattering studies, especially the beautiful experiment and analysis of the CERN Munich Group, described by Manner[15] at this conference;

— the K-π scattering analyses from LBL,[16] CERN[17] and now in progress at SLAC.[18]

These analyses have brought out the dominant elastic meson-meson structure very clearly, and the continuing work on the π-π data[15] is now beginning to see exciting signs of underlying structure below the ρ, f and g mesons.

— the inelastic scattering analysis package of Ascoli,[19] which has been successfully used in several studies of $\pi \rightarrow 3\pi$ and K $\rightarrow$ K$\pi\pi$;

— the development of the LBL-SLAC s-channel isobar model program to deal with production reactions, reported by Lasinski[20] at this meeting, and being applied to $\pi \rightarrow 3\pi$ studies at LBL,[21] and to K $\rightarrow$ K$\pi\pi$ at SLAC.[22]

It is good to have these two independent partial wave analysis programs based on different assumptions and using quite different fitting methods. Ascoli's package fits the density matrix elements, ρ_{ij}, while the program described by Lasinski attempts to fit the scattering amplitudes.

— new insights on these analyses of inelastic scattering in production reactions should come from the analysis of a Cal-Tech— LBL-SLAC[23] experiment studying;

$$\pi^\pm p \rightarrow \pi^\pm_f (p\pi^+\pi^-) \qquad \text{at 14 GeV/c}$$

using a hybrid spark chamber — bubble chamber set-up. A comparison of the N$\pi\pi$ amplitudes from this experiment with those obtained in the s–channel SLAC-LBL experiment[6,7] described above, should bring to light any unusual, unexpected t-channel effects for these production amplitude analyses. (See Fig. 8.)

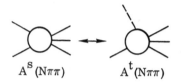

$A^s(N\pi\pi)$ $A^t(N\pi\pi)$

FIG. 8--The comparison of the amplitude for $N^* \rightarrow N\pi\pi$ derived from s-channel studies (see Refs. 6, 7) and from t-channel studies (see Ref. 23).

B. Classification.

The talk by Gilman[4] discussed the quark model and some aspects of classification. Let me quickly recap the rules of the model:

— quarks belong to the 3 group of SU(3), and are the elementary units of all SU(3) representations;

— mesons are made up of q$\bar{\text{q}}$ pairs, which implies that they will be found in the 8, and 1 SU(3) representations and only in those groups;

— the quarks have spin 1/2, which implies that mesons will have integral spin;

— the q$\bar{\text{q}}$ system undergoes orbital excitation, such that the parity of a system, $P = (-1)^{L+1}$ and the total angular momentum, $J = \underline{L} + \underline{S}$.

Thus, within the quark model, we expect supermultiplets with L=0,1,2, — where L is the orbital angular momentum in the q$\bar{\text{q}}$ system. The resulting multiplet structure is summarized in Fig. 9, where we have also tried to assign known meson states to the various J^{PC} multiplets. The table also gives an indication of whether the I=0 members of the various multiplets might be expected to be unmixed (like the η, η'), magically mixed (like the ω, ϕ) or complicated (ϵ, S* situation). The magically mixed are marked (●) on the right hand side. Remember, magically mixed means that the "ϕ–like meson" has only strange quarks and the "ω–like meson" has no strange quarks, i.e.

$$"\phi" \sim \lambda\bar{\lambda}$$

$$"\omega" \sim \frac{\bar{pp} + \bar{nn}}{\sqrt{2}} \quad .$$

FIG. 9--The L=0,1,2 supermultiplets of the quark model.

L=0	0^{-+}, 1^{--}
L=1	0^{++}, 1^{++}, 1^{+-}, 2^{++}
L=2	1^{--}, 2^{--}, 2^{-+}, 3^{--}

L=0

| 0^{-+} | π | η | η' | K |
| 1^{--} | ρ | ω | ϕ | K* |

(●)

L=1

0^{++}	δ	ϵ	S*	κ
1^{++}	A_1	D	?	Q^A
1^{+-}	B	h^ω	h^ϕ	Q^B
2^{++}	A_2	f	f'	$K^{*'}$

(●)
(●)
(●)

L=2

1^{--}	ρ'	?	?	?
2^{--}	F_1	?	?	?
2^{-+}	A_3	?	?	L
3^{--}	g	ω'	ϕ'	$K^{*''}$

(●)
(●)

A few comments are in order:

— there are candidates for each of the J^{PC} multiplets, for each of the quark model supermultiplets; the L = 0 supermultiplet is complete, the L = 1 is almost complete (although there are some reservations about the diffractively produced states in the 1^{++} and 1^{+-} nonets), while the L = 2 system has only the bare bones showing at present.

— the I = 0 states are a real "disaster area", and need a lot of work. For the magically mixed states (see Fig. 9: 1^{--}, 1^{+-}, 2^{++}, 2^{--}, 3^{--}), the "ϕ" meson may be expected to couple more strongly to K⁻-induced reactions than to π-induced reactions, in analogy to the observed ϕ production.[24] (See Fig. 10.) In this case the K⁻ cross section is ~50 times greater than the π cross section. These "ϕ-like" states will decay to K$\overline{\text{K}}$ (except for 1^{++}, 1^{+-}, 2^{--}, 2^{-+} which are forbidden), $\overline{\text{K}}$K* and K*K*. The non-strange quark, I = 0 states should couple to channels like $\pi^0\rho^0$, $\pi\delta$, $\pi\omega$, πA_2.

— the A_1 and Q states are another problem area, as we heard in the talks by Ascoli,[19] Lasinski[20] and Jones.[25] We expect mesons with the quantum numbers and the masses of these diffraction produced enhancements to complete the quark model classification scheme discussed above. However, no resonant-like behavior of the 1^+ phase is observed for either the (3π)[19] or $(K\pi\pi)$[26] systems in the various amplitude studies which have been reported. The usual plea is to study these states in non-diffractive processes;

$$\pi^+p \rightarrow \pi^+\pi^-\pi^0\Delta^{++}$$

$$\pi^-p \rightarrow K^0\pi^+\pi^-\Lambda$$

$$K^+p \rightarrow K^0\pi^+\pi^-\Delta^{++}$$

$$K^-p \rightarrow \pi^+\pi^-\pi^0\Lambda$$

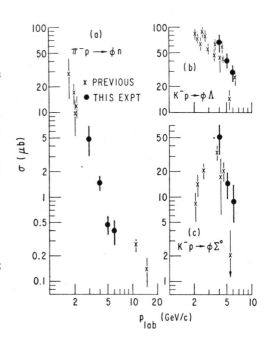

FIG. 10--The cross sections for ϕ-meson production by π^-p and K$^-$p collisions.

However, the high multiplicity of these final states and the small cross sections make them non-trivial experiments. I will comment below on two other reactions which may be favorable for the study of A or Q production,

$$\gamma p \rightarrow A_1 p$$
$$e^+e^- \rightarrow \pi A_1$$
$$\rightarrow \overline{K}Q$$

All of these experiments become more accessible with the new tools we will discuss below.

The typical reactions to be studied in searching for the missing states of in Fig. 9 are listed in Table III. Many of the reactions involve 4, 5, or 6 charged particles which would be an impossible situation for the present generation of forward magnet spectrometers,[27] which have been so successful for the two and three body final state analysis. The new multiparticle spectrometers to be discussed below should be capable of handling these inelastic reactions. There is a whole body of other reactions involving two or more neutrals (e.g. $\pi^0 n$, $\omega\eta$ etc.) which will require additional photon and neutron detectors — but the present instrumentation for the new spectrometers is sufficient to open up a wide vista of important and interesting new channels for PWA.

As the study of meson states progresses to higher masses, several difficulties befall the conventional experimental tools:

— the cross section for production of the states becomes smaller and smaller;

— the total width becomes larger, and the elastic width becomes smaller; (this makes it more essential to use partial wave analysis techniques to search out the resonant structure, and eventually forces the study of inelastic channels to obtain information on these high mass states — see Fig. 11;

— the spin of the leading particle states becomes higher, resulting in more and more

Table III Typical inelastic hadron reactions to be studied with new multiparticle spectrometers.

Decay Mode	Reaction		
$\pi\pi$	$\pi^- p \to \pi^+ \pi^- n$	S I M I L A R	
$K\bar{K}$	$\to K^+ K^- n$		
$\pi^0 \rho^0$	$\pi^- p \to \pi^+ \pi^- \pi^0 n$		
	$\pi^+ p \to \pi^+ \pi^- \pi^0 \Delta^{++}$	M O D E S	
$\pi\omega$	$\pi^- p \to \pi^- \pi^+ \pi^- \pi^0 p$		
$\pi\phi$	$\pi^- p \to \pi^- K^+ K^- p$	F O R	
$\rho\rho$	$\pi^- p \to \pi^+ \pi^- \pi^+ \pi^- n$		K^*
	$\pi^+ p \to \pi^+ \pi^- \pi^+ \pi^- \Delta^{++}$		
			$K\pi$
			$K\phi$
$\pi\pi N$ πB	$\pi^+ p \to \pi^+ \pi^- \pi^+ \pi^- \pi^0 \Delta^{++}$		$K\rho$
			$K\omega$
	$\pi^+ p \to \pi^+ \pi^- \pi^+ \pi^- \Delta^{++}$		$K\pi N$
πA_2			KA_2
	$\pi^- p \to \pi^+ \pi^- \pi^+ \pi^- n$		$K^*\pi$
			$K^*\omega$
$K\bar{K}^*$	$\pi^+ p \to K^+ \bar{K}^0 \pi^- \Delta^{++}$		$\bar{\Lambda}p$
	$\to K^0 K^- \pi^+ \Delta^{++}$		
$K^* \bar{K}^*$	$\pi^- p \to K^+ \pi^- K^- \pi^+ n$		
	$\to K^0 \pi^+ \bar{K}^0 \pi^- n$		

of the production cross section peaking in the very forward (backward) direction — just the region in which the acceptance of the conventional forward spectrometers is poorest;[28]

— the multiplicity of the decay becomes larger, causing problems for the efficient detection of all the final state particles, and also for the design of a good, selective trigger.

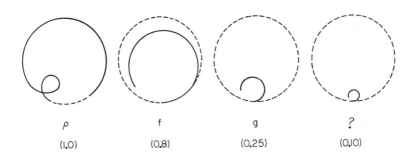

ρ f   g   ?

(1.0)   (0.8)   (0.25)   (0.10)

FIG. 11--Argand plots for the elastic scattering amplitude for four resonances with $x_{el} = 1.0, 0.8, 0.25$ and 0.10.

Finally, a comment on the statistics required for experiments attempting an amplitude (or partial wave) analysis of inelastic reactions. From the LBL-SLAC s-channel[6] N* studies we know that ~ 10,000 events per mass bin are required to find a unique solution for a problem with 60 partial waves. As the statistics are reduced ambiguous solutions appear and typically for an experiment with ~2000 events per bin, two or three different solutions were found. We use this lower estimate of the number of events required to set a scale for the new experiments, and hope that continuity constraints will allow selection of the unique fit from the several ambiguous alternatives.

Suppose now we want to study the process $\pi \rightarrow 3\pi$ over a 2000 MeV region of (3π) mass, in 20 MeV bins and with 2-3 bins in momentum transfer (to see the production mechanism); this study will require an experiment of ~1000 events/μb for a diffractive process, or ~7-10,000 events/μb for a non-diffractive reaction. A similar experiment is shown in Fig. 12, where the

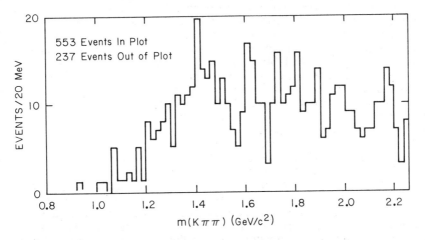

FIG. 12--The effective mass spectrum of (K$\pi\pi$) from the reaction $K^+p \rightarrow (K\pi\pi)\Delta^{++}$ at 10 GeV/c.

mass spectrum for charge exchange $K \to K\pi\pi$ is shown from a 10 GeV/c K^+p HBC experiment.[29] This experiment has a sensitivity of ~20 events/μb; so to obtain our 2000 events/20 MeV mass bin and with 2-3 t bins, we require a K^+p experiment of ~10,000 events/μb sensitivity to successfully perform an amplitude study of the non-diffractive ($K \to K\pi\pi$) scattering.

To summarize, new progress in understanding meson spectroscopy requires:

- systematic, high statistics experiments;
- partial wave analysis in 2-body and multi-body scattering to identify the full resonance structure and determine the properties (e.g. mass, width, spin, parity, isospin, coupling signs and branching ratios);
- typical decay modes to be studied — $\pi\delta$, $\pi\rho$, $\pi\omega$, πA_2, $\overline{K}K^*$, (i.e. multiparticle final states);
- sensitivity required — 1000 events/μb for diffractive processes 10,000 events/μb for non-diffractive processes;
- need detectors with good geometrical acceptance, capable of handling high multiplicity final states and with high data rate capacity to accomplish 10^4 events/μb experiments in a reasonable amount of calendar and accelerator time.

C. The New Tools.

In this section I want to briefly describe several new areas of opportunity in experimental meson spectroscopy and the physics they should impact — (a) new detectors, (b) photon beams, (c) e^+e^- storage rings.

(a) New detectors. There is a new generation of multiparticle spectrometers becoming available — the MPS system at BNL, OMEGA at CERN and LASS at SLAC. The MPS and OMEGA set-ups have been described in detail before,[30] so I will only briefly remind you of them. The general layout of these spectrometers is shown in Fig. 13. Both are large aperture magnets with ~1.5 meters gap and large field volume for the detection and measurement of the secondary tracks. In both cases, the sides of the magnet (the iron flux return path) may be moved and restacked in a flexible geometry to allow specific particle orbits to exit the system, or to insert special purpose instrumentation. The liquid hydrogen target in both systems is positioned within the field volume and is surrounded by spark chambers. The BNL spark chamber system is completely digitized, while the Omega has T.V. camera readout of optical spark chambers. Both are high data rate devices, capable of logging (30-50) events/sec. and of being triggered on highly selective event samples. The types of triggers available at present are slow proton and neutron recoils, fast forward p, $\overline{p}$ or $K^\pm$ and forward produced V^0. Multiwire proportional chambers are being built to provide even more flexible triggers. Particle identification is provided by large Cerenkov counters positioned downstream. Both spectrometers are capable of doing 10^3-10^4 event/μb experiment in a typical 10-day accelerator period.

A new multiparticle spectrometer is currently being assembled at SLAC— LASS — it is shown schematically in Fig. 14. Since it is new and the basic philosophy is somewhat different from the above spectrometers, I will describe it more fully. It is a two magnet system with a conventional forward magnet spectrometer and a target vertex detector. The forward system is the classic field-free "spark chamber — large aperture magnet — spark chamber" geometry, which has proved so powerful (in terms of data rate and resolution) and so economical (in terms of computer time) in the $\pi\pi \to \pi\pi$ and $K\pi \to K\pi$ experiments at CERN, ANL and SLAC. The magnet has a 2 meter $\times$ 1 meter gap

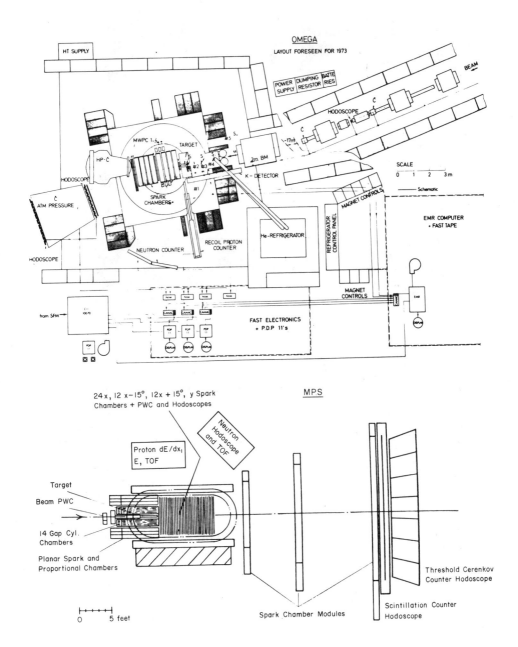

FIG. 13--The layout of the (a) OMEGA, (b) MPS spectrometer facilities.

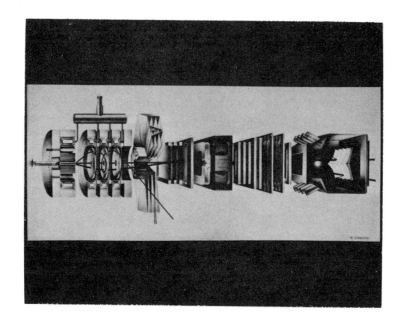

FIG. 14--A schematic of the SLAC LASS system.

with 30 kG m analyzing field integral. The design of the vertex detector was
chosen to be as non-interfering with the forward system, (i. e. as orthogonal
in measurement space), as possible. The magnet is a superconducting solenoid, 2 meters in diameter and 3-1/2 meters long with the 25 kG magnetic
field parallel to the beam. This configuration does not interfere with the
small angle fast tracks, (the peripheral particles), which are measured downstream (since the momentum vector is essentially parallel to the magnetic
field) but measures well the large angle tracks and the slow tracks missed by
the forward system.

The particle identification is provided by two threshold Cerenkov
counters — an atmospheric pressure counter, with 37 separate cells, mounted
at the end of the solenoid, and a 3 atmosphere pressure counter with 8 cells
positioned at the end of the downstream system. The detectors are a varied
mixture of techniques — downstream they are magneto-strictive spark chambers, while in the solenoid and between the two magnets, they are capacitive
read-out spark chambers and proportional wire chambers. This system,
which is directly on-line to an IBM 370/168 computer, is designed for high
data rates, and is able to handle up to 100 events/sec.

The LASS system may be triggered on the forward multiplicity, recoil
proton or forward produced V^0. In addition, two novel triggers are being prepared which deserve further description: (1) the solenoid-axial field causes
the secondary particles to sweep out a helical trajectory, whose diameter is
related to P_T and whose pitch is related to P_L. P_T and P_L are directly measured using the induced signal from three proportional wire chamber wheels,
where the high voltage planes have been etched to give polar coordinate, (R, θ),

information (i. e. the pitch and radius of the trajectory is available digitally). Thus, trigger arrangements may be devised to select specific regions of P_T or P_L or both; (2) the data acquisition system has an inbuilt pause of (3-10) msec. before committing an event to tape, during which time the full arithmetic power of the IBM 370/168 may be brought to bear on making decisions to accept or reject the event, or to set a flag for the control of subsequent analysis. One feature of this setup which should provide a powerful trigger is that effective masses of recoil particles may be calculated in real time, using the digital information from the proportional chambers surrounding the target and immediately downstream. For example, the effective mass of a recoil $\Lambda(\rightarrow p\pi^-)$ or $\Delta^{++}(\rightarrow p\pi^+)$ can be calculated within the 10 msec pause, with an accuracy of ~100 MeV.

The performance and general characteristics of these new devices is summarised in Table IV.

Table IV Summary of MPS, OMEGA and LASS characteristics

MPS (operate in summer 1974)	OMEGA (currently operating)	LASS (operate in fall 1974)
Target in Field Volume and Surrounded by Spark Chambers		
45 kG meters of analysis	54 kG meters of analysis	30 kG meters + 87 kG meters of longitudinal field
35 events/sec	50 events/sec	(40 – 100) events/sec
$\leq$ 9 GeV/c separated	$\leq$ 15 GeV/c unseparated	$\leq$ 16 GeV/c separated
Forward Particle Identification by Downstream Cerenkov Counters		
Triggers: ● Recoil Proton ● Forward V^O ● Forward Multiplicity ● $2V^O$ ● Recoil Neutron	● Recoil Proton ● Recoil Neutron ● Fast V^O ● Forward K, p	● Recoil Proton ● Recoil Δ^{++} ● Forward Multiplicity ● Forward V^O ● P_T, P_L

These new devices represent new opportunities for meson spectroscopy with
— almost 4π acceptance spectrometers,
— good multiparticle detection efficiency,
— high data rate capability (making 10^3-10^4 event/μb experiments possible),
— good measurement resolution,
— particle identification,
— a variety of trigger options
  ● recoil p, n, Λ, Δ^{++}
  ● forward $K^{\pm}$, $p^{\pm}$
  ● forward multiplicity
  ● P_T, P_L

(b) <u>Photon Interactions</u> Photon beams should be an important tool in meson spectroscopy, since the photon is the only stable meson with spin 1. The techniques for making polarized, and even monochromatic photon beams are now well established.[31] One may even study photoproduction with longitudinally polarized photons, using the virtual exchange in inelastic lepton scattering. We may also probe subnuclear structure using the Primakoff effect, which should be a very characteristic signal and quite accessible at high energies; these possibilities were reviewed by Rosen[32] at this meeting. The photon couplings can also be investigated in the study of e^+e^- collisions, which are dominated by the one photon exchange intermediate state; I will talk more of this in a moment.

Historically, the photoproduction experiments[33] have taught us a great deal of reaction dynamics, from the beautiful studies of ρ, ω, ϕ production. However, these experiments, large as they were, have not been sensitive enough to probe new meson structure; typical sensitivity has been ~ 500 events/μb while non-diffractive meson production is expected to have a cross section $\sim 0.1\,\mu$b at energies $\sim$10 GeV. Typical reactions to be studied are listed in Table V; from the quark model calculations of Gilman and Karliner,[34] the cross sections for all these processes are predicted to be approximately equal.

Table V   Some interesting photoproduction reactions

$\gamma p \rightarrow A_2^+ \Delta^0$	$\rightarrow$	$\pi^+ \pi^+ \pi^- p \pi^-$
$(\gamma n \rightarrow A_2^- p$	$\rightarrow$	$\pi^- \pi^+ \pi^- p)$
$\gamma p \rightarrow A_1^+ \Delta^0$	$\rightarrow$	$\pi^+ \pi^+ \pi^- p \pi^-$
$\gamma p \rightarrow B^+ \Delta^0$	$\rightarrow$	$\pi^+ \pi^+ \pi^- \pi^0 p \pi^-$
$\gamma p \rightarrow h^0 p$	$\rightarrow$	$\pi^+ \pi^- \pi^0 p$
$\gamma p \rightarrow \delta^+ \Delta^0$	$\rightarrow$	$\pi^+ \pi^+ \pi^- \pi^0 p \pi^-$
$\gamma p \rightarrow f p$	$\rightarrow$	$\left(\begin{array}{c} \pi^+ \pi^- \\ \pi^+ \pi^- \pi^+ \pi^- \\ K^+ K^- \end{array} \right) p$
$\gamma p \rightarrow f' p$	$\rightarrow$	$K^+ K^- p$
$\gamma p \rightarrow D^0 p$	$\rightarrow$	$\pi^+ \pi^- \pi^+ \pi^- \pi^0 p$

Typical Cross Section at 10 GeV/c $\sim 0.1\,\mu$b

The OMEGA and LASS spectrometers will both embark on experimental programs in photon beams during the next few years. At CERN, the OMEGA facility will use a tagged, polarized photon beam from the SPS.[35] The beam will allow study of photoproduction reactions in the range (10-60) GeV, with useful polarization information in the (30-50) GeV range, and with sensitivities of a few 10^4 events/μb. The beam parameters are summarized in Fig. 15.

LASS has a unique property for photon and electron beam experiments. In the past the main problem for these experiments has been that the electromagnetic background is many orders of magnitude larger than the hadronic cross sections. This background is characterized by having $q^2 = 0$, which means in the solenoid there is no net bending, and therefore, all the background is maintained in a tight bundle on the beam axis. As this background emerges from the solenoid we will perhaps have to "drain" the downstream system using a superconducting tube to exclude the magnetic field from the beam region similar to the one described at the last meson conference by Martin.[36] In these conditions we expect to be able to study photoproduction processes in the (5-20) GeV range, with useful polarization around 15 GeV, with a sensitivity of $\sim 10^4$ events/μb/day.

Another interesting study of meson production with photons concerns the helicity structure in the $\rho\pi$, $\omega\pi$ decays. For example, we know the decay

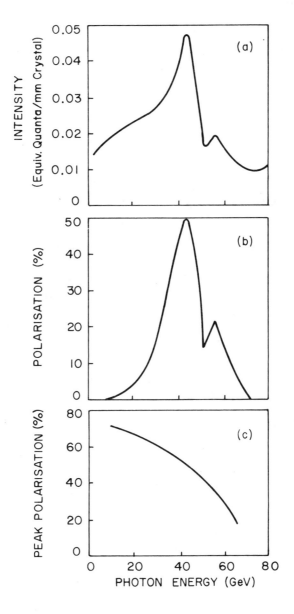

FIG. 15--The intensity and polarization parameters
from the proposed photon beam for the
OMEGA experiments at the CERN SPS.

418

$A_1 \rightarrow \rho\pi$ has dominantly longitudinal ρ's, and $B \rightarrow \omega\pi$ produces dominantly transverse ω's. Thus, B mesons should be strongly observed with real photons while A_1 production should be enhanced using longitudinal photons from inelastic electron scattering. (See Fig. 16.)

We expect many higher mass mesons to couple to the vector mesons, and this trick of using real or virtual photons may be important in their production and will certainly be interesting for studying the helicity structure of their decays.

(c) Electron-Positron Storage Rings

In the last few years we have seen a new area of physics opened up with the effective operation of electron-positron colliding beam facilities. These new tools allow the study of meson spectroscopy through three different pro-

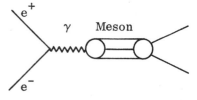

FIG. 16--Diagram for inelastic electroproduction of A_1 meson.

cesses — (i) the s-channel formation of meson states in e^+e^- collisions, (ii) the production of meson states in e^+e^- multiparticle collisions, and (iii) the utilization of the two photon process to form mesonic states in s-channel γ-γ collisions. The characteristics of operating (and proposed) storage ring machines are shown in Fig. 17, where the total energy in the center of mass of the e^+e^- collision is plotted against the luminosity of the accelerator.

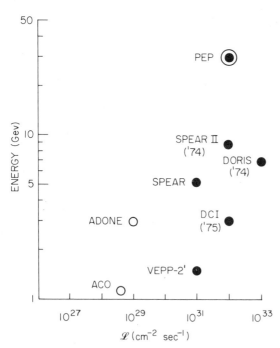

FIG. 17--The present and proposed e^+e^- colliding beam facilities.

Let us consider each of the three e^+e^- processes in turn.

(i) The old generation of storage ring machines (viz. ACO, ADONE, and VEPP) have performed extremely elegant experiments on the study of s-channel formation of the leading vector mesons — ρ, ω, and ϕ — through the one photon annihilation process —

FIG. 18--Formation of meson resonance in e^+e^- s-channel process.

As the total center of mass energy is varied by changing the energy of the stored e^+e^- beams, the experiments scan a range of meson mass and look for an enhancement in the cross section of a given final state. Such an enhancement is interpreted as evidence for the formation of a resonant state, with the same quantum numbers as the photon, (viz. $J^{PC} = 1^{--}$). Evidence of other vector mesons, one at 1600 MeV (the ρ' (1600)), and possible signs of another structure at 1250 MeV, have been reported by groups at ADONE,[37] but all three of these older machines are very limited by their luminosities (i.e. the flux of colliding electrons). The new generation of accelerators — VEPP 2', and DCI facility in Paris — should allow for a detailed scan for new 1^- states up to masses of 3000 MeV, with 10^2-10^3 times the sensitivity. In addition to studying $\pi\pi$ and $K\bar{K}$ channels it should be interesting to look for new vector mesons coupling to inelastic channels like $\pi\rho$, $\pi\omega$, πA_2, $\bar{K}K^*$, πB, etc.

(ii) The production of meson resonances ought to be a very profitable hunting ground for the new e^+e^- facilities — SPEAR I, II and DORIS. (See Fig. 17). We know many states which couple strongly to vector mesons, (see Table VI), and would expect to see such states produced in e^+e^- collisions through the following diagram:

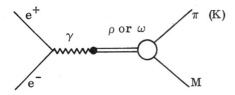

FIG. 19--Production of a meson resonance in e^+e^- multiparticle process.

Such a reaction would be an interesting place to study the A_1, Q question, as they should be seen in

$$e^+e^- \to \text{"}\rho\text{"} \to \pi A_1$$
$$\to \bar{K}Q .$$

Table VI Examples of Mesons decaying to Vector Meson

$\omega \to \rho\pi$
$\pi_N \to \rho\pi$
$\phi \to \rho\pi$
$A_1 \to \rho\pi$
$B \to \omega\pi$
$A_2 \to \rho\pi$
$g \to \omega\pi$
Also $Q \to K\rho$
$K_{1420} \to K\rho$

and the supposed $J=3,4$ K^* also couple to $K\rho$.

Since we know that the decay chain $\omega \to \rho\pi$ provides a good description of $\omega \to 3\pi$ decay characteristics, we may take the $e^+e^- \to \text{"}\rho\text{"} \to \omega\pi$ cross section as setting a scale for the other processes listed in Table VI. Taking this cross section ($\sigma(\omega\pi) \sim 2.10^{-32}$ cm^2 at threshold), and going (500-1000) MeV above threshold to kinematically separate the recoiling particles and assuming an E^{-6} fall-off of the exclusive cross section still leaves a yield of many thousands of events per day at DORIS. Both DORIS and SPEAR have large, nearly-4π magnetic detectors which should be ideal for these studies.

(iii) Although the single photon intermediate state dominates e^+e^- processes, there is considerable interest in studying the two photon exchange process.

The amplitude for this process is down by α^2 compared to the single photon channel, but at high energies it

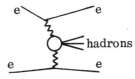

FIG. 20--Formation of a meson state in s-channel γ-γ collision.

begins to compete and becomes an attractive tool in its own right, since the cross section is enhanced by factors of $\ln^2(E/m_e)$. The cross section for the process shown in Fig. 20 has been calculated,[38] and predicts reasonable event rates for SPEAR II, and especially PEP energies (see Fig. 17). The cross sections are shown for three meson mass regions in Table VII.

Table VII Two-Photon Cross Sections

Cross Section (10^{-33}cm^2)		
Mass	$(2 \times 3$ GeV$)$	$(2 \times 15$ GeV$)$
(300-1000)	11	30
(1000-2000)	1.7	6.9
(2000-6000)	0.6	5.1

For $\mathscr{L} \sim 10^{32}$, expect yield of
$1000 < M < 2000$ hadrons, to be
$\sim 6,000$/day at SPEAR II
and $\sim 25,000$/day at PEP energies.

These are basically experiments on γ-γ scattering and therefore allow study of mesons with C = +1. The two photon initial state couples to spin-parity multiplets $0^{++}, 2^{++}, 2^{-+}$. (Since the two photons are almost real, they cannot couple to J = 1 systems.) Such experiments would allow further study of the confusing δ, ϵ, S* states in the L = 1, 0^{++} multiplet, and clarify their relationship to the ϵ', the other s-wave π-π state under the f-meson. In addition, these experiments should allow investigation of the diffractive enhancement called the A_3 (— the Regge recurrence of the pion), in a non-diffractive reaction.

These experiments are very difficult in that the recoil electrons must be detected in addition to the final hadronic products. As so often is the case in storage ring experiments, this requires a close coupling between the design and operation of the storage ring and the experiment itself. Both DORIS and PEP have been designed with such experimental programs in mind. It should be an interesting and profitable area of experimental meson spectroscopy.

Experience from SPEAR is not too encouraging, as far as meson spectroscopy is concerned, as we heard from Chinowsky[39] — the exclusive cross sections fall very fast and there is little resonance structure seen. However, at present the mass resolution is not very good (~ 40 MeV), and the statistics are very poor. Many of you may remember the first results of high energy pp $\rightarrow$ many π's were thought to be completely phase-space-like. Only when much work had been done on these channels and very high statistics were finally achieved did it become clear that the final states were resonance dominated. The resonant signal had been swamped by the diluting effect of the large number of combinations per event in the mass plots. I think for the moment the data is too sparse and the promise too great not to continue to look in e^+e^- annihilation for more information on meson resonances.

TOWERS OF MESONS

In this section I would like to point out that the empirical evidence for the existence of towers of meson states is strengthening. In Fig. 21, the Chew-Frautschi plot for the natural parity mesons is displayed, showing the ρ, f, g-meson leading trajectory, and below it I have circled several fairly established states — ϵ, δ, ρ', and others somewhat more speculative — ϵ', the s-wave π-π effect under the f-meson;[40] an fπ effect around a mass of 1750 MeV seen in the 2^+ wave in the Ascoli analysis of $(\pi \rightarrow 3\pi)$ scattering;[19] a 1^- effect

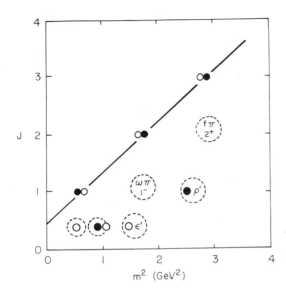

FIG. 21--Chew-Frautschi plot for the natural parity mesons.

near the f-meson coupling to $\omega\pi$, which has been suggested in an ADONE experiment reported at this meeting,[37] and the SLAC-LBL back-scattered laser beam experiment.[41]

Such a tower structure receives tentative support from recent results on high mass π-π phase shifts from the CERN-MUNICH group, discussed by Manner.[15] One of their four possible solutions shows structure in the s-, and p-waves below the f-meson and in the s-, p- and d-waves below the g-meson (see Fig. 22).

Similar regularities have been expected in various theories. The Veneziano model predicts many daughters of alternating I-spin and parity, which were originally introduced to kill the non-observed backward peak in π-π scattering in the region of the ρ- and f-mesons. Table VIII shows the kind of structure which would be expected in this model, and marked as circles are the possible daughter candidates. An alternative view is offered in the quark model, where in addition to the L=0, 1,2 --, supermultiplets discussed in Section III above, one expects radial excitations to appear. Evidence for radial excitations in the baryon system is growing with the reporting of the observation of the first and second radial excitations of the proton.[7] Such a possibility for the mesons is outlined in Fig. 23. It is not our purpose to decide which picture is correct, but to point out that an interesting structure of lower lying states is emerging from our meson studies.

Further support for these suggestions comes from studies of $\pi\pi \to \overline{N}N$ and $N\overline{N} \to \pi\pi$. In Fig. 24, the results of an analysis by Hyams et al[42] on $\pi^- p \to \overline{p}pn$ are presented. They tried to fit the complicated structure in the $\overline{p}p$ angular distributions by allowing a resonance plus background in each partial wave up to $\ell = 6$. The "resonances" resulting from this fit are shown on the Chew-Frautschi plot together with results from analysis of $\overline{N}N \to \pi\pi$. The purpose is not to claim any specific spectroscopic states from these analyses, but to draw attention to the possible existence of structure below the leading trajectory and the formation of towers of states.

One final comment on the influence of the angular momentum barrier on studies of low lying states such as those discussed above. Figure 25 shows a Chew-Frautschi plot with the leading trajectory, and the curve $J \propto k \cdot R$. Here J is the total angular momentum of the π-π state and is taken as being proportional to the product of the center of mass momentum of the pion times their spatial separation, R, which is known to be ~ 1 fermi for non-diffractive hadronic amplitudes. We are all familiar with the inhibition of high mass leading particles coupling to the elastic channel due to the angular momentum

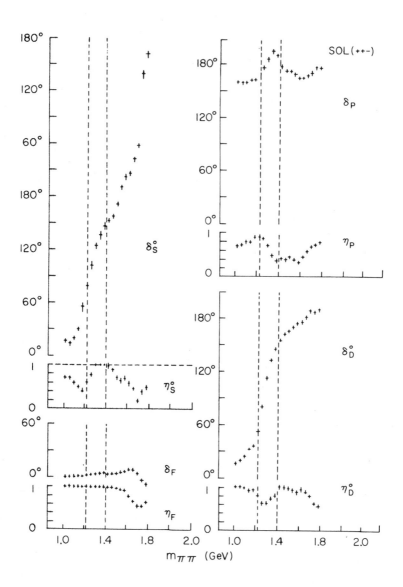

FIG. 22--Preliminary π-π phase shift — one of several
possible solutions. See Ref. (15) for details.

Table VIII Daughter structure in natural parity mesons.

	J^{PG}, I		J^{PG}, I
ρ	$1^{-+}, 1$	ω	$1^{--}, 0$
ϵ	$0^{++}, 0$	δ	$0^{+-}, 1$
A_2	$2^{+-}, 1$	f	$2^{++}, 0$
	$1^{--}, 0$	$\omega\pi$	$1^{-+}, 1$
	$0^{+-}, 1$	ϵ'	$0^{++}, 0$
g	$3^{-+}, 1$	ω'	$3^{--}, 0$
	$2^{++}, 0$	πf	$2^{+-}, 1$
ρ'	$1^{-+}, 1$		$1^{--}, 0$
	$0^{++}, 0$		$0^{+-}, 1$

barrier, but, conversely, low lying states with a given J and k will imply a value of R much smaller than 1 fermi and should be suppressed due to their non-peripherality. Perhaps the small coupling of the ρ' to $\pi\pi$ is a sign of such a suppression. Study of these low lying states will certainly be important in our full understanding of the properties in angular momentum space, of the non-diffractive $\pi\pi$ elastic scattering amplitude. At any rate, as we push forward with studies of these low lying states — and it seems interesting to do so — we are again forced to consider seriously, the inelastic channels.

EXOTICS

Cohen[43] has reviewed the status of experiments searching for exotic mesons at this meeting. The sensitivity of current experiments is ~50-100 events/μb. Since the existence of non q$\bar{\text{q}}$ meson states is such a stringent test of the quark model theories, it is important to push the limit on the existence of exotics as far as possible. The new multiparticle spectrometers — MPS, OMEGA, LASS offer a major increase in sensitivity, making 10^3-10^4 event/μb experiments quite accessible. These experiments should be done.

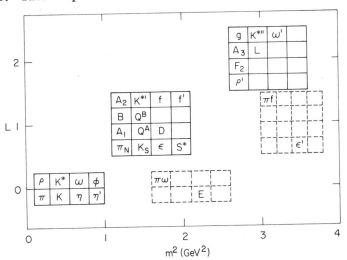

FIG. 23--Radial excitations of L=0 and L quark model supermultiplets.

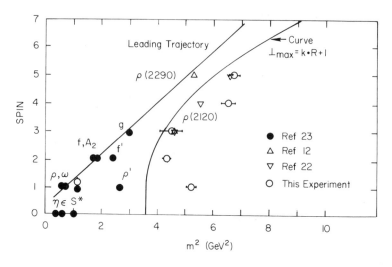

FIG. 24--Chew-Frautschi plot showing possible low lying NN̄ states. See text and Ref. (42) for details.

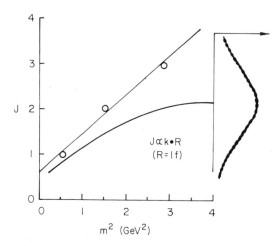

FIG. 25--Chew-Frautschi plot, showing the curve $J \propto k \cdot R$, with $R = 1$ fermi. In addition to angular momentum barrier suppression of leading particles coupling to the π-π channel, one should also expect this coupling of the low lying states to be suppressed due to their "non-peripherality."

HIGH ENERGY FOR DEMOCRACY

Finally I would like to comment on the possible use of very high energy beams to study meson spectroscopy. In all the reactions we have discussed so far, mesons are produced in t-channel exchange processes and so the ones we see are those that couple to healthy exchange amplitudes. This leads to the strong hierarchy we know in the meson world.

Perhaps, large P_T collisions are a way of producing mesons more democratically.

Let me make two observations on this: (a) we know that the cross section for exclusive processes becomes roughly equal at large momentum transfer. In Fig. 26, the differential cross section for several exclusive reactions involving pomeron, π, K, K^* and A_2 exchange, are shown for beam momenta of 4.0 and 5.0 GeV/c.[44] This is an indication of a rather democratic production mechanism dominating the large t region. (b) Let us suppose that the constituent interchange model[45] is a good guide to what is going

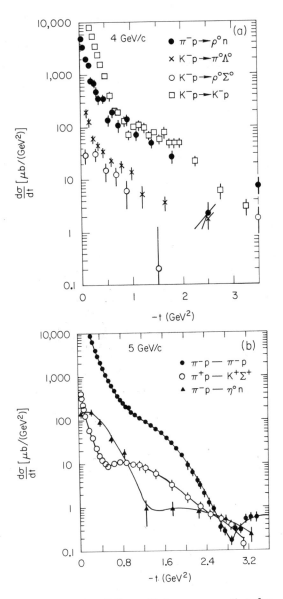

FIG. 26--The differential cross section for several exclusive channels at (a) 40 GeV/c, (b) 5.0 GeV/c, showing the near equality of all the reactions at large t.

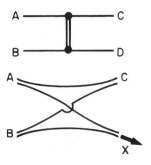

FIG. 27--Schematic diagrams for resonance formation in (a) peripheral collisions, (b) large P_T collisions.

on at large P_T. Then meson production may be thought to proceed as in Fig.
27. For small momentum transfers the exchange is dominated by some Regge-
like mechanism, while for large P_T the reaction proceeds with the interchange
of some fundamental constituent — say, the quarks. The meson final state
"C" is produced when the quark and anti-quark come together to form mesonic
matter. This $q\bar{q}$ state may now project itself onto all the available meson
states, and these states will have an equal probability of being formed, up to a
statistical factor $(2 L + 1)$, where L is the $q\bar{q}$ relative angular momentum. We
then have a situation where we may produce all mesons — ρ's and δ's, A_2's
and D's — with equal probability — an interesting and intriguing possibility.
Such experiments should become possible in the near future at NAL.

CONCLUSIONS

The following points serve to summarize this talk:

- It is important to have data on the meson states and their couplings to
  test the models of constituent interactions and resonance classification;

- from baryons we have learned that this requires systematic high
  statistics work on amplitude analysis for elastic and inelastic reactions;

- new tools are becoming available to help in these tasks

  high data rate spectrometers with good multiparticle acceptance,

  highly tuned partial wave analysis programs,

  new capability with photon beams,

  e^+e^- facilities with high luminosity and 4π magnetic detectors,

- maybe high energy and large P_T will provide a new view of the meson
  world.

ACKNOWLEDGMENTS

I would like to thank my colleagues at SLAC for many helpful discussions
on the subject of this review. Special thanks go to R. K. Carnegie, M. Davier,
F. Gilman and R. Schwitters.

REFERENCES

1. S. Almehed and C. Lovelace Nucl. Phys. B40, 157 (1972).
2. R. Ayed and P. Baregre, Paper presented to the IInd Aix-en-Provence International Conference on Elementary Particles (1973).
3. R.D. Tripp, Paper presented at Berkeley APS Meeting on Particle Physics, p. 188, Berkeley, August 1973.
4. F.J. Gilman, Stanford Linear Accelerator Center Report No. SLAC-PUB-1320, Lectures presented at the Scottish University Summer School, 1973; and Invited talk at IVth International Conference of Experimental Meson Spectrometry, Boston, April 1974.
5. A.D. Brody et al., Phys. Letters 34B, 665 (1971); V. Mehtani et al., Phys. Rev. Letters 29, 1634 (1972); A. Kernan et al., Proceedings of Baryon Resonance Conference, p. 113, Purdue, 1973.
6. D.J. Herndon et al., Stanford Linear Accelerator Center Report No. SLAC-PUB-1108 (Rev.) and LBL 1065 (Rev.) submitted to Phys. Rev. (1974). A.H. Rosenfeld et al., Stanford Linear Accelerator Center Report No. SLAC-PUB-1386 and LBL 2633, submitted to Phys. Rev. Letters (1974). R.J. Cashmore et al., Stanford Linear Accelerator Center Report No. SLAC-PUB-1387 and LBL 2634, submitted to Phys. Rev. Letters (1974).
7. R.S. Longacre et al., Stanford Linear Accelerator Center Report No. SLAC-PUB-1390 and LBL 2637, submitted to Phys. Rev. Letters (1974). R.S. Longacre et al., Stanford Linear Accelerator Center Report No. SLAC-PUB-1389 and LBL 2636, to be submitted to Phys. Rev. R.S. Longacre (Ph. D. thesis) LBL-948 (1973).
8. D.J. Herndon, P. Soding and R.J. Cashmore LBL 543 (1973), submitted to Phys. Rev.
9. The notation used to describe our partial wave amplitudes is described fully in Ref. (6). The four character description gives the incident relative angular momentum, the outgoing relative angular momentum, 2 × the I-spin of the channel and 2 × the total spin, $(L_{inc}, L_{out}, 2I, 2J)$. In addition there is an indication of the sub-particle state being considered $\pi\Delta$, ρN, ϵN: since the spins of the ρ-meson and the nucleus may be combined to form total spin 1/2 or 3/2 there are two amplitudes denoted $\rho_1 N$ or $P_3 N$ respectively.
10. The Argand plots have been made using the "Baryon First" convention. A more detailed discussion is given in Ref. (6).
11. J. Rosner, Invited talk at the Berkeley APS Meeting on Particle Physics, p. 130, Berkeley, 1973 and Stanford Linear Accelerator Center Report No. SLAC-PUB-1391, submitted to Physics Reports.
12. F.J. Gilman, M. Kugler, S. Meshkov, Phys. Letters 45B, 481 (1973) and Stanford Linear Accelerator Center Report No. SLAC-PUB-1286, submitted to Phys. Rev. D. Faiman and J. Rosner, Phys. Letters 45B, 357 (1973).
13. D. Bowen et al., IIIrd International Conference on Experimental Meson Spectroscopy, p. 215 (1972), and Phys. Rev. Letters 29, 890 (1972).
14. R. Huff and J. Kirz, LRL Physics Note 474 (unpublished).
15. W. Manner, talk in these proceedings.
16. A. Barbaro-Galtieri et al., Meeting on π-π Scattering, p. 1 (Florida 1973).

17. Aachen-Berlin-CERN-London-Vienna Collaboration. Paper submitted to Symposium on Multiparticle Dynamics, Leipzig, 1974.
18. G. Brandenburg, R. K. Carnegie, R. J. Cashmore, M. Davier, D. W. G. S. Leith, J. A. J. Matthews, P. Walden, S. Williams, F. Winkelman (private communication).
19. G. Ascoli, talk in these proceedings. Yu. M. Antipov et al., Contribution to IIIrd International Conference on Meson Spectroscopy, p. 164, 1972. G. Ascoli et al., Contributed paper to the 16th International Conference on High Energy Physics, Batavia, 1972.
20. T. Lasinski, talk in these proceedings.
21. T. Lasinski, et al., private communication.
22. R. K. Carnegie et al., private communication.
23. California Institute of Technology-LBL-SLAC Collaboration, private communication.
24. D. S. Ayres et al., Argonne Report ANL-HEP-7410, submitted to Phys. Rev. Letters.
25. L. Jones, talk in these proceedings.
26. Aachen-Berlin-CERN-London-Vienna Collaboration, CERN Reports — CERN/D. Ph II/74-1 and 74-10.
27. CERN-Munich Spectrometer — CERN, G. Grayer et al.,
    Lindebaum-Ozaki Spectrometer — BNL
    K. J. Foley et al., Nucl. Inst. and Meth. 108, 33 (1973).
    Effective Mass Spectrometer — ANL.
    D. S. Ayres, Argonne Report ANL/HEP 7314.
    SLAC Spark Chamber Spectrometer — SLAC
    R. K. Carnegie et al., (private communication)
28. G. Luste and D. W. G. S. Leith, IInd International Conference on Experimental Meson Spectroscopy, p. 593 (1970).
29. K. W. Barnham et al., Nucl. Phys. B25, 49 (1970).
30. A. Michelini, Invited talk presented at the International Conference on Instrumentation for High Energy Physics, Frascati, 1973.
31. Photon beam literature:
    Annihilation beams:
    J. Ballam et al., Nucl. Inst. and Meth. 73, 53 (1968).
    Coherent Diamond beams:
    G. Dianibrini Palazzi, Rev. Mod. Phys. 40, 611 (1968).
    Ash et al., Contribution to the Bonn Conference on Electron-Photon Physics, 1973.
    Polarized Bremsstrahlung by attenuation:
    N. Cabibbo et al., Phys. Rev. Letters 9, 270 (1962)
    Eisele et al., Nuc. Inst. and Meth. 113, 489 (1973).
    Back-scattered laser beam:
    R. Milburn, Phys. Rev. Letters 10, 75 (1963).
    C. K. Sinclair et al., IEEE Trans Nucl. Sci., 16, 1065 (1969).
    Tagged photon beams:
    G. R. Brookes et al., Nucl. Inst. and Meth. 85, 125 (1970).
    D. Caldwell et al., Phys. Rev. D7, 1362 (1973).
32. J. Rosen, talk in these proceedings.
33. K. Moffeit et al., Stanford Linear Accelerator Center Report No. SLAC-PUB-1004, submitted to Phys. Rev. M. Davier et al., Stanford Linear Accelerator Center Report No. SLAC-PUB-1205.

34. F. Gilman and I. Karliner, Stanford Linear Accelerator Center Report No. SLAC-PUB-1382 (1974).

35. T. Armstrong et al., CERN Report CERN/SPSC/74-29.

36. F. Martin, IIIrd International Conference on Meson Spectroscopy, p. 429 (1972).

37. ρ' (1650):
    G. Barbarino et al., Lettere al Nuovo Cimento, 3, 689 (1972)
    M. Grille et al., Nuovo Cimento, 13 A, 593 (1973).
    F. Ceradini et al., Phys. Letters 43B, 341 (1973).
    ρ'(1250):
    M. Conversi et al., Paper presented at this meeting.

38. J. Rosner, Review of Resonances, Stanford Linear Accelerator Center Report No. SLAC-PUB-1391, submitted to Physics Reports (1974).

39. W. Chinowsky, talk in these proceedings.

40. Particle Data Group Tables, T. Lasinski et al., submitted to Rev. Mod. Phys. (1974).

41. J. Ballam et al., Stanford Linear Accelerator Center Report No., SLAC-PUB-1373, submitted to Phys. Rev. (1974).

42. B. Hyams et al., CERN Report, submitted to Nucl. Phys. (1974).

43. K. Cohen, talk in these proceedings.

44. J.A.J. Matthews, talk presented at Berkeley APS Meeting on Particle Physics, Berkeley, 1973.

45. R. Blankenbecler, S. Brodsky and R. Gunion, Phys. Letters 39B, 649 (1972); Phys. Rev. D6, 2652 (1972).

REVIEW OF PARTICLE PROPERTIES

Particle Data Group

Vladimir CHALOUPKA and Claude BRICMAN
CERN, 1211 Genève 23, Switzerland

Angela BARBARO-GALTIERI, Denyse M. CHEW, Robert L. KELLY,
Thomas A. LASINSKI, Alan RITTENBERG, Arthur H. ROSENFELD,
Thomas G. TRIPPE and Fumiyo UCHIYAMA
Lawrence Berkeley Laboratory, University of California,
Berkeley, California 94720, USA

Naomi BARASH-SCHMIDT
Brandeis University, Waltham, Massachusetts 02154, USA

Paul SÖDING
DESY, 2000 Hamburg 52, Germany

Matts ROOS
Department of Nuclear Physics, University of Helsinki,
Helsinki 17, Finland

April 1974

(next edition, April 1975, to be published in Reviews of Modern Physics)

Reprinted from Physics Letters, Vol. 50B, No. 1, April 1974.

Printed at Lawrence Berkeley Laboratory, Berkeley

Meson Table

April 1974

In addition to the entries in the Meson Table, the Meson Data Card Listings contain all
substantial claims for meson resonances. See Contents of Meson Data Card Listings[1].

Quantities in italics have changed by more than one (old) standard deviation since April 1973.

Name $\frac{I\ 0\ 1}{-\ \omega/\phi\ \pi}$ $\frac{}{+\ \eta\ \rho}$ ⊢—⊣estab.	$I^G(J^P)C_n$	Mass M (MeV)	Full Width Γ (MeV)	M^2 $\pm\Gamma M^{(a)}$ (GeV)^2	Mode	Partial decay mode Fraction (%) [Upper limits are 1σ (%)]		p or Pmax[b] (MeV/c)
$\pi^\pm(140)$ $\pi^0(135)$	$1^-(0^-)+$	139.57 134.96	0.0 7.8 eV ±.9 eV	0.019483 0.018217	See Stable Particle Table			
$\eta(549)$	$0^+(0^-)+$	548.8 ±0.6	2.63 keV ±.58 keV	0.301 ±.000	All neutral $\pi^+\pi^-\pi^0 + \pi^+\pi^-\gamma$	71 29	See Stable Particle Table	
ε	$0^+(0^+)+$	≲ 700[c]	≳ 600[c]		$\pi\pi$			
Existence of pole not established. See note on $\pi\pi$ S wave¶.								
$\rho(770)$	$1^+(1^-)-$	770§ ±10§	150§ ±10§	0.593 ±.116	$\pi\pi$ e^+e^- $\mu^+\mu^-$ For upper limits, see footnote (e)	≈100 0.0043±.0005 (d) 0.0067±.0012 (d)		359 585 370
$\omega(783)$	$0^-(1^-)-$	782.7§ ±0.6§	10.0 ±.4	0.613 ±.008	$\pi^+\pi^-\pi^0$ $\pi^+\pi^-$ $\pi^0\gamma$ e^+e^- For upper limits, see footnote (g)	90.0±0.6 1.3±0.3 8.7±0.5 0.0076±.0017	S=1.2* S=1.5* S=1.9*	327 366 380 391
$\eta'(958)$ or X^0	$0^+(\ ^-)+$ J=0 or 2	957.6 ±0.3	< 1	0.917 <.001	$\eta\pi\pi$ $\rho^0\gamma$ $\gamma\gamma$ For upper limits, see footnote (h)	70.6±2.5 27.4±2.2 1.9±0.3	S=1.4* S=1.6*	234 458 479
$\delta(970)$	$1^-(0^+)+$	976§ ±10§	50§ ±20§	0.953 ±.049	$\eta\pi$ $\rho\pi$	seen < 25		315 139
Possibly a virtual bound state of the I = 1 $K\bar{K}$ system¶.								
$S^*(993)$	$0^+(0^+)+$	∿ 993[c] ±5	40[c] ±8	0.986 ±.040	$K\bar{K}$ $\pi\pi$		near threshold	479
See notes on $\pi\pi$ and $K\bar{K}$ S wave¶.								
$\phi(1019)$	$0^-(1^-)-$	1019.7 ±0.3 S=1.9*	4.2 ±.2	1.040 ±.004	K^+K^- $K_L K_S$ $\pi^+\pi^-\pi^0$ (incl. $\rho\pi$) $\eta\gamma$ e^+e^- $\mu^+\mu^-$ For upper limits, see footnote (i)	46.6±2.5 34.6±2.2 15.8±1.5 3.0±1.1 .032±.002 .025±.003	S=1.6* S=1.6* S=1.2* S=1.6* S=1.4*	127 111 462 362 510 499
$A_1(1100)$	$1^-(1^+)+$	∿ 1100	∿ 300	1.21 ±.33	$\rho\pi$	∿ 100		253
Broad enhancement in the $J^P=1^+$ $\rho\pi$ partial wave; not a Breit-Wigner resonance¶.								
$B(1235)$	$1^+(1^+)-$	1237§ ±10§	120§ ±20§	1.53 ±.12	$\omega\pi$ [D/S amplitude ratio = .24±.06] For upper limits, see footnote (j)	only mode seen		352

Meson Table *(cont'd)*

Name	$I^G(J^P)C_n$ estab.	Mass M (MeV)	Full Width Γ (MeV)	M² ±ΓM[(a)] (GeV)²	Partial decay mode		
					Mode	Fraction (%) [Upper limits are 1σ (%)]	p or Pmax[(b)] (MeV/c)
f(1270)	$0^+(2^+)+$	1270§ ±10§	170§ ±30§	1.61 ±.22	$\pi\pi$ $2\pi^+2\pi^-$ $K\bar{K}$	83±5§ 4±1§ 4±3 S=1.5*	619 556 394
					For upper limits, see footnote (f)		
D(1285)	$0^+(A)\pm$	1286§ ±10§	30§ ±20§	1.65 ±.03	$K\bar{K}\pi$ $\eta\pi\pi$ †[$\delta(970)\pi$ $2\pi^+2\pi^-$ (prob. $\rho^0\pi^+\pi^-$)	seen seen seen] seen	305 484 245 565
	$J^P = 0^-$, 1^+, 2^-, with 1^+ favoured						
A₂(1310)	$1^-(2^+)+$	1310§ ±10§	100§ ±10§	1.72 ±.13	$\rho\pi$ $\eta\pi$ $\omega\pi\pi$ $K\bar{K}$ $\eta'(958)\pi$	71.5±1.8 S=1.2* 15.2±1.2 8.6±1.8 S=1.3* 4.7±0.6 <1	413 529 354 428 280
E(1420)	$0^+(A)+$	1416§ ±10§	60§ ±20§	2.01 ±.08	$K\bar{K}\pi$ †[$K^*\bar{K} + \bar{K}^*K$ $\eta\pi\pi$ †[$\delta(970)\pi$	∿ 40 ∿ 20] ∿ 60 possibly seen]	421 130 564 352
f'(1514)	$0^+(2^+)+$	1516 ±3	40 ±10	2.29 ±.06	$K\bar{K}$	only mode seen	572
					For upper limits, see footnote (k)		
F₁(1540)	$1(A)$	1540 ±5	40 ±15	2.37 ±.06	$K^*\bar{K} + \bar{K}^*K$	only mode seen	321
	Evidence based on only one experiment						
ρ'(1600)	$1^+(1^-)-$	∿ 1600	∿ 400	2.56 ±.64	4π †[$\rho\pi^+\pi^-$ $\pi\pi$ $K\bar{K}$	dominant seen with $\pi^+\pi^-$ in S-wave] possibly seen < 8	738 575 788 629
A₃(1640)	$1^-(2^-)+$	∿ 1640	∿ 300	2.69 ±.49	$f\pi$		305
	Broad enhancement in the $J^P = 2^-$ $f\pi$ partial wave; not a Breit-Wigner resonance.¶						
ω(1675)	$0^-(N)-$	1666§ ±10§	142§ ±20§	2.78 ±.24	$\rho\pi$ 3π 5π †[$\omega\pi\pi$	seen possibly seen possibly seen possibly seen]	647 805 778 614
g(1680)	$1^+(3^-)-$	1686§ ±20§	180§ ±30§	2.84 ±.30	2π 4π (incl. $\pi\pi\rho,\rho\rho,A_2\pi,\omega\pi$) $K\bar{K}$ $K\bar{K}\pi$ (incl. $K^*\bar{K}$)	26±5§ ∿ 70 (ℓ) ∿ 2 ∿ 3	831 784 680 621
	J^P, M and Γ from the 2π mode[(ℓ)].						
	See note (1) for possible heavier states.						
K^+(494) K^0(498)	$1/2(0^-)$	493.71 497.70		0.244 0.248	See Stable Particle Table		
K^*(892)	$1/2(1^-)$	892.2 ±0.5	49.8 ±1.1	0.796 ±.044	$K\pi$ $K\pi\pi$ $K\gamma$	≈ 100 < 0.2 < 0.16	288 216 310
		(Charged mode; $m^0 - m^\pm = 6.1±1.5$ MeV)					

Meson Table *(cont'd)*

Name			Mass M (MeV)	Full Width Γ (MeV)	$M^2 \pm \Gamma M^{(a)}$ (GeV)^2	Partial decay mode		
$\frac{G \mid I \mid 0 \mid 1}{\frac{-\mid \omega/\phi\mid \pi}{+\mid \eta \mid \rho}}$	$I^G(J^P)C_n$ estab.					Mode	Fraction (%) [Upper limits are 1σ (%)]	p or $P_{max}^{(b)}$ (MeV/c)

κ	$1/2(0^+)$			δ_0^1 goes slowly through 90° near 1300 MeV.				

See note on Kπ S wave¶.

Q	$K_A(1240)1/2(1^+)$ or C	1242 ±10 127 ±25 seen in p̄p at rest		1.54 ±.16		Kππ	only mode seen	
						$+[K^*\pi$	large]	
	$K_A(1280\ 1/2(1^+)$ to 1400)	1280 to 1400				$+[K\rho$	seen]	
	See note (m).					$+[K(\pi\pi)_{\ell=0}$	possibly seen]	

$K^*(1420)$	$1/2(2^+)$	1421_6 ±5§	100_6 ±10§	2.02 ±.14		Kπ	55.0±2.7	616
				See note (n).		$K^*\pi$	29.5±2.5	414
						Kρ	9.2±2.4	319
						Kω	4.4±1.7	306
						Kη	2.0±2.0	482

L(1770)	$1/2(A)$	1765_6 ±10§	140_6 ±50§	3.11 ±.25		Kππ	dominant	788
						Kπππ	seen	757
						$+[K^*(1420)\pi$ and other subreactions¶]		

$J^P=2^-$ favoured, 1^+ and 3^+ not excluded.

See note (1) for possible heavier states.

(1) Contents of Meson Data Card Listings

	Non-strange (Y = 0)					Strange (|Y| = 1)	
entry	$I^G(J^P)C_n$	entry	$I^G(J^P)C_n$	entry	$I^G(J^P)C_n$	entry	$I\ (J^P)$
π (140)	$1^-(0^-)+$	→ η_N (1080)	$0^+(N)+$	ρ' (1600)	$1^+(1^-)-$	K (494)	$1/2(0^-)$
η (549)	$0^+(0^-)^+$	A_1 (1100)	$1^-(1^+)+$	A_3 (1640)	$1^-(2^-)+$	K^* (892)	$1/2(1^-)$
ε (600)	$0^+(0^+)+$	→ M (1150)		ω (1675)	$0^-(N)-$	κ	$1/2(0^+)$
ρ (770)	$1^+(1^-)-$	→ $A_{1,5}$ (1170)	1^-	g (1680)	$1^+(3^-)-$	Q	$1/2(1^+)$
ω (783)	$0^-(1^-)-$	B · (1235)	$1^+(1^+)-$	→ X (1690)	1^-	K^* (1420)	$1/2(2^+)$
→ M (940)		→ ρ' (1250)	$1^+(1^-)-$	→ X (1795)	1	→ K_N(1660)	$1/2$
→ M (953)	$^+$	f (1270)	$0^+(2^+)+$	→ S (1930)	1	→ K_N(1760)	$1/2$
η' (958)	$0^+(0^-)+$	D (1285)	$0^+(A)+$	→ A_4 (1960)	1^-	L (1770)	$1/2(A)$
δ (970)	$1^-(0^+)+$	A_2 (1310)	$1^-(2^+)+$	→ ρ (2100)	1^+	→ K^*_N(1850)	
→ H (990)	$0^-(A)-$	E (1420)	$0^+(A)+$	→ T (2200)	1	→ K^*(2200)	
S^* (993)	$0^+(0^+)+$	→ X (1430)	0	→ ρ (2275)	1^+	→ K^*(2800)	
φ (1019)	$0^-(1^-)-$	→ X (1440)	1	→ U (2360)	1		
→ M (1033)		f' (1514)	$0^+(2^+)+$	→ N̄N (2375)	0	→ Exotics	
→ B_1(1040)	1^+	F_1 (1540)	1 (A)	→ X(2500-3600)			

Meson Table *(cont'd)*

→ indicates an entry in Meson Data Card Listings not entered in the Meson Table. We do not regard these as established resonances.

¶ See Meson Data Card Listings.

* Quoted error includes scale factor $S = \sqrt{\chi^2/(N-1)}$. See footnote to Stable Particle Table.

† Square brackets indicate a subreaction of the previous (unbracketed) decay mode(s).

§ This is only an educated guess; the error given is larger than the error of the average of the published values. (See Meson Data Card Listings for the latter.)

(a) ΓM is approximately the half-width of the resonance when plotted against M^2.

(b) For decay modes into ≥ 3 particles, p_{max} is the maximum momentum that any of the particles in the final state can have. The momenta have been calculated by using the averaged central mass values, without taking into account the widths of the resonances.

(c) From pole position $(M - i\Gamma/2)$. For both ϵ and S^* the pole is on Riemann Sheet 2.

(d) The e^+e^- branching ratio is from $e^+e^- \to \pi^+\pi^-$ experiments only. The $\omega\rho$ interference is then due to $\omega\rho$ mixing only, and is expected to be small. See note in Meson Data Card Listings. The $\mu^+\mu^-$ branching ratio is compiled from 3 experiments; each possibly with substantial $\omega\rho$ interference. The error reflects this uncertainty; see notes in Meson Data Card Listings. If $e\mu$ universality holds, $\Gamma(\rho^0 \to \mu^+\mu^-) = \Gamma(\rho^0 \to e^+e^-) \times$ phase space correction.

(e) Empirical limits on fractions for other decay modes of $\rho(765)$ are $\pi^\pm\gamma < 0.5\%$, $\pi^\pm\eta < 0.8\%$, $\pi^+\pi^+\pi^-\pi^- < 0.15\%$, $\pi^\pm\pi^+\pi^-\pi^0 < 0.2\%$.

(f) Empirical limits on fractions for other decay modes of $f(1270)$ are $\eta\pi\pi < 15\%$; $K^0K^-\pi^+ + c.c. < 6\%$.

(g) Empirical limits on fractions for other decay modes of $\omega(783)$ are $\pi^+\pi^-\gamma < 5\%$, $\pi^0\pi^0\gamma < 1\%$, η + neutral(s) $< 1.5\%$, $\mu^+\mu^- < 0.02\%$, $\pi^0\mu^+\mu^- < 0.2\%$, $\eta\gamma < 0.5\%$.

(h) Empirical limits on fractions for other decay modes of $\eta'(958)$: $\pi^+\pi^- < 2\%$, $\pi^+\pi^-\pi^0 < 5\%$, $\pi^+\pi^+\pi^-\pi^- < 1\%$, $\pi^+\pi^+\pi^-\pi^-\pi^0 < 1\%$, $6\pi < 1\%$, $\pi^+\pi^-e^+e^- < 0.6\%$, $\pi^0e^+e^- < 1.3\%$, $\eta e^+e^- < 1.1\%$, $\pi^0\rho^0 < 4\%$, $\pi^0\omega + \gamma\omega < 8\%$.

(i) Empirical limits on fractions for other decay modes of $\phi(1019)$ are $\pi^+\pi^- < 0.03\%$, $\pi^+\pi^-\gamma < 0.7\%$, $\omega\gamma < 5\%$, $\rho\gamma < 2\%$, $\pi^0\gamma < 0.35\%$, $2\pi^+2\pi^-\pi^0 < 1\%$.

(j) Empirical limits on fractions for other decay modes of $B(1235)$: $\pi\pi < 15\%$, $K\bar{K} < 2\%$, $4\pi < 50\%$, $\phi\pi < 1.5\%$, $\eta\pi < 25\%$, $(\bar{K}K)^\pm\pi^0 < 8\%$, $K_SK_S\pi^\pm < 2\%$, $K_SK_L\pi^\pm < 6\%$.

(k) Empirical limits on fractions for other decay modes of $f'(1514)$ are $\pi^+\pi^- < 20\%$, $\eta\eta < 50\%$, $\eta\pi\pi < 30\%$, $K\bar{K}\pi + K^*\bar{K} < 35\%$, $2\pi^+2\pi^- < 32\%$.

(ℓ) We assume as a working hypothesis that peaks with $I^G = 1^+$ observed around 1.7 GeV all come from g(1680). For indications to the contrary see Meson Data Card Listings.

(m) See Q-region note in Meson Data Card Listings. Some investigators see a broad enhancement in mass $(K\pi\pi)$ from 1250-1400 MeV (the Q region), and others see structure. The $K\eta$, $K\omega$, and $K\pi$ modes are less than a few percent.

(n) The tabulated mass of 1421 MeV comes from the $K\pi$ mode; the $K\pi\pi$ mode can be contaminated with diffractively produced $Q^\pm$.

Established Nonets, and octet-singlet mixing angles from Appendix IIB, Eq. (2'). Of the two isosinglets, the "mainly octet" one is written first, followed by a semicolon.

$(J^P)C_n$	Nonet members	$\theta_{lin.}$	$\theta_{quadr.}$
$(0^-)+$ or: $(0^-)+$	π, K, η; η' π, K, η; E	$24 \pm 1°$ $16 \pm 1°$	$10 \pm 1°$ $6 \pm 1°$
$(1^-)-$	ρ, K^*, ϕ; ω	$36 \pm 1°$	$39 \pm 1°$
$(2^+)+$	A_2, $K^*(1420)$, f'; f	$29 \pm 2°$	$31 \pm 2°$

OTHER PAPERS SUBMITTED TO EMS '74

Below is a list of papers submitted to EMS '74 other than those which appear in the proceedings. In some cases these papers were not submitted by the authors themselves but by a member of the program committee or others who felt they should be brought to the attention of the conference. The numbers are those assigned by EMS '74. To obtain a copy of an item on this list, please write directly to the authors; their addresses can be found in the list of preprint source addresses published by "Preprints in Particles and Fields", available from: SLAC-PPF, P. O. Box 4349, Stanford, CA 94305, U.S.A.

5. Differential Cross-Sections for $\bar{p}p \to \pi^-\pi^+$, K^-K^+ Between 0.8 and 2.4 GeV/c
   E. Eisenhandler, W.R. Gibson, C. Hojvat, P.I.P. Kalmus, L.C.Y. Lee Chi Kwong, T.W. Pritchard, E.C. Usher, D.T. Williams -- Queen Mary College; M. Harrison, W.H. Range -- Liverpool University; M.A.R. Kemp, A.D. Rush, J.N. Woulds -- Daresbury; G.T.J. Arnison, A. Astbury, D.P. Jones, A.S.L. Parsons -- Rutherford

6. Interpretations of the Differential Cross-Sections for $\bar{p}p \to \pi^-\pi^+$
   E. Eisenhandler, W.R. Gibson, C. Hojvat, P.I.P. Kalmus, L.C.Y. Lee Chi Kwong, T.W. Pritchard, E.C. Usher, D.T. Williams -- Queen Mary College; M. Harrison, W. H. Range -- Liverpool University; M.A.R. Kemp, A.D. Rush, J.N. Woulds -- Daresbury; G.T.J. Arnison, A. Astbury, D.P. Jones, A.S.L. Parsons -- Rutherford

8. Determination of 3π Nucleon Cross Section in $J^P = 0^-$ State and Possible ω Exchange in 3π Coherent Production on Nuclei
   G. Bellini, M. de Corato, E. Meroni, F. Palombo, P.G. Rancoita, G. Vegni -- INFN, Milan

11. Search for Neutral Mesons near 1 GeV/c
    D.M. Binnie, J. Carr, N.C. Debenham, A. Duane, D.A. Garbutt, W.G. Jones, J. Keyne, I. Siotis -- Imperial College; J.G. McEwen -- Southampton University

12. A Study of the Reaction $\pi^-p \to \omega n$ near Threshold
    D.M. Binnie, J. Carr, N.C. Debenham, A. Duane, D.A. Garbutt, W.G. Jones, J. Keyne, I. Siotis -- Imperial College; J.G. McEwen -- Southampton University

14. On Spin Anisotropies in the Production and Decay Correlations of $X^o(960)$
    R. Lednicky, V.I. Ogievetsky, A.N. Zaslavsky -- JINR, Dubna

15. Amplitude Analysis of $\pi^- p \to \rho N$
     J. Bouchez, J. Mallet -- Saclay

17. A Two Resonance Analysis of the $Q(K\pi\pi)$ Enhancement
     M.G. Bowler, J.B. Dainton, A. Kaddoura -- Oxford
     University, N.P.L.; I.J.R. Aitchison -- Oxford University

21. Partial Wave Analysis of $K\omega$ System
     S.D. Protopopescu, S.U. Chung, R.L. Eisner, N.I. Samios,
     R.C. Strand -- Brookhaven National Laboratory

22. Study of the Decay Distributions of the η' Meson
     C. Baltay, D. Cohen, S. Csorna, M. Habibi, M. Kalelkar --
     Columbia University; W. Smith, N. Yeh -- SUNY, Binghamton

23. Antiproton-Proton Charge-Exchange Between 1 and 3 GeV/c
     D. Cutts, M.L. Good, P.D. Grannis, D. Green, Y.Y. Lee,
     R. Pittman, J. Storer -- SUNY, Stony Brook: A. Benvenuti,
     G.C. Fischer, D.D. Reeder -- University of Wisconsin,
     Madison

24. Summary of Status of Approved Omega Experiments and Summary of
     New Proposals with Tentative Schedule for 1974
     J.D. Dowell -- CERN

25. The Transverse Momentum Distribution and Statistical Models
     J.-J. Dumont -- LIBHE-ULB, Brussels; L. Heiko --
     Universite Catholique de Louvain

26. Review of Meson Spectroscopy
     R.L. Eisner -- Brookhaven National Laboratory

27. Experimental Test of the Charge Independence Principle in
     Strong Interactions by the $d + d \to He^4 + \pi^\circ$ Reaction
     J. Banaigs, J. Berger, L. Goldzahl, L. Vu Hai -- Saclay;
     M. Cottereau, C. Le Brun -- Universite de Caen;
     F. L. Fabbri, P. Picozza -- Frascati

32. Study of the R Meson Produced in 13 GeV/c $\pi^+ p$ Interactions
     G. Thompson, J.A. Gaidos, R.L. McIlwain, D.H. Miller,
     T.A. Mulera, R.B. Willmann -- Purdue University

33. A Study of the Reactions $\pi^+ p \to (p^\circ, \omega)\Delta^{++}$ at 13 GeV/c
     J.A. Gaidos, A.A. Hirata, R.J. DeBonte, T.A. Mulera,
     G. Thompson, R.B. Willmann -- Purdue University

34. Partial Wave Analysis of the $(\pi^+\pi^+\pi^-)$ System Through the
     Region of the A_3 Meson
     G. Thompson, R.C. Badewitz, J.A. Gaidos, R.L. McIlwain,
     K. Paler, R.B. Willmann -- Purdue University

35. Partial-Wave Analysis of the Low-Mass $(\pi^+\pi^+\pi^-)$ System
    Produced by Incident π^+ Mesons at 13 GeV/c
        G. Thompson, J.A. Gaidos, R.L. McIlwain, R.B. Willmann --
        Purdue University

37. Pion-Deuteron Elastic Scattering at Intermediate Energies
        K. Gabathuler -- CERN

39. Backward $\bar{p}p$ Charge Exchange From 1 to 3 GeV/c
        J. Storer, D. Cutts, M.L. Good, P.D. Grannis, D. Green,
        Y.Y. Lee, R. Pittman -- SUNY, Stony Brook; A. Benvenuti,
        G.C. Fischer, D.D. Reeder -- University of Wisconsin,
        Madison

43. Vector Mesons and (Virtual) Nuclear Optics
        M. Greco -- Frascati; Y.N. Srivastava -- Northeastern
        University

44. A Finite Dispersion Relations Approach to $\eta \rightarrow \pi^+\pi^-\gamma$ Decay
        A. Bramon, M. Greco -- Frascati

45. Quark Model Predictions for Radiative Decays of Mesons
        A. Bramon, M. Greco -- Frascati

48. Formalism and Assumptions Involved in Partial Wave Analysis
    of Three-Meson Systems
        J.D. Hansen, G.T. Jones, G. Otter, G. Rudolph -- CERN

53. Peripheral $p\bar{p}$ Production and Decay Angular Distribution
    in the Reaction $\pi^-p \rightarrow p\bar{p}n$ at 18.8 and 9.8 GeV
        B. Hyams, C. Jones, P. Weilhammer -- CERN; W. Blum,
        H. Dietl, G. Grayer, W. Koch, E. Lorenz, G. Lütjens,
        W. Männer, J. Maissburger, U. Stierlin -- Max-Planck-
        Institut

55. High Statistics Study of the Reaction $\pi^-p \rightarrow \pi^-\pi^+n$: Apparatus,
    Method of Analysis, and General Features of Results at 17 GeV/c
        G. Grayer, B. Hyams, C. Jones, P. Schlein, P. Weilhammer
        -- CERN; W. Blum, H. Dietl, W. Koch, E. Lorenz,
        G. Lütjens, W. Männer, J. Meissburger, W. Ochs,
        U. Stierlin -- Max-Planck-Institut

58. A Deck Model Calculation of $\pi^-p \rightarrow \pi^-\pi^+\pi^-p$
        G. Ascoli, R. Cutler, L.M. Jones, U. Kruse, T. Roberts
        B. Weinstein, H.W. Wyld, Jr. -- University of Illinois,
        Urbana

59. Partial Wave Analysis of the Deck Amplitude for $\pi N \rightarrow \pi\pi\pi N$
        G. Ascoli, L.M. Jones, B. Weinstein, H.W. Wyld, Jr. --
        University of Illinois, Urbana

60. A Calorimeter for 300 GeV Neutrons and Protons
    L.W. Jones, J.P. Chanowski, H.R. Gustafson, M.J. Longo,
    P.L. Skubic, J.L. Stone -- University of Michigan;
    B. Cork -- Argonne National Laboratory

62. Comments on the $\eta'(958)$
    G.R. Kalbfleisch -- Brookhaven National Laboratory

65. Direct Channel $N\bar{N}$ Phenomena
    T.E. Kalogeropoulos -- Syracuse University

66. Structure in the Reaction $\pi^+ p \to \Delta^{++} \rho^0$ at 5 GeV/c
    Y. Eisenberg, B. Haber, U. Karshon, J. Mikenberg,
    S. Pitluck, E.E. Ronat, A. Shapira, G. Yekutieli --
    Weizmann Institute

67. Production and Decay Mechanism of the B Meson and the $\omega\pi$
    System in $\pi^+ p$ Interactions at 5 GeV/c
    U. Karshon, G. Mikenberg, Y. Eisenberg, S. Pitluck,
    E.E. Ronat, A. Shapira, G. Yekutieli -- Weizmann
    Institute

68. Structure in the $\omega\pi\pi$ System at the A_2 Mass Region
    U. Karshon, G. Mikenberg, S. Pitluck, Y. Eisenberg,
    E.E. Ronat, A. Shapira, G. Yekutieli -- Weizmann
    Institute

72. Coherent Pion Dissociation into Three Pions on Heavy Nuclei
    U.E. Kruse, T.J. Roberts -- University of Illinois, Urbana;
    R.M. Edelstein, E.J. Makuchowski, C.M. Meltzer,
    E.L. Miller, J.S. Russ -- Carnegie-Mellon University;
    B. Gobbi, J.L. Rosen -- Northwestern University;
    H.A. Scott, S.L. Shapiro, L. Strawczynski -- University
    of Rochester

75. Is the Quark Mass as Small as 5 MeV?
    H. Leutwyler -- Bern University

80. Partial Wave Analysis of the 3π System Produced in the Reaction
    $\pi^+ p \to (\pi^+ \pi^+ \pi^-)p$ at 8, 16 and 23 GeV/c
    Aachen-Berlin-Bonn-CERN-Heidelberg Collaboration

85. Effective Angular-Momentum Barrier in the SU(3) Test of
    Reactions Involving Pseudoscalar Mesons
    E. Takasugi, S. Oneda -- University of Maryland

86. Can the $J^{PC} = 1^{++}$ Mesons Form an Ideal Nonet?
    T. Laankan, S. Oneda -- University of Maryland

117. t-Dependence and Production Mechanisms of the ρ, f and g
Resonances from $\pi^-p \to \pi^-\pi^+n$ at 17.2 GeV
    B. Hyams, C. Jones and P. Weilhammer -- CERN; W. Blum,
    H. Dietl, G. Grayer, E. Lorenz, G. Lütjens, W. Männer,
    J. Meissburger, W. Ochs, U. Stierlin -- Max-Planck-
    Institut

118. Investigation of the $\pi \pm p \to \pi^\pm \pi^+ n$ Reactions Near Threshold
    S.A. Bunyatov -- JINR, Dubna

119. Spin-Parity Analysis of the B Meson
    S.U. Chung, S.D. Protopopescu -- Brookhaven National
    Laboratory; G.R. Lynch, M. Alston-Garnjost, A. Barbaro-
    Galtieri, J.H. Friedman, R.L. Ott, M.S. Rabin,
    F.T. Solmitz -- Lawrence Berkeley Laboratory; S.M. Flatté
    -- University of California, Santa Cruz

121. Quasi-Two Body Reactions and Properties of Resonances
    G.L. Kane - University of Michigan

123. Dual Property of Diffractive Resonances from Semilocal
Factorisation
    R.G. Roberts, D.P. Roy -- Rutherford

125. $\rho-\omega$ Interference in $\pi^-p \to \pi^-\pi^+n$ at 17.2 GeV/c
    P. Estabrooks, B. Hyams, C. Jones, A.D. Martin,
    P. Weilhammer -- CERN; W. Blum, H. Dietl, G. Grayer,
    W. Koch, E. Lorenz, G. Lütjens, W. Männer,
    J. Meissburger, U. Stierlin -- Max-Planck-Institut

126. Astrophysics and Tachyons
    R. Mignani, E. Recami -- Universita di Catania, Italy

127. Determination of the Degree of the $\Delta T = 1/2$ Rule Violation
and the Estimation of the Pion-Pion Scattering Lengths from
the Data on the $K \to 3\pi$ Decays
    S.A. Bunyatov, P.E. Volkovitsky, H.R. Gulkanyan --
    JINR, Dubna

128. Some Aspects of Semi-Inclusive Reactions in Four-Prong π^-p
Interactions at 5 GeV/c
    C. Besliu, A. Constantinescu, F. Cotorobai,
    V.N. Emei'yanenko, V.V. Clagolev, E.S. Kuznetsova,
    R.M. Lebedev, M. Sabau, I.S. Saitov, G. Sharkhu,
    L.I. Zhuravleva, J. Hlavacova, I. Michalcak, I. Patocka
    -- JINR, Dubna

129. The A_2-A_1 Interference Phase
    G. Ascoli, L.M. Jones, R. Klanner, U.E. Kruse, H.W. Wyld
    -- University of Illinois, Urbana

132. Observation of an A_3-Amplitude Phase Change
    G. Thompson, R.C. Badewitz, J.A. Gaidos, R.L. McIlwain,
    K. Paler, R.B. Willmann -- Purdue University

133. The Reaction $K^+d \to K^0\pi^+d$ at 4.6 GeV/c and the Effective
    Exchanged Trajectory for the Reactions $K^\pm d \to K^{*\pm}(892)d$
    G. Charriere, W. Dunwoodie, A. Grant, Y. Goldschmidt-
    Clermont, F. Muller, J. Quinquard -- CERN; G. De Jongh,
    S. Tavernier -- LIBHE-ULB, Brussels; P. Cornet,
    P. Dufour, F. Grard, V.P. Henri, J. Schlesinger,
    R. Windmolders -- Universite de l'Etat, Mons; G. Dehm,
    G. Göbel, W. Wittek, G. Wolf -- Max-Planck-Institut

134. Spin Parity Analysis of the $\omega\pi$ System in the B Meson Mass
    Region
    V. Chaloupka, A. Ferrando, M.J. Losty, L. Montanet --
    CERN; J. Alitti, B. Gandois, J. Louie -- Saclay

135. Status of $\bar{p}p \to \pi(2190) \to \rho^0\rho^0\pi^0$
    G.R. Kalbfleisch, R.C. Strand, V. VanderBurg --
    Brookhaven National Laboratory; J.W. Chapman --
    University of Michigan

136. Structures in the $\pi^+\pi^+\pi^-\pi^-$ System at 1.57 and 1.77 GeV
    F.A. DiBianca, W.J. Fickinger, J.A. Malko, J.F. Owens,
    D.K. Robinson -- Case Western Reserve University;
    S. Dado, A. Engler, G. Keyes, R.W. Kraemer -- Carnegie-
    Mellon University

137. On the Possible Existence of a Vector Meson $\rho'(1250)$
    M. Conversi, L. Paoluzi -- INFN-Universita di Roma;
    F. Ceradini, S. d'Angelo, M.L. Ferrer, R. Santonico --
    Universita di Roma; M. Grilli, P. Spillantini,
    V. Valente -- Frascati

138. A Measurement of $K^*(1420)$ Decay Branching Ratios
    G. Dehm, G. Gobel, W. Wittek, G. Wolf -- Max-Planck-
    Institut; G. De Jongh, S. Tavernier -- LIBHE-ULB,
    Brussels; P. Cornet, P. Dufour, F. Grard, V.P. Henri,
    R. Windmolders -- Universite de l'Etat, Mons;
    G. Charriere, W. Dunwoodie, A. Grant, Y. Goldschmidt-
    Clermont, F. Muller, J. Quinquard -- CERN

139. Backward Production of K^* (892) in the Reactions $K^+N \to K\pi N$ in Hydrogen and Deuterium for the Incident Momentum Range 3-5 GeV/c
        G. Charriere, W. Dunwoodie, C. Ferro-Fontan, Y. Goldschmidt-Clermont, F. Muller, M. Nikolic, J. Quinquard -- CERN; G. Dehm, W. Geist, G. Göbel, W. Wittek, G. Wolf -- Max-Planck-Institut; P. Cornet, P. Dufour, F. Grard, V.P. Henri, R. Windmolders -- Universite de l'Etat, Mons; G. De Jongh, S. Tavernier -- LIBHE-ULB, Brussels

140. Meson Production in $\pi^- p$ and $K^- p$ Interactions from 3 to 6 GeV/c
        D.S. Ayres, R. Diebold, A.F. Greene, S.L. Kramer, J.S. Levine, A.J. Pawlicki, A.B. Wickland -- Argonne National Laboratory

141. Multi-Hadrons Production Through the Photon-Photon Interaction
        G. Barbiellini, F. Ceradini, M.L. Ferrer, S. Orito, L. Paoluzi, R. Santonico, T. Tsuru -- Frascati

142. Experimental Determination of $\pi\pi$ and $K\pi$ S-Wave Cross Sections from the Reactions $\pi^+ p \to \pi^+\pi^-\Delta^{++}$ and $K^+ p \to K^+\pi^-\Delta^{++}$
        F. Wagner -- Lawrence Berkeley Laboratory

143. s- and t- Channel Helicity Conservation of Pion Induced Nucleon Diffraction Dissociation at 3.9 GeV/c and the Implications of a Two Component Model
        G. Berlad, B. Haber, M.F. Hodous, R.I. Hulsizer, V. Kistiakowsky, A. Levy, I.A. Pless, R.A. Singer, J. Wolfson, R.K. Yamamoto -- M.I.T.

146. Inclusive Study of ρ^0 Production in $\pi^- p$ Interactions at 8 GeV/c
        T. Kitagaki, J. Abe, K. Hasegawa, A. Yamaguchi, T. Nozaki, K. Temai, R. Sugahara -- Tohoku University

LIST OF PARTICIPANTS

BELGIUM

    Universite de l'Etat , Mons   V. P. Henri

CANADA

    McGill University   C. Hojvat
                                              P. M. Patel

    University of Toronto   A. W. Key
                                              G. J. Luste

DENMARK

    Niels Bohr Institute   J. D. Hansen

ENGLAND

    University of Birmingham   M. F. Votruba

    Durham University   A. D. Martin

    Imperial College   D. M. Binnie
                                            I. Butterworth

    University of Liverpool   D. N. Edwards

    Oxford University   M. G. Bowler
                                          R. H. Dalitz

    Queen Mary College   D. V. Bugg

    Rutherford High Energy Laboratory   M. R. Jane
                                            M. L. Mallary
                                          D. H. Saxon
                                          S. N. Tovey

FINLAND

    University of Helsinki   M. G. W. Roos

FRANCE

    Universite de Caen   G. Y. Bizard

    Laboratoire de l'Accelerateur   J. M. Buon
        Lineaire, Orsay

    Universite de Paris VI   J. T. Laberrigue
                                          T. P. Yiou

FRANCE (Cont'd)

Centre d'Etudes Nucleaires, Saclay H. A. Blumenfeld
J. M. Bouchez
J. Ernwein
L. Goldzahl

Centre de Recherches Nucleaires, A. Fridman
Strasbourg

FEDERAL REPUBLIC OF GERMANY

I. Physikaliches Institut der Tech. K. Lubelsmeyer
Hochschule, Aachen

DESY G. Wolf

Max-Planck-Institut G. G. Lutz
H. M. Dietl
G. H. Grayer
W. E. Manner

ISRAEL

University of Tel-Aviv A. Levy

Weizmann Institute E. E. Ronat

ITALY

Universita di Bologna A. Minguzzi-Ranzi

Laboratori Nazionali di Frascati F. L. Fabbri
C. Mencuccini
M. Greco

Universita di Genova D. Teodoro

Universita di Padova M. Cresti

Universita di Roma R. Bizzarri
L. Paoluzi

JAPAN

Tohoku University T. Kitagaki

THE NETHERLANDS

Universiteit van Amsterdam W. Hoogland

SCOTLAND

    University of Glasgow   I. S. Hughes

SPAIN

    Junta de Energia Nuclear, Madrid   J. B. Diaz

SWITZERLAND

    Universite de Geneve   M. D. Martin

    CERN   A. Birman
                                      V. Chaloupka
                                      D. R. Morrison
                                      P. M. Weilhammer

UNITED STATES OF AMERICA

    Argonne National Laboratory   R. E. Diebold
                                        T. H. Fields
                                        J. J. Phelan
                                        A. B. Wicklund

    Brandeis University   J. R. Bensinger
                                        L. E. Kirsch
                                        H. J. Schnitzer

    Brookhaven National Laboratory   S. U. Chung
                                          R. L. Eisner
                                        K. J. Foley
                                        H. A. Gordon
                                        G. R. Kalbfleisch
                                        S. J. Lindenbaum
                                        S. Ozaki
                                        S. D. Protopopescu
                                      I. Stumer

    Brown University   G. T. Y. Chen
                                        R. E. Lanou
                                        A. M. Shapiro
                                        M. Widgoff

    California Institute of Technology W. Ochs

    University of California, Berkeley   H. H. Bingham
                                          W. Chinowsky
                                        A. H. Rosenfeld

    University of California, Davis   W. T. Ko

    University of California, Irvine   G. Shaw

UNITED STATES OF AMERICA (Cont'd)

University of California, Los Angeles	H. K. Ticho
University of California, Riverside	B. C. Shen
University of California, Santa Cruz	S. M. Flatte
Carnegie-Mellon University	S. Dado
	R. M. Edelstein
	J. S. Russ
Case Western Reserve University	W. J. Fickinger
	D. K. Robinson
University of Cincinnati	B. T. Meadows
Columbia University	D. Cohen
	M. S. Kalelkar
Cornell University	B. Gittelman
	D. H. White
Duke University	J. S. Loos
	W. D. Walker
Florida State University	J. R. Albright
	S. L. Hagopian
	V. Hagopian
	C. P. Horne
	J. D. Kimel
	P. K. Williams
Harvard University	S. Glashow
University of Hawaii	M. W. Peters
University of Illinois	G. Ascoli
	L. M. Jones
	T. B. W. Kirk
	R. Klanner
	U. E. Kruse
	D. A. Rhines
	R. D. Sard
Indiana University	D. B. Lichtenberg
Iowa State University	K. E. Lassila
	F. C. Peterson
Johns Hopkins University	C. Y. Chien
	R. A. Zdanis

UNITED STATES OF AMERIA (Cont'd)

University of Kansas	R. G. Ammar
Lawrence Berkeley Laboratory	G. S. Abrams
	D. M. Chew
	G. Gidal
	G. Goldhaber
	T. A. Lasinski
	P. J. Oddone
	F. Wagner
	F. C. Winkelmann
Massachusetts Institue of Technology	D. S. Barton
	A. Bodek
	E. S. Hafen
	R. M. O'Donnell
	I. A. Pless
	J. Wolfson
University of Massachusetts, Amherst	J. Button-Shafer
	M. S. Z. Rabin
Southeastern Massachusetts University	Z. H. Bar-Yam
	W. G. E. Kern
	J. J. Russell
University of Michigan	G. L. Kane
National Bureau of Standards	S. Meshkov
National Accelerator Laboratory	V. A. Ashford
	P. F. M. Koehler
City College of New York	M. A. Kramer
State University of New York at Albany	C. R. Sun
State University of New York at Binghamton	N. K. Yeh
State University of New York at Stony Brook	C. Quigg
Northeastern University	R. Aaron
	R. Arnowitt
	G. J. Blanar
	C. F. Boyer
	D. Earles
	W. L. Faissler
	M. H. Friedman
	D. A. Garelick

UNITED STATES OF AMERICA (Cont'd)

Northeastern University (Cont'd)

M. W. Gettner
M. J. Glaubman
B. Gottschalk
H. Johnstad
M. Jordan
H. Lee
B. Malenka
P. Nath
E. L. Pothier
D. Potter
M. T. Ronan
E. J. Saletan
G. Srivastava
Y. Srivastava
M. F. Tautz
M. Vaughn
E. von Goeler
R. Weinstein

Northwestern University

M. M. Block
J. L. Rosen

University of Notre Dame

N. M. Cason

Oak Ridge National Laboratory

H. O. Cohn

University of Pennsylvania

H. M. Brody
L. A. Kroger
W. Selove

Purdue University

N. V. Baggett
R. H. Capps
J. A. Gaidos
L. J. Gutay
D. H. Miller
E. I. Shibata
Y. W. Tang
R. B. Willmann

University of Rochester

D. E. Andrews
W. C. Carithers
T. Ferbel
C. A. Nelson

Rockefeller University

H. R. Pagels

Rutgers University

J. L. Alspector
K. J. Cohen
W. C. Harrison
M. D. Jones

UNITED STATES OF AMERICA (Cont'd)

Rutgers University (Cont'd)	R. J. Plano
	P. Yamin
Stanford Linear Accelerator Laboratory	J. Ballam
	G. W. Brandenburg
	R. K. Carnegie
	J. T. Carroll
	G. J. Feldman
	F. J. Gilman
	D. W. G. S. Leith
	D. J. Linglin
	J. A. J. Matthews
	K. C. Moffeit
	A. C. Odian
	T. L. Schalk
	A. Wetsch
Stanford University	D. G. Hitlin
Syracuse University	M. Goldberg
	T. Kalogeropoulos
	G. Moneti
University of Tennessee	W. M. Bugg
University of Texas	A. Bohm
Tufts University	G. Wolsky
Upsala College	J. M. Oostens
Vanderbilt University	S. L. Stone
University of Washington	R. M. Harris
	W. J. Podolsky
University of Wisconsin	R. Prepost
Yale University	T. W. Ludlam

UNION OF SOVIET SOCIALIST REPUBLICS

Joint Institute for Nuclear Research, Dubna	V. A. Belyakov
	S. A. Bunyatov

COMPLETE AUTHOR INDEX

		L.C. Number	ISBN
No. 1	Feedback and Dynamic Control of Plasmas (Princeton 1970)	70-141596	0-88318-100-2
No. 2	Particles and Fields - 1971 (Rochester)	71-184662	0-88318-101-0
No. 3	Thermal Expansion - 1971 (Corning)	72-76970	0-88318-102-9
No. 4	Superconductivity in d- and f-Band Metals (Rochester 1971)	74-188879	0-88318-103-7
No. 5	Magnetism and Magnetic Materials - 1971 (2 parts) (Chicago)	59-2468	0-88318-104-5
No. 6	Particle Physics (Irvine 1971)	72-81239	0-88318-105-3
No. 7	Exploring the History of Nuclear Physics (Brookline 1967, 1969)	72-81883	0-88318-106-1
No. 8	Experimental Meson Spectroscopy - 1972 (Philadelphia)	72-88226	0-88318-107-X
No. 9	Cyclotrons - 1972 (Vancouver)	72-92798	0-88318-108-8
No.10	Magnetism and Magnetic Materials - 1972 (2 parts) (Denver)	72-623469	0-88318-109-6
No.11	Transport Phenomena - 1973 (Brown University Conference)	73-80682	0-88318-110-X
No.12	Experiments on High Energy Particle Collisions - 1973 (Vanderbilt Conference)	73-81705	0-88318-111-8
No.13	π-π Scattering - 1973 (Tallahassee Conference)	73-81704	0-88318-112-6
No.14	Particles and Fields - 1973 (APS/DPF Berkeley)	73-91923	0-88318-113-4
No.15	High Energy Collisions - 1973 (Stony Brook)	73-92324	0-88318-114-2
No.16	Causality and Physical Theories (Wayne State University, 1973)	73-93420	0-88318-115-0
No.17	Thermal Expansion - 1973 (Lake of the Ozarks)	73-94415	0-88318-116-9
No.18	Magnetism and Magnetic Materials - 1973 (2 parts) (Boston)	59-2468	0-88318-117-7
No.19	Physics and the Energy Problem - 1974 (APS Chicago)	73-94416	0-88318-118-5
No.20	Tetrahedrally Bonded Amorphous Semiconductors (Yorktown Heights, 1974)	74-80145	0-88318-119-3
No.21	Experimental Meson Spectroscopy - 1974 (Boston)	74-82628	0-88318-120-1
No.22	Neutrinos - 1974 (Philadelphia)	74-82413	0-88318-121-X